Lecture Notes in Mechanical Engineering

Lecture Notes in Mechanical Engineering (LNME) publishes the latest developments in Mechanical Engineering - quickly, informally and with high quality. Original research reported in proceedings and post-proceedings represents the core of LNME. Volumes published in LNME embrace all aspects, subfields and new challenges of mechanical engineering. Topics in the series include:

- Engineering Design
- Machinery and Machine Elements
- Mechanical Structures and Stress Analysis
- Automotive Engineering
- Engine Technology
- Aerospace Technology and Astronautics
- Nanotechnology and Microengineering
- Control, Robotics, Mechatronics
- MEMS
- Theoretical and Applied Mechanics
- Dynamical Systems, Control
- Fluid Mechanics
- Engineering Thermodynamics, Heat and Mass Transfer
- Manufacturing
- Precision Engineering, Instrumentation, Measurement
- Materials Engineering
- Tribology and Surface Technology

To submit a proposal or request further information, please contact the Springer Editor of your location:

China: Dr. Mengchu Huang at mengchu.huang@springer.com
India: Priya Vyas at priya.vyas@springer.com
Rest of Asia, Australia, New Zealand: Swati Meherishi at swati.meherishi@springer.com
All other countries: Dr. Leontina Di Cecco at Leontina.dicecco@springer.com

To submit a proposal for a monograph, please check our Springer Tracts in Mechanical Engineering at http://www.springer.com/series/11693 or contact Leontina.dicecco@springer.com

Indexed by SCOPUS. The books of the series are submitted for indexing to Web of Science.

More information about this series at http://www.springer.com/series/11236

D. A. Indeitsev · A. M. Krivtsov
Editors

Advanced Problems in Mechanics

Proceedings of the XLVII International Summer School-Conference "Advanced Problems in Mechanics", June 24–29, 2019, St. Petersburg, Russia

Springer

Editors
D. A. Indeitsev
IPME RAS, Peter the Great St. Petersburg Polytechnic University
Saint-Petersburg, Russia

A. M. Krivtsov
IPME RAS, Peter the Great St. Petersburg Polytechnic University
Saint-Petersburg, Russia

ISSN 2195-4356 ISSN 2195-4364 (electronic)
Lecture Notes in Mechanical Engineering
ISBN 978-3-030-49881-8 ISBN 978-3-030-49882-5 (eBook)
https://doi.org/10.1007/978-3-030-49882-5

This Springer imprint is published by the registered company Springer Nature Switzerland AG
The registered company address is: Gewerbestrasse 11, 6330 Cham, Switzerland

Contents

Influence of the Sperm Velocity on Fertilization Capacity in the Oscillatory Model of Mouse Zona Pellucida

Andjelka Hedrih[1(✉)], Ana Mitrovic-Jovanovic[2,3], and Mihailo Lazarevic[4]

[1] Department of Mechanics, Mathematical Institute of Serbian Academy of Sciences and Arts, Belgrade, Serbia
handjelka@gmail.com
[2] School of Medicine, University of Belgrade, Belgrade, Serbia
[3] Clinics for Gynecology and Obstetrics "Narodni Front" Belgrade, Belgrade, Serbia
[4] Faculty of Mechanical Engineering, University of Belgrade, Belgrade, Serbia

Abstract. Considering the fertilization process as an oscillatory phenomenon, based on mechanotransduction theory of sperm–oocyte interaction, influence of sperm velocity and their specific arrangement on outer surface of oocyte- Zona pellucida (ZP) relative to the oscillatory behavior of ZP was studied using discrete continuum oscillatory spherical net model of mouse ZP. For the calculated favorable impact angles of spermatozoa by using generalized Lissajous curves, a parametric frequency analysis of oscillatory behavior of the knot molecules in the mouse ZP spherical net model is conducted. In order to mimic successful fertilization in physiological conditions in this numerical experiment, velocities of the progressive and hyperactivated spermatozoa were used. The resultant trajectories of knot molecules in mouse ZP (mZP) spherical net model, in the form of generalized Lissajous curves, are presented. Influences of the sperm velocity and its arrangement on the resultant trajectory of the corresponding knot molecules are discussed. Component displacements in the meridian and circular directions of the knot molecules of ZP are in the form of multi-frequency oscillations. Symmetrical arrangements of spermatozoa having effective velocities are more favorable for achieving a favorable oscillatory multi-frequency state of mZP for a successful fertilization. Determining the optimal parameters of spermatozoa impact that will induce a ZP favorable oscillatory state opens the possibilities for more complete explanation of the fertilization process.

Keywords: Zona pellucida · Discrete oscillatory spherical net model · Oscillations · Lissajous curves · Sperm velocity · Sperm arrangement

D. A. Indeitsev and A. M. Krivtsov (Eds.): APM 2019, LNME, pp. 1–21, 2020.
https://doi.org/10.1007/978-3-030-49882-5_1

1 Introduction

This study utilizes the oscillatory model proposed for fertilization by [1, 2]. A cellular response to a mechanical stimulus could be evaluated by cell-stretching device [3] as well as by atomic force microscope [4] and micropipette aspiration technique [5] and micro-tactile sensor [6].

Velocity of spermatozoa and their distribution on the ZP surface significantly affect mice's zona pellucida (ZP) oscillatory behaviour. Furthermore, since velocity is directly related with success of fertilization, swimming velocity and parameters having influence on it are considered in this study. A basic description of the ZP oscillatory mechanical net model is presented in Sect. 1. along with a brief explanation on why distribution of spermatozoa has effect on oscillatory behaviour of ZP.

1.1 Importance of Sperm Velocity for Fertilization Success

Mouse spermatozoa are very vulnerable cells. After ejaculation, their life span in female reproductive tract is limited. During their chemotactical, rheotactical, and thermotactical [7] movement in female reproductive tract, some of them will pass through complex metabolic and structural changes, becoming fertilization-competent [8]. As ejaculate contains spermatozoa of different stage of maturation, only a small fraction (on average, about 10%) is chemotactically responsive and represents capacitated (mature) spermatozoa [9] "which maximum rate matched exactly the male gonadosomatic index of the species" [10].

Sperm competition thus favors a larger population of spermatozoa that are competent to fertilize, and spermatozoa that are more sensitive to the signals emitted by the ovum and undergo the acrosome reaction during fertilization [11]. Centola et al. in [12] confirm the declining tendency in sperm concentration, total count, and total motile count in human semen over the then year period in sperm donors with stable or improved demographic and lifestyle factors [12].

Capacitation process consists of several biochemical and functional modifications of sperm cells that enable the sperm to undergo both the acrosome reaction and hyperactivation (HA). HA is a functional change in the sperm movement pattern in a form of relatively progressive motion with high-amplitude flagellar bending [13]. In hyperactivated spermatozoa linearity, amplitude of lateral head displacement beat cross frequency and average wavelength gradually change with time [14]. HA movement pattern gives direct mechanical advantage to spermatozoa, permitting their penetration through a viscous medium. HA also facilitates penetration of spermatozoa through the zona pellucida (ZP) [10, 13] and could be influenced by different factors: cumulus cells that enter the ampulla with the oocyte [15], progesterone (a dose-dependent stimulation of HA) [13]. Acrosome reaction must be timed precisely, because both premature and delayed acrosome reactions will prevent sperm from penetrating the cumulus oophorus and adhering to the ZP [10].

The oviduct appears to influence the extracellular matrix properties of the spermatozoa as well as the Cumulus Oophorus Complex that is of interest in ZP-cumulus-sperm interaction [16]. Spermatozoa that reach the ZP first will have a chance to interact with ZP molecules and one of them will have a chance to fertilize the oocyte.

Sperm swimming velocity is a key determinant of male fertilization success [17]. Sperm velocity is correlated with the ability to fuse with the oolemma [18, 19]. Many factors are positively associated with maximum sperm swimming velocity like: mating style, total sperm length [20–22], relative testis weight and sperm midpiece volume [23], increase in the length of the sperm head and hydrodynamic shape of the sperm head [21], shape of the head and the proportions between the components of the sperm flagellum [17], content of functional mitochondria [24], and midpiece size [25]. Relation between sperm head shape and sperm motility is guided by genetic factors and it is seem to be poligenetcly determined. Using both in mice and human models of infertility, many gens, which mutations have repercussions on sperm motility as well as sperm head shape were identified [26].

The ratio between head length and total flagellum length was significantly associated with straight-line velocity so that as the proportion of the flagellum length in relation to head length decreased, sperm velocity also decreased.

Kinetic parameters of spermatozoa are used for evaluation of fertility and differ for healthy and unhealthy sperm cells [17]. The rotational speed of sperm cells from infertile men is observed to be significantly less as compared to controls [27]. Hyperbaric oxygen treatment could improve vigility of spermatozoa from subfertile man [28]. There are strain-dependent differences in sperm motility profiles [29]: at early time points of capacitation, sperm motility patterns have largely progressive, but in later time points tracks shift to more varied patterns of movement, including hyperactivation [29]. For identifying sperm motility parameters by claster analysis a Bayesian network could be used [30].

1.2 Bio-Mechanical Models of Sperm-Oocyte Interaction

There are attempts to explain the fertilization process in mammals from the biomechanical point of view. The purpose of these models is to identify biomechanical parameters of importance for the fertilization process that are related with both ovum and spermatozoa. The next step would be to interconnect biomechanical and biochemical models of fertilization into a unique integrated model. One of the assumptions is that the spermatozoa generating a higher contact pressure upon ZP surface would have a better chance of penetration through ZP [31, 32]. That assumption was discussed in [2]. Kozlovsky and Gefen [32] show via numerical simulations that spermatozoa with sharper head generate a higher contact pressure on the ZP. Numerical simulation carried out in [31] showed that contact pressure is more sensitive to spermatozoa's' motility than to spermatozoa's' head stiffness.

Although many molecules on sperm surface involved in sperm-oocyte interaction have been determined, the precise molecular mechanism of mammalian fertilization is still not known. An alternative theory for successful sperm–ZP binding events and post-fertilization inhibition of sperm binding was proposed by [33] based on modifications of ZP2 and ZP3 and sperm binding affinity for these molecules. There are several candidates for complementary molecules on the mammalian sperm surface involved in the early interactions with the ZP [34].

In the oscillatory model of fertilization, it is not essential which individual molecules are being involved in fertilization, but the way in which a specific structure having certain biomechanical characteristics, oscillates.

In order to model the effect of spermatozoa impact on ZP, an oscillatory net model of mouse ZP has been developed.

1.3 The Oscillatory Model of Fertilization – Basic Concept

It is well known that mechanical loading may change biological microstructure [35]. Theory of oscillations can be applied to biological systems at the macro (biorhythms, oscillations in blood pressure, hormone level, and nervous system activities), micro (cell), and nano (subcellular structures, e.g. movement of a molecule in a spermatozoid flagellum) levels. In the course of accomplishing their own function, molecules of a cell are moving-oscillating periodically (e.g. movement of myosin at muscular contraction) or aperiodically. Each cell and all its structures exhibit oscillatory movement. The characteristics of oscillatory movement (set of oscillatory frequencies, amplitude of movement, etc.) of a cell will be dependent on its functional state (degree of metabolic activity, healthy/unhealthy cell) and also on cell type (type of cell, size, structure …). Oscillatory movement of a cell varies with its contact with other cells. From the biomechanical and electrical points of view, receptor recognition is based on this oscillatory movements. An oocyte exposed to surface actions of spermatozoa could be considered as an oscillatory system oscillating in a forced regime under the influence of an external force. The external force is generated by spermatozoa. Owing to the character of movement of spermatozoa, it could be considered that this external force is of periodic character [36]. In [36] is established that movement of spermatozoa had oscillatory character. According to the experiments in [36], mouse spermatozoa have twisted planar type of 3D waveform movement, clockwise roll direction, and clockwise circling direction of movement.

The oscillatory movement of an oocyte depends also on the characteristics of the external force (above all, on frequency). Since ZP is the first structure of the oocyte to interact with spermatozoa (in in vitro conditions), our interest is to characterize ZP oscillatory behaviour under the action of spermatozoa having different velocities.

Under the action of external force or initial perturbations of the ZP due to pulse action of spermatozoa, molecules of the ZP will move-oscillate in certain manner. This movement can be in the form of a straight line, periodic, aperiodic, resemble a stochastic movement, or look like chaotic movement. Since straight line movement is predictable, we assume that if a large number of ZP molecules (or of one compact part of it) moves in this way, this will facilitate penetration (from the biomechanical point of view) of spermatozoa through the ZP consequently resulting in fertilization. We also assume, on the basis of theory of oscillations, that for initial penetration through the ZP it is required that spermatozoa oscillate in a multi-frequency regime having at least one part of the frequency range coinciding with that of the oscillations of the oocyte (or at least the part of the ZP being acted upon). Then, the conditions exist for a resonance to appear over the time interval where this ratio of frequencies is maintained, when structures of the ZP and spermatozoa oscillate at the same frequency; then, theoretically, amplitudes of molecular movement tend towards infinity, i.e. they increase during the time of existence of this regime which, from the biomechanical point of view, could be the condition for breaking bonds of the ZP molecules and consequent initial penetration of spermatozoa through the ZP. Aperiodic or chaotic movements are hard to predict and certainly do not facilitate penetration through ZP's net structure. For this reason, it is important to know under which conditions ZP molecules would move in a predictable way compatible with fertilization. Oocyte is a dynamic structure changing in time through all the phases of its maturing, thus changing the mechanical properties (e.g. outer shell of the ZP, cytoplasm) and structure. As already mentioned, this causes changes of its oscillatory state [37].

1.4 Sperm Arrangement on ZP Surface and ZP Oscillations

In the course of fertilization in both in vivo and in vitro conditions, oocyte is in contact with spermatozoa of different quality (morphology, velocity, acrosome status). ZP is mechanosensitive structure and response on mechanical stimulation that come from sperm on its surface.

Acting upon the surface of ZP all these spermatozoa, each in its own way, contribute to the change of oscillatory behaviour of the ZP and egg cell as a whole. Their joint action contributes to the final result – fertilization. Action of spermatozoa having different characteristics can be considered as action of external periodic forces having different characteristics.

Different distributions of spermatozoa having the same/different kinetic parameters result in different distributions of external forces acting on the ZP surface. Each individual spermatozoid generates certain force which acts upon the ZP surface [38] and their joint action will give specific distribution of force on the ZP surface. From what has been said so far, it turns out that a symmetric or asymmetric distribution of force produced by action of spermatozoa upon ZP surface will cause different oscillatory states of the ZP. Theoretically, there are combinations of the distribution of this force resulting in the same or similar oscillatory states of the ZP. Examples of different distributions of spermatozoa on the ZP surface are shown in Figs. 1A, B, and C. Since it is universally acknowledge that the success of fertilization is measured by the quality of spermatozoa, primarily by the percentage of progressively moving spermatozoa, it is important to investigate on how the velocity of spermatozoa influences the oscillatory behaviour of ZP molecules. For the analysis of oscillatory behaviour of ZP molecules, we applied the model of oscillatory spherical net of mouse ZP. Detailed description of the model is given in [2]. For the analysis only a small segment of the ZP model retaining the molar ratio of mouse molecule ZP is used, Figs. 1D and E. The segment consists of four crisscrossed orthogonal chains consisting of 11 ZP molecules, each distributed in a specific way (molar ratio retained). Masses of the ZP molecules are included in the model. At this stage, we are interested to know what will be the behaviour of ZP1 molecules which represent nodal molecules in the model. They have been selected for two reasons – their masses are the largest thus their contribution to the oscillatory behaviour will be significant; upon fertilization, by creating disulphide bridges between ZP1 molecules on one side and ZP2 and ZP3 on the other, followed by a change of ZP structure, Young modulus of elasticity increases [5] and the conditions for a polyspermy block are created. In addition, according to the modified Wassarman model of ZP [39], without the ZP1 bonding molecules the net structure of ZP does not exist [40, 41].

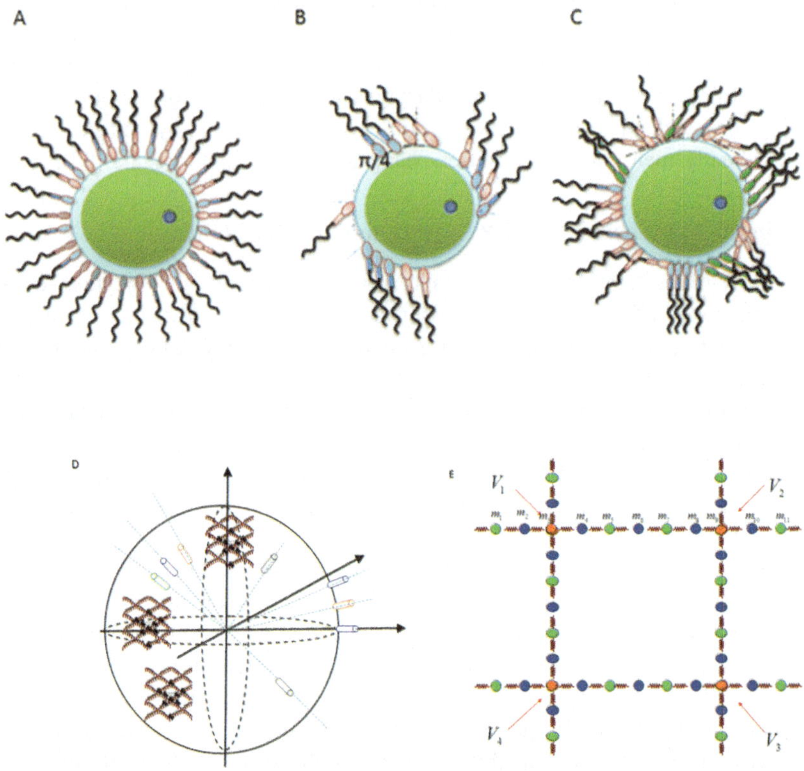

Fig. 1. Hypothetical arrangement of spermatozoa on ZP surface: **A** Symmetrical arrangement of spermatozoa having two different swimming velocities. **B** Asymmetrical arrangement of spermatozoa having two different swimming velocities but the same impact angle. **C** Asymmetrical arrangement of spermatozoa having three different swimming velocities and different impact angles arbitrarily arranged. Different colors of spermatozoa (pink, blue or green) denote their different swimming velocities. **D** Model of *ZP* spherical surface that shows radial direction of axis of constructive elements of the model - *ZP* proteins, **E** Part of the *mZP* network on a part of the sphere (oocyte). Orange (*ZP*1), blue (*ZP*2) and green (*ZP*3) represent molecules of *ZP* proteins. Chains of spherical net are identical in the circular and meridian directions. Each *ZP* protein is connected to the sphere with elastic springs that can oscillate in radial direction. V1–V4 are sperm cell effective velocities. Red arrows denote sperm cell impact with a knot molecule.

1.5 Oscillatory Net Model of Mouse Zona Pellucida

Zona pellucida (ZP) is an acellular 3D matrix that surrounds all mammalian oocytes, ovulated eggs, and embryos up to the early blastocyst stage of development is very important structure in the process of oogenesis, fertilization, and preimplantation development and dynamically changes its mechanical properties during this processes [5, 42]. In mice, ZP consists of 3 sulphated glycoproteins: ZP1, ZP2, and ZP3. ZP2 and ZP3 have fibrillary structure. They are interchangeably connected making fibril structures that are cross-linked with ZP1 [39].

In order to describe the oscillatory behaviour of ZP structure, an oscillatory spherical net model of mouse ZP was created [1, 43–46] by using discrete continuum method [47–50]. Discretization method is used in solving other problems in reproductive biology and medicine. Time discretization method was also used in the mathematical model in pattern formation during zebrafish embryo development [51]. Method of continuum mechanics is used for modeling the tumor growth [52], continuum modeling of growth and remodeling of living matter [35].

According to the model [1, 43, 45], the system of ZP oscillatory net oscillates in a free regime after ovulation without presence of spermatozoa. If there is only an initial perturbation by the kinetic and potential energy given to oscillatory structures, only free vibration regimes of a vibrating discrete structure appear.

Under numerous vibro-impacts between spermatozoa and *ZP*, the ZP oscillatory net starts to oscillate in a forced regime. The application of one or multi-frequency external excitation forces to a *ZP* discrete net that oscillates in a free regime results in multi-frequency forced regime oscillations. External excitation is presented as impact of spermatozoa. As spermatozoa movement have oscillatory character [36], their impact can be considered as repeated external forced oscillations. The molar ratio and the masses of mouse ZP (mZP) glycoproteins [39] were included into the model. Material particles in the ZP oscillatory spherical net model correspond to the *ZP* glycoproteins (see Figs. 1D and 1E). Knot material particles correspond to the ZP1 glycoproteins and are assigned as knot molecules. A detailed description of the ZP oscillatory model is given in [2, 43, 45]. In Fig. 1B a part of the spherical net ZP model is presented. For the purpose of making frequency and trajectory analysis of knot mass oscillations along trajectory in the forced regime, we consider a small representative part of the discrete spherical net that preserves the ratio ZP1:ZP2–ZP3 as 1:5 [39], Fig. 1E [1, 2, 43–46].

The oscillatory spherical net model of mZP allows analyzing movements of ZP molecules from the biomechanical point of view. These movements will be affected by sperm's velocity, impact angle, arrangement or impact position. The present model can assess to what extent ZP motion depends on each of the above mentioned factors.

The main goal of this study is to understand if different sperms' effective velocities affect the resultant trajectories of knot molecules. Effective velocities of progressive and hyperactivated spermatozoa impacting the mZP net at certain favourable angles [2] will be considered. The analysis will focus on the resulting movement of the knot molecules in the plane tangential to the model surface. Resultant trajectories of knot molecules, expressed as generalized Lissajous curves, have two displacement components, circular and meridian displacements. These component trajectories drive the open or closed paths of spermatozoa through the ZP.

The second goal of the study is to analyze if arrangements and ratio of sperm cells with effective velocity affect the resultant trajectory motion of mZP knot molecules. According to the shape of resultant trajectory motion of mZP knot molecules in the oscillatory model of mZP, the following conclusions can be drawn:

- If the resultant motion of knot molecules is synchronized, it will be easier for spermatozoa to pass through ZP net. This ZP oscillatory state favours penetration and fertilization.
- If trajectories are chaotic or stochastic like, probability that one of many spermatozoa which are periodically "attacking" ZP will penetrate through it is very low. This counteracts penetration and fertilization.

2 Methods

The analytical method used for modeling oscillatory behaviour of the mZP net [2, 45] is based on the vibration theory of mechanical chain systems [53]. Chain systems are hypothesized to be heterogeneous and formed by elements with ideal elastic bonds. A small portion of the mZP spherical surface net with 11 molecules in the chain which periodically repeat preserving the molar ratio of mZP glycoproteins (see Fig. 1E) is considered. Therefore, molecules/mass particles of the chains are characterized by 11 free vibration modes [45]. In Figs. 1D and 1E, the knot molecules ZP1 are colored in orange. Pulse action of spermatozoa gives some initial velocity to model knot molecules. Here, we know the masses of ZP molecules (expressed in terms of mass units), stiffness of springs (calculated with Eq. (1)), the γ angle between the direction of spermatozoa's movements and the tangent plane to spherical net, the θ angle between direction of spermatozoa's movements and the meridian direction of the tangent plane, and distribution of spermatozoa in terms of effective velocity. Initial conditions are as following: angles γ and θ are set equal to favourable values according to [2], the initial movements of all points of the model are zero (as prior to action of spermatozoa), the initial velocities of all points also are zero, except those of knot points (points 3 and 9 in the chain) to which spermatozoa transfer an initial velocity after the impact. Pulse forces have high intensity but short interaction. In order to capture the effect of different distributions of spermatozoa's impact forces on the resultant trajectories of knot molecules and cover all knot points in ZP model, the following cases where four spermatozoids with effective velocity impact against four different neighboring knot molecules are considered.

The following analyses of different combinations of spermatozoa's effective velocities were done:

- with four hyperactivated spermatozoa,
- with four progressive spermatozoa,
- with three hyperactivated and one progressive spermatozoa,
- with two hyperactivated and two progressive spermatozoa when the spermatozoa of the same velocity are arranged in the diagonal up position,
- with two hyperactivated spermatozoa and two progressive when spermatozoa with same velocity has impact on the neighboring knot molecule, and
- with one hyperactivated and three progressive spermatozoa.

The position of the hyperactivated spermatozoa is varied to evaluate if trajectories of knot molecules are sensitive to this factor.

Multi-frequency oscillatory motions of knot molecules when each of the four sperm cells with effective velocity simultaneously impact with different neighboring knot molecule were analyzed. See Fig. 1E.

Rigidities (C_m) of the massless elastic elements that interconnect the ZP molecules are assumed to be equal and calculated according to the oscillatory motion formula:

$$c_m = \frac{E(R^2 - r^2)\pi}{2R} \tag{1}$$

where E-is Young modulus of elasticity, R is radius of the mouse oocyte, r-is radius of the oocyte minus approximate thickness of mZP. Data were taken from [6]: $E = 17.5$ kPa, $2R = 56.5$ μm-average diameter of the oocyte from [6], $\delta = 4.94$ μm-approximate thickness of the oocyte, $C_m = 0.253$ N/m. Molecular masses of the corresponding ZP glicoproteins from kDa were transformed into standard units (kg). Data for the mouse spermatozoa having an effective velocity were taken from [29].

Progressive and hyperactivated sperm cells were considered as sperm cells having an effective velocity. Hyperactivation of sperm cells is necessary for capacitating the process whereby sperm cells become competent to fertilize the oocyte. In physiological conditions, these two types of sperm cells will be capable of reaching the oocyte (progressive) and responding to specific stimuli from female reproductive tract (to be hyperactivated in the process of capacitation). Slow and weakly motile sperm cells would not survive in the mouse female reproductive tract; they are not capable of reaching the oocyte, thus they were excluded from the analysis. For the numerical experiment, the authors did not take into account the sperm cells of intermediate velocities because agreement between visual and model-assigned tracks for this group was 52.1% [29], due to the character of movement of spermatozoa of intermediate velocities. For this numerical experiments the following *VAP* - average path velocity in μm/s were taken: for progressive spermatozoa $V_p = 146.9$ μm/s and for hyperactivated spermatozoa $V_{hp} = 171.1$ μm/s. According to the data for mass of bull spermatozoa (1,82 × 10^{-14} kg) [54], the approximated mass for a mouse spermatozoon was taken as 10^{-14} kg.

Velocity that spermatozoa transfer to the knot molecule at the moment of impact was calculated according to the formula: $V = V_{hp} \cdot m_{sp}/m_3$. Here, m_{sp} is the approximated mass of mouse spermatozoa; m_3 is mass of ZP1 glycoprotein-knot molecule.

In order to find the resultant motion of knot molecules, it is necessary to determine and then superpose displacement components in the circular and meridian directions in the tangent plane to the spherical ZP net. Frequency analysis of mZP knot oscillation trajectories for the case of four knot molecules impacted by spermatozoa with effective velocities entails the following steps (for details see [2, 45]):

A. Define the system of eleventh ordinary differential equation of chains dynamics with 11 material particles motion for each chain;
B. Add 22 initial conditions for each system of differential equations that describe molecular chain dynamics. In the considered cases, for each chain 11 initial conditions are related to initial displacements, and other 11 initial conditions are related to initial velocities of the molecules;

C. Each vibration mode is characterized by one frequency and unknown amplitudes for each molecule. By introducing this solution into the system of differential equations, a system of eleven homogenous algebraic equations in the eigen amplitudes of mass particles oscillatory motion with coefficients depending on the unknown/assumed frequency;
D. Obtain frequency equations. There is a system of eleven homogeneous algebraic equations which admits the non-zero solution only if the determinant of this system is equal to zero. Since the system contains 11 unknown amplitudes and one unknown frequency, we obtain a polynomial equation of the eleventh order with respect to the square of frequency. This equation has 11 roots each of which corresponds to the square of one eigen circular frequency
E. Determination of the system of determinants and characteristic functions and the characteristic numbers or the frequency equation of the system oscillatory behaviour;
F. Determination of a set of eleven eigen circular frequencies as roots of the frequency equation;
G. Determination of cofactors which correspond to each of the eleven eigen frequencies and the corresponding eigen amplitudes which correspond to each of the eleven eigen frequencies and eigen modes;
H. Analytical expressions for each of the eleven mass particles, eleventh frequency oscillatory displacements in the chain direction and the corresponding integral constants for the corresponding particular and defined initial conditions;
I. Numerical multi-parametric analysis of the system dynamics for certain initial conditions.

When a sperm cell periodically impacts against a knot molecule, it gives an initial velocity to the knot molecule perturbing its state. This generates the 11-modes oscillatory motion of chain molecules. According to the present model [2, 45], the resultant displacements of each knot molecule in the circular $u(t)$ and meridian $v(t)$ directions are obtained by summing the effect of three sperm cells on the neighbouring knot molecules. It is possible to change velocity and sperm impact angle for each observed spermatozoon.

Equations for displacement in the circular and meridian directions of the first knot molecule are presented by formulas:

$$u_1(t) = u_{3,N_1}(t) + u_{9,N_2}(t) \tag{2}$$

$$v_1(t) = v_{3,N_1}(t) + v_{9,N_4}(t) \tag{3}$$

where: N_1–N_4 denotes positions of the knot molecules in the corresponding chain in representative part of the spherical net; indexes 3 and 9 denote order of molecules in the chain (see Fig. 1E). We introduced this notation so that different initial conditions could be applied to each knot molecule.

Expressions for the second, third, and fourth knot molecule have the same form - summing the effect of the neighbouring knot molecules. These formulas are generated from the model.

Displacements in the circular and meridian directions caused by an impact of the sperm cell at the first knot molecule at initial moment, giving initial velocity to knot molecule, are calculated according to the formulas:

$$u_{3,N_1}(t) = V_1 \sum_{s=1}^{11} K_{3s} C_{1s} \cos(\omega_s t + \alpha_{1s}) \quad (4)$$

$$v_{3,N_1}(t) = V_1 \sum_{s=1}^{11} K_{3s} D_{1s} \cos(\omega_s t + \beta_{1s}) \quad (5)$$

$$u_{9,N_1}(t) = V_1 \sum_{s=1}^{11} K_{9s} C_{1s} \cos(\omega_s t + \alpha_{1s}) \quad (6)$$

$$v_{9,N_1}(t) = V_1 \sum_{s=1}^{11} K_{9s} D_{1s} \cos(\omega_s t + \beta_{1s}) \quad (7)$$

where V_1 is ratio between the sperm effective velocity (velocity of the hyperactivated or progressive mouse spermatozoa, taken from [29], and the velocity of the hyperactivated sperms previously set as $V_{hp} = 171.1$ μm/s. C_{1s} and D_{1s} are integral constants determined in step (E), from initial conditions (when the velocity is transferred only to the third or ninth molecule in the chain in the equilibrium state and all other molecules did not have any velocity at the initial moment after the spermatozoa's impact), α_s and β_s are the phase and integral constants determined from initial conditions, ω_s $s = 1, 2, 3\ldots, 11$ is the corresponding eigen circular frequency determined in steps (D) and (F), K_{3s} and K_{9s} are cofactors of the system determinant defining the characteristic (frequency) system equation for the corresponding value of s-th circular frequency determined in steps (G) and (E). The above equations are derived from the theory of oscillations of chain systems [53, 55]. Similar equations for displacements of the corresponding knot molecules caused by the second, third, and fourth sperm cells can be derived. The integral constants C_s and D_s are determined from initial conditions as follows:

$$D_s = \frac{V}{L} |\Delta_s| \cos(\gamma) \sin(\theta) \quad (8)$$

$$C_s = \frac{V}{L} |\Delta_s| \cos(\gamma) \cos(\theta) \quad (9)$$

where γ is an angle between direction of spermatozoa's movements and the tangential plane, and θ is an angle between direction of spermatozoa's movements and meridian direction of the tangential plane, V is velocity that spermatozoa transfer to the knot molecule at the moment of impact, L is determinant of the system and is specific determinant.

System determinant s was determined in step (C).

Eigen circular frequencies ω_s, s = 1, 2, 3, …, 11, were determined using the appropriate software and the graphical method to represent the characteristic function for each frequency equation by the formula: $\omega_s(x_s) = \sqrt{x_s}$, $x_s = 1, 2, 3, \ldots 11$, where x_s is reduced chain eigen characteristic number (determined in steps (D) and (E) to a no dimensional value of square of eigen circular frequency, read from the graphic. 10^6 is factor of correction of roots obtained from the characteristic frequency equations. We use the correction factor for two reasons: to harmonize the unit measures and because we multiplied masses of molecules in MatCad by 10^{12}.

The system of homogenous algebraic equations in terms of eigen amplitudes of mass particles oscillatory motion was made by means of the matrix of inertia and matrix of coefficient of elasticity in step (C). The system of differential equations of chains dynamics with 11 material particles motion for each chain was defined in step A according to the literature as well as by assuming their solutions [1].

By using appropriate software and graphical method for presenting the characteristic function from frequency equations (the 11 frequencies equations) of the system, 11 zeros (roots) of the system characteristic function were determined. Eleven zeros (roots) of the system characteristic function are the squares of the corresponding eigen circular frequencies of the representative part of the discrete spherical surface net model. All eleven material particles in the chain oscillate at these eleven circular frequencies. These frequencies are characteristic of the system and its structure and do not depend on initial conditions. They depend only on molecular masses and rigidity of the spring elements in the system.

The analysis were done for certain favourable angles γ and θ ($\pi/4$ and $\pi/4$, $\pi/6$ and $\pi/4$ and $\pi/12$ and $\pi/4$ respectively) established in studies of [2] and for favourable angle of $5/12\pi$ and $\pi/4$.

3 Results

Knot molecules of the ZP net representative model excited by impacting spermatozoa can oscillate in three directions: there will be eleven natural frequencies in the meridian and circular directions and two forced vibrations in the radial direction. Component displacements of knot molecules in circular and meridian directions are obtained for each combination of impact angles. Resultant displacements of knot molecules are presented on the same graph in the form of generalized Lissajous curves obtained by composing two orthogonal eleven frequency component vibrations in the plane tangential to the oscillatory spherical net model of mZP.

The resultant motion of the knot molecules in the tangential plane to the spherical net are obtained for the initial moment as perturbation of the equilibrium state of the representative part of the oscillatory spherical net model of mZP.

Knot molecule trajectories obtained for the case of four hyperactivated spermatozoids $V_1 = V_1 = V_2 = V_3 = V_4 = V_{hp}$. impacting against four neighbouring knot molecules under different favorable angles γ and θ ($\pi/4$ and $\pi/4$, $\pi/6$ and $\pi/4$; $\pi/12$ and $\pi/4$ and $5/12\pi$ and $\pi/4$ respectively) established in studies of [2] are shown in Fig. 2. It can be seen that all trajectories are straight lines. This case corresponds to the full synchronization of the component displacements of the knot molecules in the circular and meridian directions.

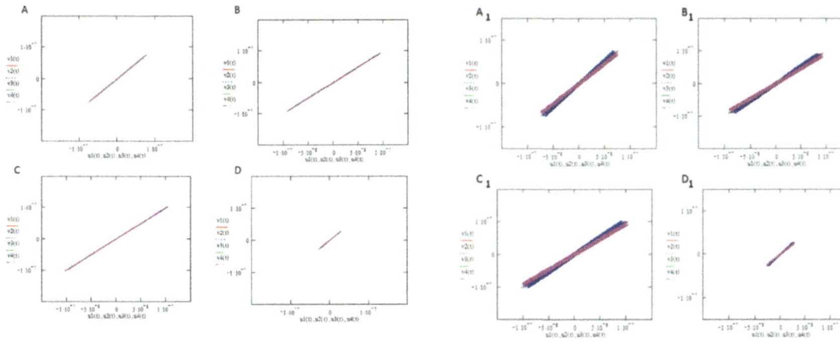

Fig. 2. Resultant trajectories for all four knot mass particles in the corresponding tangent plane to the spherical net when impact angles γ and θ: A and A1. $\gamma_1 = \gamma_2 = \gamma_3 = \gamma_4 = \pi/4$ and $\theta_1 = \theta_2 = \theta_3 = \theta_4 = \pi/4$. B and B1. $\gamma_1 = \gamma_2 = \gamma_3 = \gamma_4 = \pi/6$ and $\theta_1 = \theta_2 = \theta_3 = \theta_4 = \pi/4$. C and C1 $\gamma_1 = \gamma_2 = \gamma_3 = \gamma_4 = \pi/12$ and $\theta_1 = \theta_2 = \theta_3 = \theta_4 = \pi/4$. D and D1. $\gamma_1 = \gamma_2 = \gamma_3 = \gamma_4 = 5\pi/12$ and $\theta_1 = \theta_2 = \theta_3 = \theta_4 = \pi/4$. When four hyperactivated spermatozoa at the initial moment, impact on the knot molecules with velocities $V_{hp} = 171.1$ µm/s (A, B, C and D) and when all three spermatozoa at the initial moment, impact on the knot molecules have velocity of $V_{hp} = 171.1$ µm/s and one (impacts first knot molecule) V1 has velocity of $V_p = 146.9$ µm/s (A1, B1, C1 and D1).

Trajectories of knot molecules computed for the case of four progressive spermatozoids V1 = V2 = V3 = V4 = Vp impacting against the ZP membrane under different favourable angles γ and θ ($\pi/4$ and $\pi/4$, $\pi/6$ and $\pi/4$ and $\pi/12$ and $\pi/4$, and $5/12\pi$ and $\pi/4$ respectively) are straight lines similar to that shown in Fig. 2(A, B, C, D). However, amplitudes and inclination of trajectories for different sperm impact angles are slightly different from the previous case.

If three hyperactivated and one progressive spermatozoa in position V1 (see Fig. 1c) impact against knot molecules under favourable angles, the resultant motion of the knot molecules in the form of Lissajous curves are shown in Fig. 2A1–D1. Trajectories at positions V1 and V3 are straight lines while resultant motions at positions V2 and V4 are stochastic-like.

Figure 3(B, C, D, E) shows the trajectories obtained for the case with two hyperactivated and two progressive spermatozoa arranged in the diagonal up position (A). For all combinations of impact angles, trajectories are straight lines, only amplitudes are different. Again, full synchronization of knot molecules motion is achieved.

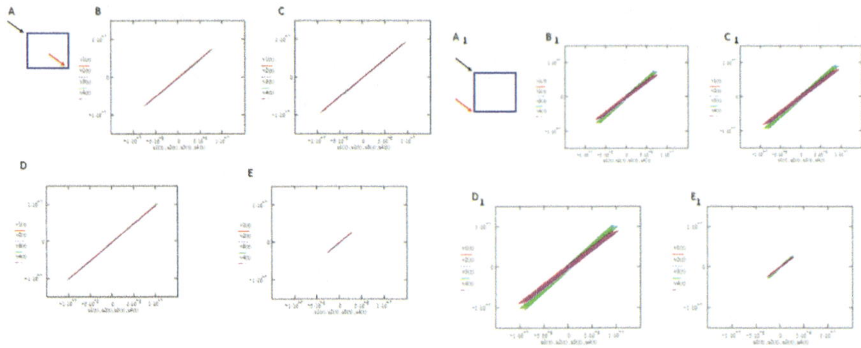

Fig. 3. Resultant trajectories for all four knot mass particles in the corresponding tangent plane to the spherical net when at the initial moment two impacting spermatozoa are hyperactivated and two have progressive velocities. Spermatozoa having the same velocities are in diagonal up position arrangement. A, B, C, D, E. And when Spermatozoa having the same velocities impact with knot molecules in the same chain A1,- E1, and impact angles are B and B1 $\gamma_1 = \gamma_2 = \gamma_3 = \gamma_4 = \pi/4$ and $\theta_1 = \theta_2 = \theta_3 = \theta_4 = \pi/4$. C and C1 $\gamma_1 = \gamma_2 = \gamma_3 = \gamma_4 = \pi/6$ and $\theta_1 = \theta_2 = \theta_3 = \theta_4 = \pi/4$. D and D1 $\gamma_1 = \gamma_2 = \gamma_3 = \gamma_4 = \pi/12$ and $\theta_1 = \theta_2 = \theta_3 = \theta_4 = \pi/4$. E and E1. $\gamma_1 = \gamma_2 = \gamma_3 = \gamma_4 = 5\pi/12$ and $\theta_1 = \theta_2 = \theta_3 = \theta_4 = \pi/4$.

The case of two hyperactivated ($V_1 = V_4 = V_{hp}$) and two progressive spermatozoa impacting the ZP net on the same chain is illustrated in Fig. 4A_1–E_1. Amplitude of Lissajous curves is smaller for the sperm impact angles $\gamma_1 = \gamma_2 = \gamma_3 = \gamma_4 = 5/12\pi$ and $\theta_1 = \theta_2 = \theta_3 = \theta_4 = \pi/4$ (see Fig. 3E and 3E_1). Knot molecule resultant trajectories in the tangent plane have stochastic-like form that is a less favourable state than when two hyperactivated and two progressive spermatozoa are arranged in the diagonal up position (see Fig. 3B–E.). This result indicates that specific arrangement of spermatozoa having effective velocities on ZP surface may generate different oscillatory states of ZP.

The resultant trajectories of knot molecules obtained for the case of one hyperactivated (for example, at node 1) and three progressive spermatozoa are shown in Fig. 4. It can be seen that trajectories of knot molecules 1 and 3 are straight lines while Lissajous curves for positions 2 and 4 have a stochastic-like shape.

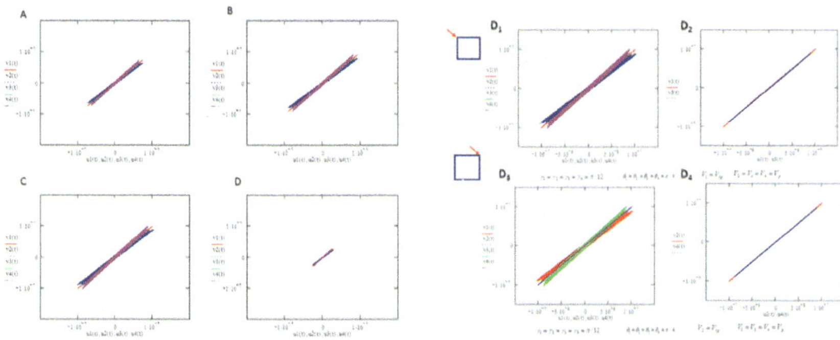

Fig. 4. Resultant trajectories for all 4 knot mass particles in the corresponding tangent plane to spherical net, at initial moment, for one hyperactivated sperm cell $V_{hp} = 171.1$ μm/s impacting ZP at nodal position 1 and three progressive $V_p = 146.9$ μm/s sperm cells impacting the other knot molecules under different favourable angles: **A.** $\gamma_1 = \gamma_2 = \gamma_3 = \gamma_4 = \pi/4$ and $\theta_1 = \theta_2 = \theta_3 = \theta_4 = \pi/4$. **B.** $\gamma_1 = \gamma_2 = \gamma_3 = \gamma_4 = \pi/6$ and $\theta_1 = \theta_2 = \theta_3 = \theta_4 = \pi/4$. **C.** $\gamma_1 = \gamma_2 = \gamma_3 = \gamma_4 = \pi/12$ and $\theta_1 = \theta_2 = \theta_3 = \theta_4 = \pi/4$. **D.** $\gamma_1 = \gamma_2 = \gamma_3 = \gamma_4 = 5\pi/12$ and $\theta_1 = \theta_2 = \theta_3 = \theta_4 = \pi/4$. For the case D: when hyperactivated sperm cell impacts knot molecule in the first position D_1 and D_2. when hyperactivated sperm cell impacts with the knot molecule in the second position D_3, D_4. D2. Knot molecules in first and third positions when hyperactivated sperm cell impacts knot molecule in the first position. D_4. Knot molecules in second and forth positions when hyperactivated sperm cell impacts with knot molecule in the second position.

In order to see if trajectories are affected by the position of the hyperactivated spermatozoa, their impact position on the ZP net is varied. If hyperactivated spermatozoa affected at position 2, the resultant trajectory of knot molecules in positions 2 and 4 are straight lines while trajectories of positions 1 and 3 are stochastic. The same behaviour is observed for all examined favourable angles (see Fig. 4D_1–D_4).

4 Discussion and Conclusions

The discrete spherical net model of mouse ZP was used in this study to carry out qualitative analyses of oscillatory behaviour of ZP knot molecules. By perturbing the initial equilibrium state with pulses representing the impact of sperm cells against the ZP membrane under certain angle, the following facts were observed:

- Component displacements in the meridian and circular directions of the knot molecules represent multi-frequency (eleventh-frequency) oscillations;
- The resultant trajectories of ZP knot molecules in the tangent plane to the sphere net are described by generalized Lissajous curves that could be of straight lines, periodic, non-periodic, or stochastic like curves. This results from summing two orthogonal multifrequency (11 modes) vibratory components;

Under physiological conditions, slow spermatozoa would never reach the oocyte, but in IVF conditions they could contribute to creation of a favourable oscillatory state of mZP.

For the cases when a symmetric distribution of spermatozoa having the same velocity exists, the resulting trajectories of the knot molecules in the plane tangential to the spherical surface have the form of a straight line, i.e. they move synchronously, only amplitudes of these movements are different depending upon the angle of impact of the spermatozoa action, Figs. 2 and 3(A, B, C, D). Larger amplitude of movement of molecules will, theoretically, cause larger deformations of the net and thus create conditions for penetration of the spermatozoa. Under specific initial conditions, it is possible that only individual knot molecules in the model move along a straight line in the plane tangential to the sphere surface, while the others do not, Figs. 2(A_1–D_1), 3 (B_1–E_1) and 4(D_1–D_4). Only some of the possible initial conditions are presented in this work. In a real biological system there are a large number of different combinations of initial conditions which could produce favourable oscillatory states of ZP which indicates that the process of fertilization could be treated as a stochastic process. On the other hand, the possibility that some knot molecules oscillate synchronously while the others do not, indicates that the fertilization phenomenon could have a local character, indicating that the position of the initial penetration of spermatozoa through ZP could be determined by local initial conditions which are of stochastic character while the structure and composition of the ZP surface could be assuming as uniform.

This numerical experiment opens new questions:

- Is, in reality, sperm velocity related to the ZP sperm impact angle?

It is known that capacitated spermatozoa are capable of attaching to ZP [16]. The oscillatory pattern is different for healthy spermatozoa and spermatozoa having morphological defects [36]. The transversal force generated by a spermatozoid is higher than the propulsive force [38, 56] indicating that kinetic parameters of sperm swimming velocity may be related to sperm impact angle. This indicates that amplitude and pattern of sperm velocity could affect the sperm impact angle.

- What is the ratio of hyperactivated sperm necessary to achieve favourable oscillatory state of mZP for successful fertilisation?

According to [29] about 22% are hyperactivated spermatozoa. It is possible to assume that the 22% rate of hyperactivated spermatozoa is consistent with the findings of the vibration analysis for the case with one hyperactivated spermatozoa (Fig. 4).

- Do sperm cells with effective velocity have stochastic like arrangements on the mZP surface in real physiological conditions?

According to sperm velocity parameters for specific sperm types [29] and the results presented in this paper, it is possible that progressive and hyperactivated sperm interacts with ZP under more favourable angles than slow or weakly motile sperm in '*in vitro*' conditions. If it is so, then weakly motile sperm could never generate Lissajous curves of knot molecules in the form of a straight line. Lisssajous curves will then be in a more chaotic-like form and with low amplitude, according to this oscillatory model. It could be that in reality specific arrangements and distribution of spermatozoa having different velocities on the mZP surface contribute to achieving the oscillatory states of ZP favourable for sperm penetration. To answer these questions, further analyses should be done.

Lefievre et al. [57] argue that the value of traditional semen parameters like concentration, motility, and morphology are not sufficient in the diagnosis and prognosis of male infertility [57] and that there is no prognostic parameter and no adequate treatment for male sub/(in) fertility. According to our oscillatory model, we are free to suggest a new type of treatment for male sub fertility: we believe that adding adequate mechanical stimulation in the medium with healthy oocyte and adequate number of spermatozoa could improve fertilization. Mechanical simulation could be in the form of a sound. The basic idea is that adding the mechanical stimulation with specific parameters (intensity and set of oscillatory frequencies) could contribute to achieving successful fertilization through creation of resonance in the oscillations between oocyte and sperm. This method could be useful in subfertility cases (when there are not enough spermatozoa with normal function and morphology to achieve fertilization in physiological conditions). Since spermatozoa have a high degree of plasticity and adaptability in their responses to the events which signal ovulation [57], we believe that they could respond to the external mechanical stimulation. Adding the external mechanical stimulation could influence the process of sperm hyperactivation. We suppose that the parameters of external mechanical stimulation for achieving successful fertilization will be species specific.

4.1 The Limitations of the Model

The present analysis had some limitations. In the first place, since the oscillatory behaviour of a finite number of molecules in the representative part of the mZP spherical net model (eleven) was analysed, a finite number of eigen circular frequencies was obtained. In a real ZP there are numerous ZP molecules, consequently numerous eigen circular frequencies. Non-linear properties of real ZP and limitations of the used analytical method of solving a system of differential non-linear equations require using numerical methods and approximations. Real ZP is a very complex oscillatory system and very selective regarding sperm penetration. This ZP selectiveness provides constant quantity of genetic material in each generation. One may speculate that in real biological conditions, each spermatozoon that impacts with the ZP contributes in a specific manner to the achievement of oscillatory state/s favourable for successful fertilization.

Furthermore, our model is one-layer model. Real ZP is a multi-layer structure with distinguishable inner and outer layers of different mechanical properties [4]. Our next step in developing mZP biomechanical model that resembles more a realistic mZP will be developing a system of a complex hybrid system dynamics which takes into account the mechanical, physico-chemical, electro- and thermal coupled fields. With this approach, it will be possible to create a spherical, multi-layered, discrete continuum model that will be more suitable for investigating the mechanisms of spermatozoa penetration through entire thickness.

Acknowledgements. Authors would like to thank to Prof. Katica (Stevanović) Hedrih from the Mathematical Institute of SANU, Belgrade, Serbia for valuable consultation and suggestions. This work was supported by the Ministry of Education, Sciences and Technology Development of the Republic of Serbia through Mathematical Institute SANU, Belgrade Grant ON174001, as well as Project III No. 41006 through Faculty of Mechanical Engineering, University of Belgrade.

References

1. Hedrih, A., Lazarevic, M., Mitrovic-Jovanovic, A.: Fertilization as a biomechanical oscillatory phenomenon in mammals. In: Maksimović, S., Igić, T., Trišović, N. (eds.) Proceedings of 4th International Congress of Serbian Society of Mechanics, Vrnjačka Banja, Serbia, 4–7 June 2013, pp. 579–584. Serbian Society of Mechanics, Belgrade (2013) (Beograd: Beotele Prom), D-01. ISBN 978-86-909973-5-0
2. Hedrih, A., Lazarevic, M., Mitrovic-Jovanovic, A.: Influence of sperm impact angle on successful fertilization through mZP oscillatory spherical net model. Comp. Biol. Med. **59**, 19–29 (2015)
3. Sato, K., Kamada, S., Minami, K.: Development of microstretching device to evaluate cell membrane strain field around sensing point of mechanical stimuli. Int. J. Mech. Sci. **52**, 251–256 (2010)
4. Martinova, Y., Petrov, M., Mollova, M., Rashev, R., Ivanova, M.: Ultrastructural study of cat zona pellucida during oocyte maturation and fertilization. Anim. Rep. Sci. **108**, 425–434 (2008)
5. Khalilian, M., Navidbakhsh, M., Valojerdi, M.R., Chizari Mand Yazdi, P.E.: Estimating Young's modulus of zona pellucida by micropipette aspiration in combination with theoretical models of ovum. J. Roy. Soc. Inter. **7**, 687–694 (2010)
6. Sun, Y., Wan, K.T., Roberts, K.P., Bischof, J.C., Nelson, B.J.: Mechanical property characterization of mouse zona pellucida. IEEE Tran Nanobiosci. **2**, 279–286 (2003)
7. Miki, K., Clapham, D.E.: Rheotaxis guides mammalian sperm. Curr. Biol. **23**, 443–452 (2013)
8. Flesch, F.M., Gadella, M.B.: Dynamics of the mammalian sperm plasma membrane in the process of fertilization. Biochim. Biophys. Acta **1469**, 197–235 (2000)
9. Eisenbach, M.: Mammalian sperm chemotaxis and its association with capacitation. Dev. Gen. **25**, 87–94 (1999)
10. Pizzari, T.: Of mice and sperm. PNAS **103**(41), 14983–14984 (2006)
11. Gomendio, M., Martin-Coello, J., Crespo, C., Magana, C., Roldan, E.R.S.: Sperm competition enhances functional capacity of mammalian spermatozoa. PNAS **103**, 15113–15117 (2006)
12. Centola, M.G., Blanchard, A., Demick, J., Li, S., Eisenberg, L.M.: Decline in sperm count and motility in young adult men from 2003 to 2013: observations from a U.S. Andrology **4** (2), 270–276 (2016). https://doi.org/10.1111/andr.12149
13. Katebi, M., Movahedin, M., Abdolvahabi, M.A., Akbari, M., Abolhassani, F., Sobhani, A., Aoki, F.: Changes in motility parameters of mouse spermatozoa in response to different doses of progesterone during course of hyperactivation. Iran. Biomed. J. **9**, 73–79 (2005)
14. Kinukawa, M., Nagata, M., Aoki, F.: Changes in flagellar bending during the course of hyperactivation in hamster spermatozoa. Reproduction **125**, 43–51 (2003)
15. Ho, H.C., Suarez, S.S.: Hyperactivation of mammalian spermatozoa: function and regulation. Reproduction **122**, 519–526 (2001)

16. Gadella, B.M.: Dynamic regulation of sperm interactions with the zona pellucida prior to and after fertilisation. Rep. Fert. Dev. **25**, 26–37 (2012)
17. Malo, A.F., Gomendio, M., Garde, J., Lang-Lenton, B., Soler, A.J., Roldan, E.R.: Sperm design and sperm function. Biol. Lett. **2**, 246–249 (2006)
18. Wolf, P., Feneux, D., Ducot, B., Rodrigues, D., Jouannet, P.: Influence of sperm movement parameters on human sperm\oolemma fusion. J. Rep. Fert. **105**, 185–192 (1995)
19. Dacheux, J.L., Dacheux, F.: New insights into epididymal function in relation to sperm maturation. Reproduction **147**, R27–R42 (2014)
20. Gomendio, M., Roldan, E.: Sperm competition influences sperm size in ammals. Proc. R. Soc. B Biol. Sci. **243**(1308), 181–185 (1991). https://doi.org/10.1098/rspb.1991.0029
21. Montoto, L.G., Sanchez, M.V., Tourmente, M., Martın-Coello, J., Luque-Larena, J.J., Gomendio, M., Roldan, E.R.S.: Sperm competition differentially affects swimming velocity and size of spermatozoa from closely related muroid rodents: head first. Reproduction **142**, 819–830 (2011)
22. Tourmente, M., Montserrat, G., Eduardo, R.R.S.: Sperm competition and the evolution of sperm design in mammals. BMC Evol. Biol. **11**, 12 (2011). https://doi.org/10.1186/1471-2148-11-12
23. Anderson, M.J., Dixson, A.F.: Sperm competition, Motility and the midpiece in primates. Nat. Brief Commun. **416**, 496 (2002)
24. Amaral, A., Lourenco¸ B., Marques, M., Ramalho-Santos, J.: Mitochondria functionality and sperm quality. Reproduction **146**, R163–R174 (2013)
25. Firman, R.C., Simmons, L.W.: Sperm midpiece length predicts sperm swimming velocity in house mice. Biol. Lett. **6**, 513–516 (2010)
26. de Boer, P., de Vries, M., Ramos, L.: A mutation study of sperm head shape and motility in the mouse: lessons for the clinic. Andrology **3**, 174–202 (2015)
27. Subramani, E., Basu, H., Thangaraju, S., Dandekar, S., Mathur, D., Chaudhury, K.: Rotational dynamics of optically trapped human spermatozoa. Sci. World. J. Article ID 154367 (2014). 7 p. http://dx.doi.org/10.1155/2014/154367
28. Mitrovic, A., Brkic, P., Jovanovic, T.: The effects of hyperbaric oxygen treatment on vigility of spermatozoids: preliminary report. Acta Phys. Hung. **98**, 84–89 (2011)
29. Goodson, S.G., Zhang, Z., Tsuruta, J.K., Wang, W., O'Brien, D.A.: Classification of mouse sperm motility patterns using an automated multiclass support vector machines. Model. Biol. Reprod. **84**, 1207–1215 (2011)
30. Anaya, G.M.C., Calle, F., Perez, C.J., Martın-Hidalgo, D., Fallola, C., Bragado, M.J., Garcıa-Marın, L.J., Oropesa, A.L.: A new Bayesian network-based approach to the analysis of sperm motility: application in the study of tench (Tincatinca) semen. Andrology **3**, 956–966 (2015)
31. Gefen, A.: The relationship between sperm velocity and pressures applied to the zona pellucida during early sperm–oocyte penetration. J. Biomech. Eng. **132**, 124501–124501-4 (2010)
32. Kozlovsky, P., Gefen, A.: Sperm penetration to the zona pellucida of an oocyte: a computational model incorporating acrosome reaction. Comp. Meth. Biomech. Biomed. Eng. **16**(10), 7 p. (2013)
33. Clark, G.F.: The molecular basis of mouse sperm–zona pellucida binding: a still unresolved issue in developmental biology. Reproduction **142**, 377–381 (2011)
34. Brewis, I., Wong, C.-H.: Gamete recognition: sperm proteins that interact with the egg zona pellucida. Rev. Reprod. **4**, 135–142 (1999)
35. Menzel, A., Kuhl, E.: Frontiers in growth and remodeling. Mech. Res. Commun. **42**, 1–14 (2012)
36. Woolley, D.M.: Motility of spermatozoa at surfaces. Reproduction **126**, 259–270 (2003)

37. Hedrih, A.: Transition in oscillatory behaviour in mouse oocyte and mouse embryo through oscillatory spherical net model of mouse Zona Pellucida. In: Awrejcewicz, J. (ed.) Applied Non-Linear Dynamical Systems, Springer Proceedings in Mathematics & Statistics, vol. 93 pp. 295–303. Springer, Cham (2014). ISBN 978-3-319-08265-3. https://doi.org/10.1007/978-3-319-08266-0_21
38. Hsiao, W.W., Liao, H.S., Lin, H.H., Ding, R.F., Huang, K.Y., Chang, C.S.: Motility measurement of a mouse sperm by atomic force microscopy. Anal. Sci. **29**, 3–8 (2013)
39. Green, D.: Three-dimensional structure of the zona pellucida. Rev Reprod. **2**, 147–156 (1997)
40. Wassarman, P.M., Litscher, E.S.: Influence of the zona pellucida of the mouse egg on folliculogenesis and fertility. Int. J. Dev. Biol. **56**, 833–839 (2012)
41. Huang, H.L., Lv, C., Zhao, Y.C., Li, W., He, X.M., Li, P., Sha, A.-G., Tian, X., Papasian, C. J., Deng, H.W., Lu, G.-X., Xiao, H.-M.: Mutant ZP1 in familial infertility. N. Engl. J. Med. **370**, 1220–1226 (2014)
42. Murayama, Y., Mizuno, J., Kamakura, H., Fueta, Y., Nakamura, H., Akaishi, K., Anzai, K., Watanabe, A., Inui, H., Omata, S.: Mouse zona pellucida dynamically changes its elasticity during oocyte maturation, fertilization and early embryo development. Hum. Cell **19**, 119–125 (2006)
43. Hedrih, A.: Modeling oscillations of zona pelucida before and after fertilization. In: Young Scientist Prize Paperawarded at the 7th European Nonlinear Dynamics Conference held in Rome, Italy, July 2011, EUROMECH Newsletter. European Mechanics Society, vol. 40, pp. 6–14 (2011). http://www.euromech.org/prizes/presentations. http://www.mi.sanu.ac.rs/projects/NL40-Euromech.pdf
44. Hedrih, A.: Frequency analysis of knot mass particles in oscillatory spherical net model of mouse zona pellucida. Lecture Session, Short Paper. In Abstract book of 23rd International Congress of Theoretical and Applied Mechanics, (IUTAM ICTAM Beijing 2012), SM01-049 Biomechanics and Biomaterials, 19–24 August 2012, Beijing, China, p. 209. ISBN 978-988-16022-3-7
45. Hedrih, A., (Stevanovic) Hedrih, K., Bugarski, B.: Oscillatory spherical net model of mouse zona pellucida. J. Appl. Math. Bioinform. **4**, 225–268 (2013)
46. Hedrih, A., Lazarevic, M., Mitrovic-Jovanovic, A.: Parametric frequency analysis of oscillatory behaviour of mouse zona pellucida spherical net model: case successful and unsuccessful fertilization. Proc. Appl. Math. Mech. PAMM **13**(1), 53–54. https://doi.org/10.1002/pamm.201310022. http://onlinelibrary.wiley.com/doi/10.1002/pamm.v13.1/issuetoc. In: 84th Annual Meeting of the International Association of Applied Mathematics and Mechanics- GAMM 2013 (Gesellschaft für Angewandte Mathematik und Mechanik), Novi Sad, Serbia, 18–22 March 2013. ISSN 1617-7061
47. Hedrih (Stevanovic), K.: Discrete continuum method, computational mechanics. In: WCCM VI in Conjunction with APCOM 04, 5–10 September 2004, Beijing, China, pp. 1–11. CD. IACAM International Association for Computational Mechanics (2004). www.iacm.info
48. Hedrih (Stevanovic), K.: Modes of the homogeneous chain dynamics. Sig. Process. **86**, 2678–2702 (2006)
49. Hedrih (Stevanović), K.: Analytical mechanics of fractional order discrete system vibrations. Adv. Nonlinear Sci. JANN **3**, 101–148 (2011). ISSN 978-86-905633-3-3
50. Hedrih (Stevanović), K.: Advances in classical and analytical mechanics: a reviews of author's results. Theoret. Appl. Mech. **40** S1, 293–383 (2012). Special Issue. https://doi.org/10.2298/tam12s1293h

51. Chen, S., Zhang, Y.-T.: Krylov implicit integration factor methods for spatial discretization on high dimensional unstructured meshes: Application to discontinuous Galerkin methods. J. Commun. Phys. **230**, 4336–4352 (2011)
52. Ramirez-Torres, A., Rodriguez-Ramos, R., Gluge, R., Bravo-Castillero, J., Guinovart-Diaz, R., Rodriguez-Sanchez, R.: Biomechanic approach of a growing tumor. Mech. Res. Commun. **51**, 32–38 (2013)
53. Hedrih (Stevanović), R.K., Hedrih, A.N.: Phenomenological mapping and dynamical absorptions in chain systems with multiple degrees of freedom. J. Vib.Conr. **22**, 18–36 (2016)
54. Bahr, G.F., Zeitler, E.: Study of bull spermatozoa. J. Cell Biol. **21**, 175–189 (1964)
55. Rašković, D.P.: Theory of Elasticity. Naučna knjiga, Beograd (1985)
56. Ishijima, S.: Dynamics of flagellar force generated by a hyperactivated spermatozoon. Reproduction **142**, 409–415 (2011)
57. Lefievre, L., Bedu-Addo, K., Conner, S.J., Machado-Oliveira, G.S.M., Chen, Y., Kirkman-Brown, J..C., Afnan, M.A., Publicover, S.J.,. For, W.C.L, Barratt, C.L.R.: Counting sperm does not add up any more: time for a new equation? Reproduction **133**, 675–684 (2007)

Parameter Determination of Metamaterials in Generalized Mechanics as a Result of Computational Homogenization

Bilen Emek Abali(✉) and Hua Yang

Institute of Mechanics, Technische Universität Berlin, Einsteinufer 5, 10587 Berlin, Germany
bilenemek@abali.org

Abstract. As the length scale starts decreasing such that the inner substructure of the material becomes dominant in material response, the well-known theory of elasticity shows inadequacies. As a remedy, generalized mechanics is proposed leading to additional, inner substructure related parameters to be determined. In order to acquire them, for a so-called metamaterial with known substructure and material response in the length scale of the substructure, we present how to apply a computational approach based on the finite element method.

Keywords: Generalized mechanics · Metamaterials · Inverse analysis · Finite element method

1 Introduction

In continuum mechanics, conventional theory of elasticity fails to model structures, where the inner substructure starts affecting the material response. An intuitive explanation for this phenomenon relies on the length scale of the geometry, *macroscale*, ratio with respect to the inner substructure, *microscale*. As this ratio approaches one and the length scales are in the same order, then the effects of the substructure shall be incorporated and we call this structure related material system *metamaterial*. This inner substructure might be simply the molecular structure. For example, in the case of crystalline materials with a lattice type substructure, the grain orientation leads to material anisotropy or change in parameters like the yield stress, these phenomena have been studied among others also in Reuss (1929); Hashin and Shtrikman (1962); Sharo and Kachanov (2000); Lebensohn et al. (2004). Such an inner substructure can be generated by adhering different materials, which is the case in composite materials and "effective" parameters read as a result of a homogenization procedure, see for example Levin (1976); Willis (1977); Kushnevsky et al. (1998); Sburlati et al. (2018). A system with inclusions like a porous material can be

D. A. Indeitsev and A. M. Krivtsov (Eds.): APM 2019, LNME, pp. 22–31, 2020.
https://doi.org/10.1007/978-3-030-49882-5_2

seen as a metamaterial, where the voids affect the material properties at the macroscale, we refer to Eshelby (1957); Mori and Tanaka (1973); Kanaun and Kudryavtseva (1986); Hashin (1991); Nazarenko (1996); Dormieux et al. (2006). Additive manufacturing—as in the case of 3D printing—is another prominent example to build up a metamaterial as applied in Kochmann and Venturini (2013); Placidi et al. (2016); Turco et al. (2017); Solyaev et al. (2018); Ganzosch et al. (2018); Yang et al. (2018). Often it is assumed that the substructure is periodic in a sense that the same cell is repeated for generating the structure at the macroscale. This so-called representative volume element is useful for an analysis of effective parameters. All these approaches are based on the assumption that the material response is modeled with the same phenomenological models at both scales.

By using the homogenization approach as in Pideri and Seppecher (1997); Bigoni and Drugan (2007); Seppecher et al. (2011); Abdoul-Anziz and Seppecher (2018); Mandadapu et al. (2018), we understand that the assumption of having the same material model can lead to inaccurate results such that a higher order theory needs to be incorporated at the macroscale as developed by Eringen and Suhubi (1964); Mindlin (1964); Eringen (1968); Steinmann (1994); Eremeyev et al. (2012); Polizzotto (2013a; 2013b); Ivanova and Vilchevskaya (2016); Abali (2018). Various times it has been observed that a generalized mechanics description is necessary for modeling mechanical response accurately as the thinner or smaller structure starts deviating from classical results as detected in Namazu et al. (2000); Lam et al. (2003); McFarland and Colton (2005); Gruber et al. (2008); Chen et al. (2010); dell'Isola et al. (2019). For a simple beam bending problem, conventional theory of elasticity fails to estimate the experimental results, as a remedy, for example the strain gradient theory in Abali and Müller (2016) is capable of capturing this effect, as applied by Abali et al. (2015), Abali et al. (2017); however, we need to know the additional parameters introduced for incorporating higher order effects.

As the inner substructure and its material response is set, a detailed model of the microscale can be used to determine the additional parameters at the macroscale. Thus, the parameter determination in generalized mechanics is not a new approach, see for example Forest et al. (1999); Pietraszkiewicz and Eremeyev (2009); Giorgio (2016) or also by using the asymptotic analysis in Bensoussan et al. (1978); Hollister and Kikuchi (1992); Chung et al. (2001); Temizer (2012) with an application in Forest et al. (2001); Li (2011); Eremeyev (2016) Barboura and Li (2018); Ganghoffer et al. (2018); Turco (2019). Often a representative volume element has been used, we remark that it is difficult to justify that the higher order theory has to inherit one, see the discussion in Rahali et al. (2015). Thus, we search for a method without implementing a representative volume element at all. In this work we briefly show the second order theory and the additional parameters occurring in this theory. Then we apply the general algorithm proposed by Abali et al. (2019) and define the parameters for a specific geometry.

2 Computational Approach

We strictly follow Abali et al. (2019) and use the equivalence of the stored energy at the microscale,

$$ {}^{\mathrm{m}}w = \frac{1}{2}\,{}^{\mathrm{m}}\varepsilon_{ij}\,{}^{\mathrm{m}}C_{ijkl}\,{}^{\mathrm{m}}\varepsilon_{kl}, \tag{1} $$

to the stored energy at the macroscale,

$$ {}^{\mathrm{M}}w = \frac{1}{2}\,{}^{\mathrm{M}}\varepsilon_{ij}\,{}^{\mathrm{M}}C_{ijkl}\,{}^{\mathrm{M}}\varepsilon_{kl} + {}^{\mathrm{M}}\varepsilon_{ij}G_{ijklm}\,{}^{\mathrm{M}}\varepsilon_{kl,m} + \frac{1}{2}\,{}^{\mathrm{M}}\varepsilon_{ij,k}D_{ijklmn}\,{}^{\mathrm{M}}\varepsilon_{lm,n}, \tag{2} $$

such that we have

$$ \begin{aligned} \int_{\mathcal{B}} {}^{\mathrm{m}}w \,\mathrm{dv} &= \int_{\mathcal{B}} {}^{\mathrm{M}}w \,\mathrm{dv}, \\ \int_{\mathcal{B}} {}^{\mathrm{m}}\varepsilon_{ij}\,{}^{\mathrm{m}}C_{ijkl}\,{}^{\mathrm{m}}\varepsilon_{kl}\,\mathrm{dv} &= {}^{\mathrm{M}}C_{ijkl}\int_{\mathcal{B}} {}^{\mathrm{M}}\varepsilon_{ij}\,{}^{\mathrm{M}}\varepsilon_{kl}\,\mathrm{dv} + 2G_{ijklm}\int_{\mathcal{B}} {}^{\mathrm{M}}\varepsilon_{ij}\,{}^{\mathrm{M}}\varepsilon_{kl,m}\,\mathrm{dv} \\ &\quad + D_{ijklmn}\int_{\mathcal{B}} {}^{\mathrm{M}}\varepsilon_{ij,k}\,{}^{\mathrm{M}}\varepsilon_{lm,n}\,\mathrm{dv}. \end{aligned} \tag{3} $$

Consider that we assume that the macroscale material properties are appropriate for an isotropic and centrosymmetric material

$$ \begin{aligned} {}^{\mathrm{M}}C_{ijkl} &= c_1\delta_{ij}\delta_{kl} + c_2(\delta_{ik}\delta_{jl} + \delta_{il}\delta_{jk}), \\ D_{ijklmn} &= c_3(\delta_{ij}\delta_{kl}\delta_{mn} + \delta_{in}\delta_{jk}\delta_{lm} + \delta_{ij}\delta_{km}\delta_{ln} + \delta_{ik}\delta_{jn}\delta_{lm}) + c_4\delta_{ij}\delta_{kn}\delta_{ml} \\ &\quad + c_5(\delta_{ik}\delta_{jl}\delta_{mn} + \delta_{im}\delta_{jk}\delta_{ln} + \delta_{ik}\delta_{jm}\delta_{ln} + \delta_{il}\delta_{jk}\delta_{mn}) \\ &\quad + c_6(\delta_{il}\delta_{jm}\delta_{kn} + \delta_{im}\delta_{jl}\delta_{kn}) \\ &\quad + c_7(\delta_{il}\delta_{jn}\delta_{mk} + \delta_{im}\delta_{jn}\delta_{lk} + \delta_{in}\delta_{jl}\delta_{km} + \delta_{in}\delta_{jm}\delta_{kl}), \\ G_{ijklm} &= 0\ , \end{aligned} \tag{4} $$

with the unknown material parameters, $\boldsymbol{c} = \{c_1, c_2, c_3, c_4, c_5, c_6, c_7\}$, which we obviously intend to determine. By simply inserting the latter into the energy equivalence and writing in a linear algebra fashion, as an example for one case denoted by the index 1 as follows:

$$ \sum_{\alpha=1}^{7} A_{1\alpha}c_\alpha = R_1, \tag{5} $$

we observe that the coefficient matrix, $\boldsymbol{A}$, as well as the right hand side, R, can be computed

$$
\begin{aligned}
A_{11} =&\delta_{ij}\delta_{kl} \int_{\mathcal{B}} {}^{\mathrm{M}}\varepsilon_{ij} \, {}^{\mathrm{M}}\varepsilon_{kl} \, \mathrm{dv} \\
A_{12} =&(\delta_{ik}\delta_{jl} + \delta_{il}\delta_{jk}) \int_{\mathcal{B}} {}^{\mathrm{M}}\varepsilon_{ij} \, {}^{\mathrm{M}}\varepsilon_{kl} \, \mathrm{dv} \\
A_{13} =&(\delta_{ij}\delta_{kl}\delta_{mn} + \delta_{in}\delta_{jk}\delta_{lm} + \delta_{ij}\delta_{km}\delta_{ln} + \delta_{ik}\delta_{jn}\delta_{lm}) \int_{\mathcal{B}} {}^{\mathrm{M}}\varepsilon_{ij,k} \, {}^{\mathrm{M}}\varepsilon_{lm,n} \, \mathrm{dv} \\
A_{14} =&\delta_{ij}\delta_{kn}\delta_{ml} \int_{\mathcal{B}} {}^{\mathrm{M}}\varepsilon_{ij,k} \, {}^{\mathrm{M}}\varepsilon_{lm,n} \, \mathrm{dv} \\
A_{15} =&(\delta_{ik}\delta_{jl}\delta_{mn} + \delta_{im}\delta_{jk}\delta_{ln} + \delta_{ik}\delta_{jm}\delta_{ln} + \delta_{il}\delta_{jk}\delta_{mn}) \int_{\mathcal{B}} {}^{\mathrm{M}}\varepsilon_{ij,k} \, {}^{\mathrm{M}}\varepsilon_{lm,n} \, \mathrm{dv} \\
A_{16} =&(\delta_{il}\delta_{jm}\delta_{kn} + \delta_{im}\delta_{jl}\delta_{kn}) \int_{\mathcal{B}} {}^{\mathrm{M}}\varepsilon_{ij,k} \, {}^{\mathrm{M}}\varepsilon_{lm,n} \, \mathrm{dv} \\
A_{17} =&(\delta_{il}\delta_{jn}\delta_{mk} + \delta_{im}\delta_{jn}\delta_{lk} + \delta_{in}\delta_{jl}\delta_{km} + \delta_{in}\delta_{jm}\delta_{kl}) \int_{\mathcal{B}} {}^{\mathrm{M}}\varepsilon_{ij,k} \, {}^{\mathrm{M}}\varepsilon_{lm,n} \, \mathrm{dv} \\
R_1 =& \int_{\mathcal{B}} {}^{\mathrm{m}}\varepsilon_{ij} \, {}^{\mathrm{m}}C_{ijkl} \, {}^{\mathrm{m}}\varepsilon_{kl} \, \mathrm{dv},
\end{aligned}
\tag{6}
$$

for a problem with given, ${}^{\mathrm{M}}\varepsilon$, and computed, ${}^{\mathrm{m}}\varepsilon$. By defining 7 distinct cases, the system, $\boldsymbol{Ac} = \boldsymbol{R}$, with $\boldsymbol{A}$ of rank 7 provides a unique determination of unknowns by $\boldsymbol{c} = \boldsymbol{A}^{-1}\boldsymbol{R}$.

These seven cases are the one of the key choices in the approach and we use the following seven cases:

$$
\begin{gathered}
\text{case1}: {}^{\mathrm{M}}\boldsymbol{u} = \Big(\tfrac{y}{2}, \tfrac{x}{2}, 0\Big) \quad \text{case2}: {}^{\mathrm{M}}\boldsymbol{u} = \Big(x, 0, 0\Big) \quad \text{case3}: {}^{\mathrm{M}}\boldsymbol{u} = \Big(-xz, 0, xy\Big) \\
\text{case4}: {}^{\mathrm{M}}\boldsymbol{u} = \Big(xz, 0, -\tfrac{x^2}{2}\Big) \quad \text{case5}: {}^{\mathrm{M}}\boldsymbol{u} = \Big(-yz, 0, xy\Big) \\
\text{case6}: {}^{\mathrm{M}}\boldsymbol{u} = \Big(0, -y, \tfrac{y^2}{2}\Big) \quad \text{case7}: {}^{\mathrm{M}}\boldsymbol{u} = \Big(0, \tfrac{y^2}{2}, 0\Big),
\end{gathered}
\tag{7}
$$

where the only necessary condition seems to be such a choice generating a rank 7 coefficient matrix. It is challenging (if even possible) to suggest experimental designs for constructing this given homogenized displacement on the structure. If we use a linear strain measure,

$$
{}^{\mathrm{M}}\varepsilon_{ij} = \frac{1}{2}\Big(\frac{\partial \, {}^{\mathrm{M}}u_i}{\partial X_j} + \frac{\partial \, {}^{\mathrm{M}}u_j}{\partial X_i}\Big) = {}^{\mathrm{M}}u_{(i,j)}, \tag{8}
$$

we can easily calculate the coefficient matrix for one of the aforementioned cases. For the right hand side, we compute ${}^{\mathrm{m}}\boldsymbol{u}$ for the detailed microscale of the continuum body, $\mathcal{B}$, by applying the boundary conditions acquired from the given ${}^{\mathrm{M}}\boldsymbol{u}$ evaluated on boundaries. Solving ${}^{\mathrm{m}}\boldsymbol{u}$ at the microscale is established by satisfying the weak form:

$$
\text{Form} = \int_{\mathcal{B}} {}^{\mathrm{m}}C_{ijkl} \, {}^{\mathrm{m}}u_{(k,l)} \delta u_{i,j} \, \mathrm{d}V, \tag{9}
$$

with the corresponding test functions, $\delta \boldsymbol{u}$, from the same HILBERTian SOBOLEV space as the unknown, ${}^{\mathrm{m}}\boldsymbol{u}$, known as the GALERKIN method,

$$\hat{\mathcal{V}} = \left\{ {}^{\mathrm{m}}\boldsymbol{u}, \delta\boldsymbol{u} \in [\mathcal{H}^n(\Omega)]^3 : {}^{\mathrm{m}}\boldsymbol{u}, \delta\boldsymbol{u} = \text{given } \forall \boldsymbol{x} \in \partial\mathcal{B} \right\}. \tag{10}$$

The construction is automatized by using open-source programs like Salome, NetGen, and FEniCS (Alnaes et al. 2009; Logg et al. 2012), by using a Python code, we refer to Abali (2017) for a standard introduction of this weak form as well as the whole implementation.

3 Application

A pantographic structure has been studied for several systems, see for example Misra et al. (2018); Turco et al. (2019); dell'Isola et al. (2018); Solyaev et al. (2018); Harrison et al. (2018); Spagnuolo and Andreaus (2019); Greco et al. (2019). We aim at determining effective parameters in a strain gradient theory by applying the procedure from the last section for the pantographic structure as shown in Fig. 1. We emphasize that no representative volume element is used, instead, we simulate only a part of the whole structure as the macroscale displacement is provided as a function applied on this part.

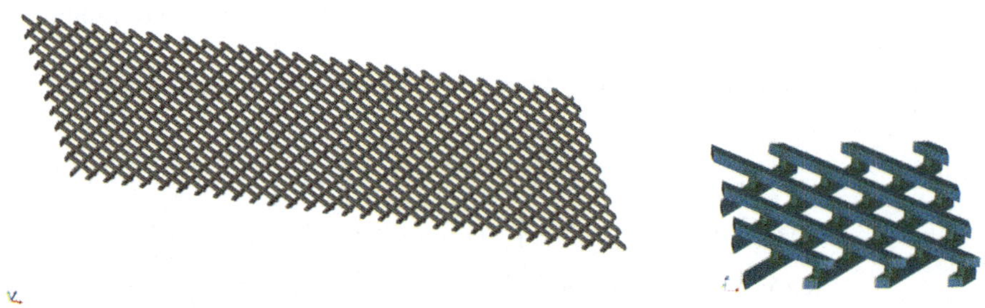

Fig. 1. Pantographic structure CAD model designed in Salome open-source platform. *Left:* the whole structure. *Right:* part of the structure used for the computation with the shown mesh generated by Netgen.

For a 3D printed pantographic structure out of ABS or PP, we may approximate a linear elastic response with YOUNG's modulus of $E = 400 \times 10^6$ Pa and POISSON's ratio of $\nu = 0.3$ leading to the following LAME parameters:

$$\lambda = \frac{E\nu}{(1+\nu)(1-2\nu)}, \quad \mu = \frac{E}{2(1+\nu)}. \tag{11}$$

They are used in the microscale material response

$${}^{\mathrm{m}}C_{ijkl} = \lambda\delta_{ij}\delta_{kl} + \mu\delta_{ik}\delta_{jl} + \mu\delta_{il}\delta_{jk}, \tag{12}$$

which is simply the HOOKE's phenomenological model in isotropic linear elasticity. We emphasize that we use this assumption for clarity and fail to know if the material response of an additively manufactured polymer material is accurately captured by this model. Especially in semi-crystalline materials like PP, fused deposition modeling 3D printers may introduce extrusion orientation dependent anisotropic response. Moreover, the polymer material may behave hyperelastic. Another model is possible for obtaining the right hand side in Eq. (6) in order to increase the accuracy. Herein we use linear elastic model for demonstrating the methodology.

After solving 7 cases subsequently, computing the coefficient matrix, we have determined the 7 material and structure related parameters as follows:

$$\begin{aligned} c_1 &= 231 \times 10^{6}\,\mathrm{Pa} \\ c_2 &= 154 \times 10^{6}\,\mathrm{Pa} \\ c_3 &= 287 \times 10^{-6}\,\mathrm{N} \\ c_4 &= 58 \times 10^{-6}\,\mathrm{N} \\ c_5 &= -264 \times 10^{-6}\,\mathrm{N} \\ c_6 &= -32 \times 10^{-6}\,\mathrm{N} \\ c_7 &= -32 \times 10^{-6}\,\mathrm{N} \end{aligned} \tag{13}$$

4 Discussion and Conclusion

A simple yet elegant computational approach has been applied for obtaining the effective parameters as a result of a homogenization procedure in space in order to reduce the complexity of the structure modeling greatly. As an expense of additional parameters, we aim at incorporating the inner substructure effects by using higher gradients in the displacement. These additional parameters have been obtained by a purely computational methodology under the following assumptions:

- At the microscale, the material model is linear elastic and isotropic.
- At the macroscale, the material model is linear strain gradient elastic and isotropic as well as centrosymmetric.

Both assumptions are difficult to verify or falsify. We use these assumptions in the modeling for simplicity, more sophisticated approaches can be implemented as well, the general methodology remains still valid. The only possible validation for a concrete structure relies on an experimental study, which is left to further research endeavors.

References

Abali, B.E.: Computational Reality, Solving Nonlinear and Coupled Problems in Continuum Mechanics. Advanced Structured Materials, vol. 55. Springer Nature, Singapore (2017)

Abali, B.E.: Revealing the physical insight of a length-scale parameter in metamaterials by exploiting the variational formulation. Continuum Mechanics and Thermodynamics, pp. 1–10 (2018)

Abali, B.E., Müller, W.H.: Numerical solution of generalized mechanics based on a variational formulation. Oberwolfach Rep. Mech. Mater. Mech. Interfaces Evolving Microstr. **17**(1), 9–12 (2016)

Abali, B.E., Müller, W.H., Eremeyev, V.A.: Strain gradient elasticity with geometric nonlinearities and its computational evaluation. Mech. Adv. Mater. Modern Process. **1**(1), 1–11 (2015)

Abali, B.E., Müller, W.H., dell'Isola, F.: Theory and computation of higher gradient elasticity theories based on action principles. Arch. Appl. Mech. **87**(9), 1495–1510 (2017)

Abali, B.E., Yang, H., Papadopoulos, P.: A computational approach for determination of parameters in generalized mechanics. In: Abali, B.E., Altenbach, H., Müller, W.H. (eds.) Higher Gradient Materials and Related Generalized Continua, chap. 1, pp. 1–18. Springer Nature, Singapore (2019)

Abdoul-Anziz, H., Seppecher, P.: Strain gradient and generalized continua obtained by homogenizing frame lattices. Math. Mech. Compl. Syst. **6**(3), 213–250 (2018)

Alnaes, M.S., Logg, A., Mardal, K.A., Skavhaug, O., Langtangen, H.P.: Unified framework for finite element assembly. Int. J. Comput. Sci. Eng. **4**(4), 231–244 (2009)

Barboura, S., Li, J.: Establishment of strain gradient constitutive relations by using asymptotic analysis and the finite element method for complex periodic microstructures. Int. J. Solids Struct. **136**, 60–76 (2018)

Bensoussan, A., Lions, J.L., Papanicolaou, G.: Asymptotic Analysis for Periodic Structures. North-Holland, Amsterdam (1978)

Bigoni, D., Drugan, W.: Analytical derivation of cosserat moduli via homogenization of heterogeneous elastic materials. J. Appl. Mech. **74**(4), 741–753 (2007)

Chen, C., Pei, Y., De Hosson, J.T.M.: Effects of size on the mechanical response of metallic glasses investigated through in situ tem bending and compression experiments. Acta Mater. **58**(1), 189–200 (2010)

Chung, P.W., Tamma, K.K., Namburu, R.R.: Asymptotic expansion homogenization for heterogeneous media: computational issues and applications. Compos. Appl. Sci. Manuf. **32**(9), 1291–1301 (2001)

dell'Isola, F., Seppecher, P., Alibert, J.J., Lekszycki, T., Grygoruk, R., Pawlikowski, M., Steigmann, D., Giorgio, I., Andreaus, U., Turco, E., et al.: Pantographic metamaterials: an example of mathematically driven design and of its technological challenges. Continuum Mechanics and Thermodynamics, pp. 1–34 (2018)

dell'Isola, F., Turco, E., Misra, A., Vangelatos, Z., Grigoropoulos, C., Melissinaki, V., Farsari, M.: Force-displacement relationship in micro-metric pantographs: experiments and numerical simulations. Comptes Rendus Mécanique **347**(5), 397–405 (2019)

Dormieux, L., Kondo, D., Ulm, F.J.: Microporomechanics. Wiley, Chichester (2006)

Eremeyev, V.A.: On effective properties of materials at the nano-and microscales considering surface effects. Acta Mech. **227**(1), 29–42 (2016)

Eremeyev, V.A., Lebedev, L.P., Altenbach, H.: Foundations of Micropolar Mechanics. Springer Science & Business Media, New York (2012)

Eringen, A.: Mechanics of micromorphic continua. In: Kröner, E. (ed.) Mechanics of Generalized Continua, pp. 18–35. Springer, Berlin (1968)

Eringen, A., Suhubi, E.: Nonlinear theory of simple micro-elastic solids. Int. J. Eng. Sci. **2**, 189–203 (1964)

Eshelby, J.D.: The determination of the elastic field of an ellipsoidal inclusion, and related problems. In: Proceedings of the Royal Society of London Series A, Mathematical and Physical Sciences, pp. 376–396 (1957)
Forest, S., Dendievel, R., Canova, G.R.: Estimating the overall properties of heterogeneous cosserat materials. Modell. Simul. Mater. Sci. Eng. **7**(5), 829 (1999)
Forest, S., Pradel, F., Sab, K.: Asymptotic analysis of heterogeneous cosserat media. Int. J. Solids Struct. **38**(26–27), 4585–4608 (2001)
Ganghoffer, J., Goda, I., Novotny, A., Rahouadj, R., Sokolowski, J.: Homogenized couple stress model of optimal auxetic microstructures computed by topology optimization. ZAMM-J. Appl. Math. Mech./Zeitschrift für Angewandte Mathematik und Mechanik **98**(5), 696–717 (2018)
Ganzosch, G., Hoschke, K., Lekszycki, T., Giorgio, I., Turco, E., Müller, W.H.: 3d-measurements of 3d-deformations of pantographic structures. Technische Mechanik **38**(3), 233–245 (2018)
Giorgio, I.: Numerical identification procedure between a micro-Cauchy model and a macro-second gradient model for planar pantographic structures. Zeitschrift für angewandte Mathematik und Physik **67**(4), 95 (2016)
Greco, L., Cuomo, M., Contrafatto, L.: Two new triangular g1-conforming finite elements with cubic edge rotation for the analysis of Kirchhoff plates. Comput. Methods Appl. Mech. Eng. **356**, 354–386 (2019)
Gruber, P.A., Böhm, J., Onuseit, F., Wanner, A., Spolenak, R., Arzt, E.: Size effects on yield strength and strain hardening for ultra-thin Cu films with and without passivation: a study by synchrotron and bulge test techniques. Acta Mater. **56**(10), 2318–2335 (2008)
Harrison, P., Taylor, E., Alsayednoor, J.: Improving the accuracy of the uniaxial bias extension test on engineering fabrics using a simple wrinkle mitigation technique. Compos. A Appl. Sci. Manuf. **108**, 53–61 (2018)
Hashin, Z.: The spherical inclusion with imperfect interface. J. Appl. Mech. **58**(2), 444–449 (1991)
Hashin, Z., Shtrikman, S.: On some variational principles in anisotropic and nonhomogeneous elasticity. J. Mech. Phys. Solids **10**(4), 335–342 (1962)
Hollister, S.J., Kikuchi, N.: A comparison of homogenization and standard mechanics analyses for periodic porous composites. Comput. Mech. **10**(2), 73–95 (1992)
Ivanova, E.A., Vilchevskaya, E.N.: Micropolar continuum in spatial description. Continuum Mech. Thermodyn. **28**(6), 1759–1780 (2016)
Kanaun, S., Kudryavtseva, L.: Spherically layered inclusions in a homogeneous elastic medium. J. Appl. Math. Mech. **50**(4), 483–491 (1986)
Kochmann, D.M., Venturini, G.N.: Homogenized mechanical properties of auxetic composite materials in finite-strain elasticity. Smart Mater. Struct. **22**(8), 084, 004 (2013)
Kushnevsky, V., Morachkovsky, O., Altenbach, H.: Identification of effective properties of particle reinforced composite materials. Comput. Mech. **22**(4), 317–325 (1998)
Lam, D.C., Yang, F., Chong, A., Wang, J., Tong, P.: Experiments and theory in strain gradient elasticity. J. Mech. Phys. Solids **51**(8), 1477–1508 (2003)
Lebensohn, R., Liu, Y., Castaneda, P.P.: On the accuracy of the self-consistent approximation for polycrystals: comparison with full-field numerical simulations. Acta Mater. **52**(18), 5347–5361 (2004)
Levin, V.: Determination of composite material elastic and thermoelastic constants. Mech. Solids **11**(6), 119–126 (1976)
Li, J.: Establishment of strain gradient constitutive relations by homogenization. Comptes Rendus Mécanique **339**(4), 235–244 (2011)

Logg, A., Mardal, K.A., Wells, G.: Automated solution of differential equations by the finite element method: The FEniCS book, vol. 84. Springer Science & Business Media (2012)

Mandadapu, K.K., Abali, B.E., Papadopoulos, P.: On the polar nature and invariance properties of a thermomechanical theory for continuum-on-continuum homogenization. arXiv preprint arXiv:180802540 (2018)

McFarland, A.W., Colton, J.S.: Role of material microstructure in plate stiffness with relevance to microcantilever sensors. J. Micromech. Microeng. **15**(5), 1060–1067 (2005)

Mindlin, R.: Micro-structure in linear elasticity. Arch. Ration. Mech. Anal. **16**(1), 51–78 (1964)

Misra, A., Lekszycki, T., Giorgio, I., Ganzosch, G., Müller, W.H., Dell'Isola, F.: Pantographic metamaterials show atypical poynting effect reversal. Mech. Res. Commun. **89**, 6–10 (2018)

Mori, T., Tanaka, K.: Average stress in matrix and average elastic energy of materials with misfitting inclusions. Acta Metall. **21**(5), 571–574 (1973)

Namazu, T., Isono, Y., Tanaka, T.: Evaluation of size effect on mechanical properties of single crystal silicon by nanoscale bending test using afm. J. Microelectromech. Syst. **9**(4), 450–459 (2000)

Nazarenko, L.: Elastic properties of materials with ellipsoidal pores. Int. Appl. Mech. **32**(1), 46–52 (1996)

Pideri, C., Seppecher, P.: A second gradient material resulting from the homogenization of an heterogeneous linear elastic medium. Continuum Mech. Thermodyn. **9**(5), 241–257 (1997)

Pietraszkiewicz, W., Eremeyev, V.: On natural strain measures of the non-linear micropolar continuum. Int. J. Solids Struct. **46**(3), 774–787 (2009)

Placidi, L., Greco, L., Bucci, S., Turco, E., Rizzi, N.L.: A second gradient formulation for a 2d fabric sheet with inextensible fibres. Zeitschrift für angewandte Mathematik und Physik **67**(5), 114 (2016)

Polizzotto, C.: A second strain gradient elasticity theory with second velocity gradient inertia-part i: Constitutive equations and quasi-static behavior. Int. J. Solids Struct. **50**(24), 3749–3765 (2013a)

Polizzotto, C.: A second strain gradient elasticity theory with second velocity gradient inertia-part ii: Dynamic behavior. Int. J. Solids Struct. **50**(24), 3766–3777 (2013b)

Rahali, Y., Giorgio, I., Ganghoffer, J., dell'Isola, F.: Homogenization à la Piola produces second gradient continuum models for linear pantographic lattices. Int. J. Eng. Sci. **97**, 148–172 (2015)

Reuss, A.: Berechnung der Fließgrenze von Mischkristallen auf grund der Plastizitätsbedingung für Einkristalle. ZAMM-Journal of Applied Mathematics and Mechanics/Zeitschrift für Angewandte Mathematik und Mechanik **9**(1), 49–58 (1929)

Sburlati, R., Cianci, R., Kashtalyan, M.: Hashin's bounds for elastic properties of particle-reinforced composites with graded interphase. Int. J. Solids Struct. **138**, 224–235 (2018)

Seppecher, P., Alibert, J.J., dell'Isola, F.: Linear elastic trusses leading to continua with exotic mechanical interactions. In: Journal of Physics: Conference Series, IOP Publishing, vol. 319, p. 012018 (2011)

Shafiro, B., Kachanov, M.: Anisotropic effective conductivity of materials with non-randomly oriented inclusions of diverse ellipsoidal shapes. J. Appl. Phys. **87**(12), 8561–8569 (2000)

Solyaev, Y., Lurie, S., Ustenko, A.: Numerical modeling of a composite auxetic metamaterials using micro-dilatation theory. Continuum Mechanics and Thermodynamics, pp. 1–9 (2018)

Spagnuolo, M., Andreaus, U.: A targeted review on large deformations of planar elastic beams: extensibility, distributed loads, buckling and post-buckling. Math. Mech. Solids **24**(1), 258–280 (2019)

Steinmann, P.: A micropolar theory of finite deformation and finite rotation multiplicative elastoplasticity. Int. J. Solids Struct. **31**(8), 1063–1084 (1994)

Temizer, I.: On the asymptotic expansion treatment of two-scale finite thermoelasticity. Int. J. Eng. Sci. **53**, 74–84 (2012)

Turco, E.: How the properties of pantographic elementary lattices determine the properties of pantographic metamaterials. In: Abali, B., Altenbach, H., dell'Isola, F., Eremeyev, V., Öchsner, A. (eds.) New Achievements in Continuum Mechanics and Thermodynamics, Advanced Structured Materials, vol. 108, pp. 489–506. Springer, Cham (2019)

Turco, E., Golaszewski, M., Giorgio, I., D'Annibale, F.: Pantographic lattices with non-orthogonal fibres: experiments and their numerical simulations. Compos. B Eng. **118**, 1–14 (2017)

Turco, E., Misra, A., Sarikaya, R., Lekszycki, T.: Quantitative analysis of deformation mechanisms in pantographic substructures: experiments and modeling. Continuum Mech. Thermodyn. **31**(1), 209–223 (2019)

Willis, J.: Bounds and self-consistent estimates for the overall properties of anisotropic composites. J. Mech. Phys. Solids **25**(3), 185–202 (1977)

Yang, H., Ganzosch, G., Giorgio, I., Abali, B.E.: Material characterization and computations of a polymeric metamaterial with a pantographic substructure. Zeitschrift für angewandte Mathematik und Physik **69**(4), 105 (2018)

A Model of Hydraulic Fractured Horizontal Well for Debit Computation of Slanged Gas and Oil

Igor Abramov[1(✉)] and Maksim Simonov[2]

[1] Peter the Great Saint Petersburg Polytechnic University, Polytechnicheskaya, 29, Saint Petersburg 195251, Russia
Igoryabra@gmail.com

[2] Gazpromneft Science and Technology Center, Moika River emb. 75-79 liter D, Saint Petersburg 190000, Russia
Simonov.MV@gazpromneft-ntc.ru

Abstract. This work presents model of liquid rate and bottomhole pressure computation in multistage hydraulic fractured horizontal well. Model is based on a solution of transient fluid flow equation in porous media with finite volume method. 3D formulation of the problem gives an opportunity to model not only axisymmetric example as it is in 2D models. Results are recorded with a second order time and space accuracy. The most innovate thing that this project can be used for is solution of inverse problem – restoring reservoir parameters on a base of known well production.

Keywords: Hydraulic fracturing · Multistage horizontal well hydraulic fracturing · Oil production · Numerical methods · Finite difference method · Alternating direction method · Finite volume method

1 Hydraulic Fracturing

At present, the priority direction of oil reserves growth in world oil production is the development and industrial application of modern integrated methods for increasing oil recovery, which can provide a synergistic effect in the development of oil fields. One of the most effective methods of increasing oil recovery and intensifying the flow of fluid and gas to wells is hydraulic fracturing. The technology consists in creating a highly conductive fracture in the target formation under the action of a proppant-fed fluid under pressure to ensure the flow of produced fluid to the bottom of the well. The fracture formed as a result of hydraulic fracturing can be horizontal or vertical.

There are a lot of mathematical models that can determine the well with hydraulic fracturing flow rate, but not all are applicable for assessing the energy state of a formation (reservoir pressure) after hydraulic fracturing. To solve this problem, 3D reservoir simulators in commercial software are used, which allow to

D. A. Indeitsev and A. M. Krivtsov (Eds.): APM 2019, LNME, pp. 32–41, 2020.
https://doi.org/10.1007/978-3-030-49882-5_3

complete tasks associated with assessing the effect of hydraulic fracturing. But the time spent on preparing a large amount of initial data for the simulator and the calculation time for these models does not allow to consider many options for selecting candidate wells for hydraulic fracturing, as well as factor analysis of hydraulic fracturing.

Thus, the goal of this work was to implement a three-dimensional model of a horizontal well with multi-stage hydraulic fracturing to predict the well fluid flow rate or bottomhole pressure after hydraulic fracturing, and also to solve the inverse problem - to restore the hydraulic fracturing and formation characteristics based on the known production dynamics.

2 Well Model Without Fracturing

As a preparatory phase a horizontal well model without fracturing is considered. For a horizontal well in a formation, it has been assumed that it is all a drain and that the same fluid volume flows into each unit of the well's volume. In Appendix A.1 a model is schematically depicted, a brown dotted line indicates a well. Taking into account that the permeability coefficients can have different values in different directions of coordinates Darcy's law gives rise to the permeability tensor, instead of the permeability coefficient. It is a tensor, on the main diagonal of which there are permeability coefficients along the corresponding axes. Thus, diffusivity equation will take the following form.

$$\varphi_0 C\frac{\partial p}{\partial t}-\frac{1}{\mu}\left(\frac{\partial}{\partial x}\left(k_x\frac{\partial p}{\partial x}\right)+\frac{\partial}{\partial y}\left(k_y\frac{\partial p}{\partial y}\right)+\frac{\partial}{\partial z}\left(k_z\frac{\partial p}{\partial z}\right)\right)=q_V\delta(M-M') \quad (1)$$

Where M is the region under consideration, M' is the region in which there is a source (sink). Bringing the equation to dimensionless form:

$$\begin{gathered} x_D=\frac{x}{L},\ y_D=\frac{y}{L},\ z_D=\frac{z}{L},\ h_D=\frac{h}{L},\ L_{Dw}=\frac{L_w}{L},\ r_{Dw}=\frac{r_w}{L}, \\ q_D=\frac{q}{q_t},\ p_D=\frac{2\pi h k_m\left(p_0-p\right)}{\mu q_t},\ t_D=\frac{k_m t}{\mu\varphi_m C_m L^2}, \\ k_{Dx}=\frac{k_x}{k_m},\ k_{Dy}=\frac{k_y}{k_m},\ k_{Dz}=\frac{k_z}{k_m} \end{gathered} \quad (2)$$

Dimensioning constants: $[L]=m$, $[qt]=\frac{m^3}{sec}$, $[k_m]=Darcy$, $[C_m]=Pa^{-1}$. In case of equality of the dimensionless constants to the reservoir characteristics, the equation takes the following form:

$$\frac{\partial p_D}{\partial t_D}-\left(k_{Dx}\frac{\partial^2 p_D}{\partial x_D^2}+k_{Dy}\frac{\partial^2 p_D}{\partial y_D^2}+k_{Dz}\frac{\partial^2 p_D}{\partial z_D^2}\right)=\frac{2\pi q_D h_D L^3}{\Delta \mathrm{V}}\delta\left(M-M'\right) \quad (3)$$

Next, a solution of the problem will be held in a dimensionless form, therefore, it is possible to discard the indices D and redesignate the right side to reduce further calculations:

$$\begin{cases} f(M-M') = \frac{2\pi q_D h_D L^3}{\Delta V}\delta(M-M'); \\ \frac{\partial p}{\partial t} - \left(k_x \frac{\partial^2 p}{\partial x^2} + k_y \frac{\partial^2 p}{\partial y^2} + k_z \frac{\partial^2 p}{\partial z^2}\right) = f\left(M-M'\right) \end{cases} \tag{4}$$

Discretize the second-order spatial derivatives at time n at the lattice position with coordinates i, j, k and introduce a special notation for them.

$$\begin{aligned} \frac{\partial^2 p^n_{i,j,k}}{\partial x^2} &= \frac{p^n_{i+1,j,k} - 2p^n_{i,j,k} + p^n_{i-1,j,k}}{(\Delta \mathrm{x})^2} + O(\Delta \mathrm{x}^2) \\ &\approx \frac{p^n_{i+1,j,k} - 2p^n_{i,j,k} + p^n_{i-1,j,k}}{(\Delta \mathrm{x})^2} = \frac{\delta^2_x p^n}{(\Delta \mathrm{x})^2} \end{aligned} \tag{5}$$

To solve equations of this type, Clark and Nicholson proposed taking the average between the explicit and implicit schemes to preserve the advantages of the implicit scheme - absolute convergence and improve accuracy. But according to this scheme, at each time step, it is necessary to find 7 unknown pressure values at neighboring points, which greatly complicates the algorithm. To solve the equation, the method of variable directions (the Douglas-Gunn approach) is used. The idea of this approach is to decompose the solution of the equation into three substeps. On each of them, a tridiagonal matrix algorithm is used along one of the coordinates.

$$\begin{aligned} I: \frac{p^{n+\frac{1}{3}} - p^n}{\Delta \mathrm{t}} &= \frac{k_x}{2\left(\Delta \mathrm{x}\right)^2}\delta^2_x\left(p^{n+\frac{1}{3}} + p^n\right) + \frac{k_y}{\left(\Delta \mathrm{y}\right)^2}\delta^2_y p^n \\ &+ \frac{k_z}{\left(\Delta \mathrm{z}\right)^2}\delta^2_z p^n + f\left(M-M'\right); \\ II: \frac{p^{n+\frac{2}{3}} - p^n}{\Delta \mathrm{t}} &= \frac{k_x}{2(\Delta \mathrm{x})^2}\delta^2_x(p^{n+\frac{1}{3}} + p^n) + \frac{k_y}{2(\Delta \mathrm{y})^2}\delta^2_y(p^{n+\frac{2}{3}} + p^n) \\ &+ \frac{k_z}{(\Delta \mathrm{z})^2}\delta^2_z p^n + f(M-M'); \\ III: \frac{p^{n+1} - p^n}{\Delta \mathrm{t}} &= \frac{k_x}{2(\Delta \mathrm{x})^2}\delta^2_x(p^{n+\frac{1}{3}} + p^n) + \frac{k_y}{2(\Delta \mathrm{y})^2}\delta^2_y(p^{n+\frac{2}{3}} + p^n) \\ &+ \frac{k_z}{2(\Delta \mathrm{z})^2}\delta^2_z(p^{n+1} + p^n) + f(M-M') \end{aligned} \tag{6}$$

3 Well Model with Hydraulic Fracturing

After the preparatory phase, a transition to modeling of a horizontal well with multi-stage hydraulic fracturing can be made. Fractures are rectangular areas in a plane perpendicular to the well. Since the volume of production through fractures is much larger than the volume of production in the well outside the fractures, in this model we consider that sink is located only at the intersection of the fractures and the well. In Appendix A.1 the model in question is schematically depicted, the brown dotted line indicates the well, and the blue one indicates the fracture.

Assuming that fluid flow in the hydraulic fractures obeys Darcy's law, constant pressure or non-flow boundary conditions can be used. The mathematical model for hydraulic fractures is given by the next equations.

$$\varphi_0 C\frac{\partial p}{\partial t} - \frac{1}{\mu}(\frac{\partial}{\partial x}(k_x\frac{\partial p}{\partial x}) + \frac{\partial}{\partial y}(k_y\frac{\partial p}{\partial y}) + \frac{\partial}{\partial z}(k_z\frac{\partial p}{\partial z})) = 0, (x, y, z) \notin fracture$$

$$\varphi_{fr0} C_{fr}\frac{\partial p}{\partial t} - \frac{1}{\mu}(\frac{\partial}{\partial x}(k_{frx}\frac{\partial p}{\partial x}) + \frac{\partial}{\partial y}(k_{fry}\frac{\partial p}{\partial y}) + \frac{\partial}{\partial z}(k_{frz}\frac{\partial p}{\partial z})) = q_V,$$
$$(x, y, z) \in fracture (7)$$

Since permeability coefficients of a fracture and a formation can differ by a factor of 10^6 the finite volume method was applied to solve this problem. In this method, the above diffusivity equation is solved in an integrated form for the formation and fractures.

$$\varphi_0 C\frac{\partial}{\partial t}\left(\int\limits_V p \, \mathrm{d}V\right) = \frac{1}{\mu}\oint\limits_S \boldsymbol{n}\cdot(\mathbf{K}\cdot\nabla \mathrm{p})\,\mathrm{d}S, (x, y, z) \notin fracture$$

$$\varphi_{fr0} C_{fr}\frac{\partial}{\partial t}\left(\int\limits_V \mathrm{p}\mathrm{d}V\right) = \frac{1}{\mu}\oint\limits_S \boldsymbol{n}\cdot(\mathbf{K}_{fr}\cdot\nabla \mathrm{p})\mathrm{d}S + q_V, (x, y, z) \in fracture \quad (8)$$

Dimensionless variables of fractured horizontal well are defined the same way it was made in first model using fracture permeability coefficients.

$$k_{frDx} = \frac{k_{frx}}{k_m}, k_{frDy} = \frac{k_{fry}}{k_m}, k_{frDz} = \frac{k_{frz}}{k_m} \quad (9)$$

Next, solution of the problem is in a dimensionless form, therefore, the indices D are discarded. In the finite volume method, the reservoir characteristics are tied not to the grid nodes, but to the cells.

$$\frac{\partial}{\partial t}\left(\int\limits_V p \, \mathrm{d}V\right) = \oint\limits_S \boldsymbol{n}\cdot(\mathbf{K}\cdot\nabla \mathrm{p})\,\mathrm{d}S, (x, y, z) \notin fracture \quad (10)$$

$$\frac{\varphi_{fr0} C_{fr}}{\varphi_0 C}\frac{\partial}{\partial t}\left(\int\limits_V \mathrm{p}\mathrm{d}V\right) = \oint\limits_S \boldsymbol{n}\cdot(\mathbf{K}_{fr}\cdot\nabla \mathrm{p})\mathrm{d}S + \frac{2\pi h q L^3}{\Delta V}, (x, y, z) \in fracture$$

A cube with the size of one cell is built around each node. Each eighth of this cube lies in a separate cell and has its own physical characteristics. The final equation for finding the value in the node lying in the cube is obtained as the sum of all the equations for the eighths of the cube.

Considering one eighth of the control volume (Fig. 1), it was supposed that it lies outside the fracture, and the equation obtained above was written for this volume element in the form of an explicit scheme.

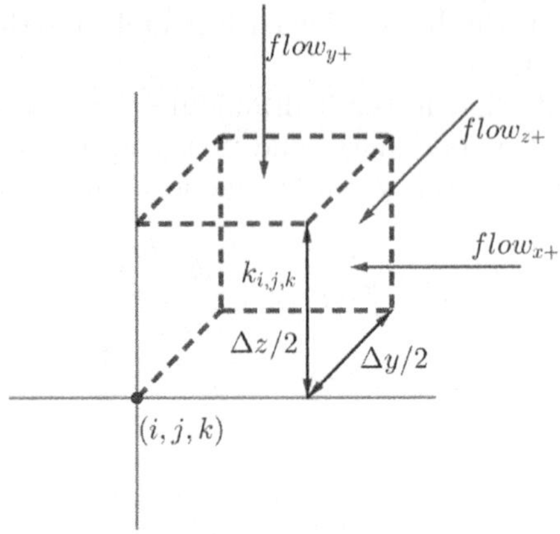

Fig. 1. One eighth of the control volume

$$\frac{\Delta x \Delta y \Delta z}{8} \frac{p_{i,j,k}^{n+1} - p_{i,j,k}^{n}}{\Delta t} = k_x \frac{\Delta y \Delta z}{4} \frac{p_{i+1,j,k}^{n} - p_{i,j,k}^{n}}{\Delta x} + k_y \frac{\Delta x \Delta z}{4} \frac{p_{i,j+1,k}^{n} - p_{i,j,k}^{n}}{\Delta y} + k_z \frac{\Delta x \Delta y}{4} \frac{p_{i,j,k+1}^{n} - p_{i,j,k}^{n}}{\Delta z} \tag{11}$$

Then, the final equation for the cell node located on the fracture boundary along the x axis, in the case of an explicit scheme, will be written in the following form.

$$\begin{aligned} &\frac{1}{2}\left(\frac{\varphi_{fr0} C_{fr}}{\varphi_0 C} + 1\right) \Delta x \Delta y \Delta z \frac{p_{i,j,k}^{n+1} - p_{i,j,k}^{n}}{\Delta t} = \\ &\quad \frac{\Delta y \Delta z}{4 \Delta x} \left(\left(p_{i+1,j,k}^{n} - p_{i,j,k}^{n}\right) k_x + \left(p_{i-1,j,k}^{n} - p_{i,j,k}^{n}\right) k_{fx}\right) \\ &+ \frac{\Delta x \Delta z}{4 \Delta y} \left(\left(p_{i,j+1,k}^{n} - p_{i,j,k}^{n}\right) \frac{1}{2}\left(k_y + k_{fy}\right) + \left(p_{i,j-1,k}^{n} - p_{i,j,k}^{n}\right) \frac{1}{2}\left(k_y + k_{fy}\right)\right) \\ &+ \frac{\Delta x \Delta y}{4 \Delta z} \left(\left(p_{i,j,k+1}^{n} - p_{i,j,k}^{n}\right) \frac{1}{2}\left(k_z + k_{fz}\right) + \left(p_{i,j,k-1}^{n} - p_{i,j,k}^{n}\right) \frac{1}{2}\left(k_y + k_{fy}\right)\right) \end{aligned} \tag{12}$$

Due to the restrictions imposed by the explicit scheme on the space and time step, it was decided to use an implicit scheme with time substeps, each of which on the right side contains flows in only one spatial direction. As a result of applying this scheme, after each time step, the pressure distribution remained stretched along the last axis of the tridiagonal matrix algorithm, which was an artifact of the selected scheme. Since such artifacts can have a strong influence on the calculation of flow rate or borehole pressure, the Douglas-Gunn scheme was tested. For the cell considered above, lying on the fracture boundary along the x axis, it is represented in the form given in Appendix A.2. Results of a computation in different time moments are presented on Fig. 2.

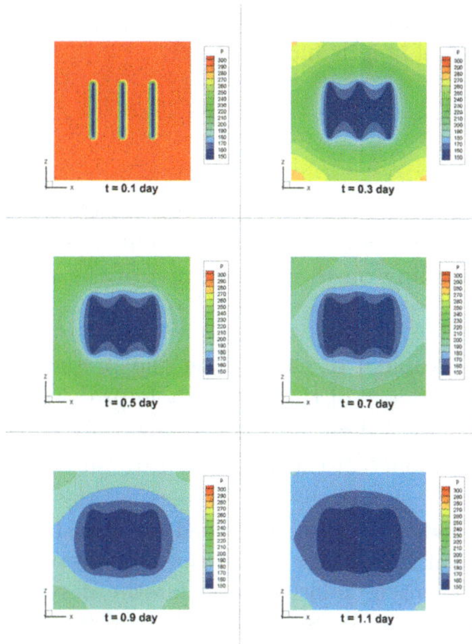

Fig. 2. The result of the algorithm - the pressure field in the slice along the y axis

4 The "Infinite Permeability" Model

During the testing of the method, it was found that the calculations with using the intersection of a well and a fracture as a drain will be stable only at time steps of the order of a tenth of a second, which does not allow to use it for industrial application. For this reason, it was decided to switch to the "infinite permeability" model, in which each unit volume of a fracture becomes a drain of a fixed power. To compare these two models in the fracture plane control points were selected, the pressure values at which are compared with a time step of the order of 10^{-2} s. The positions of this points are shown in the diagram of Appendix A.3, the boundaries of the fracture are indicated by a blue dotted line on it. The final relative deviation of the "infinite permeability" model from the model with point drain at the intersection of the fracture and the well is shown in the graph. The initial reservoir pressure is two times greater than the bottomhole pressure, so this plot begins from a mark of 50%. The boundary non-flow conditions are satisfied when using the finite volume method automatically, because streams into the boundary cells flow only from the inside of the volume under consideration. To use the boundary conditions of constant pressure instead of the equation derived from tridiagonal matrix algorithm pressure value in boundary nodes are set equal to boundary pressure. Thus, the pressure distribution differs fundamentally only in the first second of the algorithm work, which allows to use the model of "infinite permeability" if calculations with a large time step are necessary. For this model tests with different time and spatial step were made, their results are presented in Appendix A.4. Optimal time step is 15 min, optimal spatial step is half of meter.

5 Conclusions

As a result of the algorithm work the pressure distribution in the reservoir is recorded in files with a user-defined time step, which allows to make factor analysis. In addition, depending on the formulation of the problem, the calculation results are recorded:

- When solving the problem of calculating the pressure field at a given well flow rate – the dependence of the bottomhole pressure on time;
- In solving the problem of determining the flow rate at predefined bottomhole pressure – the dependence of the well flow rate on time.

Acknowledgement. This work was supported by Ministry of Science and Higher Education of the Russian Federation within the framework of the Federal Program "Research and development in priority areas for the development of the scientific and technological complex of Russia for 2014 – 2020" (activity 1.2), grant No. 14.575.21.0146 of September 26, 2017, unique identifier: RFMEFI57517X0146. The industrial partner of the grant is LLC "Gazpromneft Science & Technology Centre".

A Appendix

A.1 Appendix 1

(See Figs. 3 and 4)

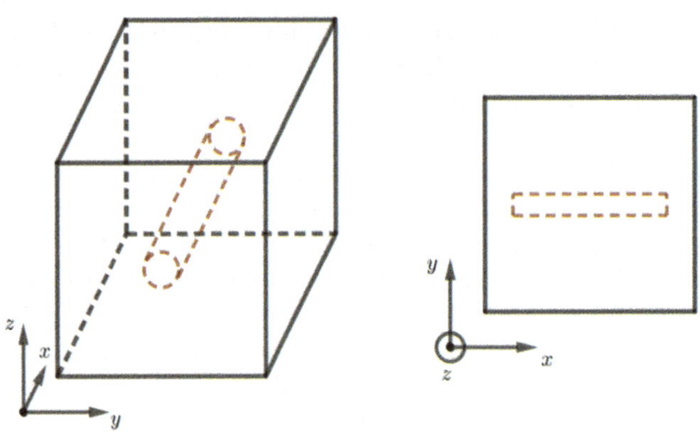

Fig. 3. Scheme of model without fracturing

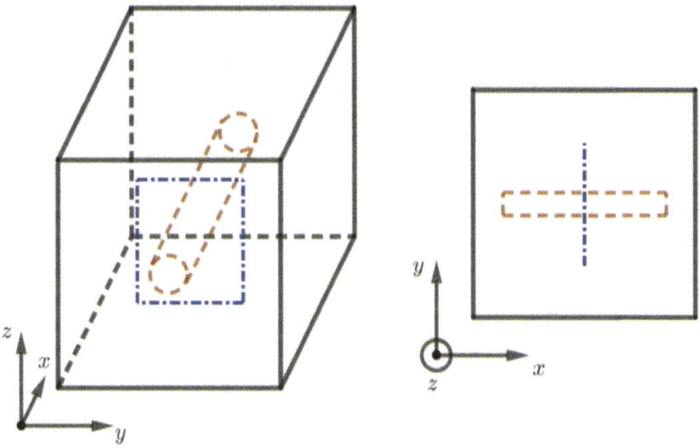

Fig. 4. Scheme of model with hydraulic fracturing

A.2 Appendix 2

$$
\begin{aligned}
I &: \tfrac{1}{2}\left(\tfrac{\varphi_{fr0}C_{fr}}{\varphi_0 C}+1\right)\Delta x\Delta y\Delta z\tfrac{p^{n+\frac{1}{3}}_{i,j,k}-p^{n}_{i,j,k}}{\Delta t}\\
&= \tfrac{\Delta y\Delta z}{8\Delta x}((p^{n}_{i+1,j,k}-p^{n}_{i,j,k})k_x+(p^{n}_{i-1,j,k}-p^{n}_{i,j,k})k_{fx}))\\
&\quad+\tfrac{\Delta y\Delta z}{8\Delta x}((p^{n+\frac{1}{3}}_{i+1,j,k}-p^{n+\frac{1}{3}}_{i,j,k})k_x+(p^{n+\frac{1}{3}}_{i-1,j,k}-p^{n+\frac{1}{3}}_{i,j,k})k_{fx})\\
&\quad+\tfrac{\Delta x\Delta z}{4\Delta y}((p^{n}_{i,j+1,k}-p^{n}_{i,j,k})\tfrac{1}{2}(k_y+k_{fy})+(p^{n}_{i,j-1,k}-p^{n}_{i,j,k})\tfrac{1}{2}(k_y+k_{fy}))\\
&\quad+\tfrac{\Delta x\Delta y}{4\Delta z}((p^{n}_{i,j,k+1}-p^{n}_{i,j,k})\tfrac{1}{2}(k_z+k_{fz})+(p^{n}_{i,j,k-1}-p^{n}_{i,j,k})\tfrac{1}{2}(k_y+k_{fy}))\\
II &: \tfrac{1}{2}\left(\tfrac{\varphi_{fr0}C_{fr}}{\varphi_0 C}+1\right)\Delta x\Delta y\Delta z\tfrac{p^{n+\frac{2}{3}}_{I,j,k}-p^{n}_{I,j,k}}{\Delta t}\\
&= \tfrac{\Delta y\Delta z}{8\Delta x}((p^{n}_{i+1,j,k}-p^{n}_{i,j,k})k_x+(p^{n}_{i-1,j,k}-p^{n}_{i,j,k})k_{fx})\\
&\quad+\tfrac{\Delta y\Delta z}{8\Delta x}((p^{n+\frac{1}{3}}_{i+1,j,k}-p^{n+\frac{1}{3}}_{i,j,k})k_x+(p^{n+\frac{1}{3}}_{i-1,j,k}-p^{n+\frac{1}{3}}_{i,j,k})k_{fx})\\
&\quad\tfrac{\Delta x\Delta z}{8\Delta y}((p^{n}_{i,j+1,k}-p^{n}_{i,j,k})\tfrac{1}{2}(k_y+k_{fy})+(p^{n}_{i,j-1,k}-p^{n}_{i,j,k})\tfrac{1}{2}(k_y+k_{fy}))\\
&\quad+\tfrac{\Delta x\Delta z}{8\Delta y}((p^{n+\frac{2}{3}}_{i,j+1,k}-p^{n+\frac{2}{3}}_{i,j,k})\tfrac{1}{2}(k_y+k_{fy})+(p^{n+\frac{2}{3}}_{i,j-1,k}-p^{n+\frac{2}{3}}_{i,j,k})\tfrac{1}{2}(k_y+k_{fy}))\\
&\quad+\tfrac{\Delta x\Delta y}{4\Delta z}((p^{n}_{i,j,k+1}-p^{n}_{i,j,k})\tfrac{1}{2}(k_z+k_{fz})+(p^{n}_{i,j,k-1}-p^{n}_{i,j,k})\tfrac{1}{2}(k_y+k_{fy}))\\
III &: \tfrac{1}{2}\left(\tfrac{\varphi_{fr0}C_{fr}}{\varphi_0 C}+1\right)\Delta x\Delta y\Delta z\tfrac{p^{n+1}_{i,j,k}-p^{n}_{i,j,k}}{\Delta t}\\
&= \tfrac{\Delta y\Delta z}{8\Delta x}((p^{n}_{i+1,j,k}-p^{n}_{i,j,k})k_x+(p^{n}_{i-1,j,k}-p^{n}_{i,j,k})k_{fx})\\
&\quad+\tfrac{\Delta y\Delta z}{8\Delta x}((p^{n+\frac{1}{3}}_{i+1,j,k}-p^{n+\frac{1}{3}}_{i,j,k})k_x+(p^{n+\frac{1}{3}}_{i-1,j,k}-p^{n+\frac{1}{3}}_{i,j,k})k_{fx})\\
&\quad+\tfrac{\Delta x\Delta z}{8\Delta y}((p^{n}_{i,j+1,k}-p^{n}_{i,j,k})\tfrac{1}{2}(k_y+k_{fy})+(p^{n}_{i,j-1,k}-p^{n}_{i,j,k})\tfrac{1}{2}(k_y+k_{fy}))\\
&\quad+\tfrac{\Delta x\Delta z}{8\Delta y}((p^{n+\frac{2}{3}}_{i,j+1,k}-p^{n+\frac{2}{3}}_{i,j,k})\tfrac{1}{2}(k_y+k_{fy})+(p^{n+\frac{2}{3}}_{i,j-1,k}-p^{n+\frac{2}{3}}_{i,j,k})\tfrac{1}{2}(k_y+k_{fy}))\\
&\quad+\tfrac{\Delta x\Delta y}{8\Delta z}((p^{n}_{i,j,k+1}-p^{n}_{i,j,k})\tfrac{1}{2}(k_z+k_{fz})+(p^{n}_{i,j,k-1}-p^{n}_{i,j,k})\tfrac{1}{2}(k_y+k_{fy}))\\
&\quad+\tfrac{\Delta x\Delta y}{8\Delta z}((p^{n+1}_{i,j,k+1}-p^{n+1}_{i,j,k})\tfrac{1}{2}(k_z+k_{fz})+(p^{n+1}_{i,j,k-1}-p^{n+1}_{i,j,k})\tfrac{1}{2}(k_y+k_{fy}))
\end{aligned}
\tag{A.1}
$$

A.3 Appendix 3

(See Figs. 5 and 6)

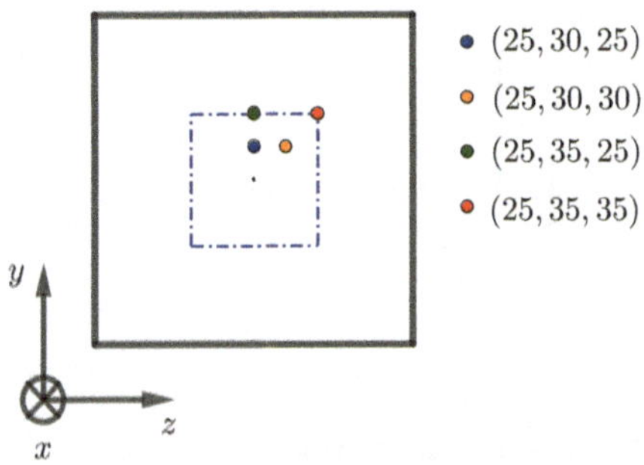

Fig. 5. Control points

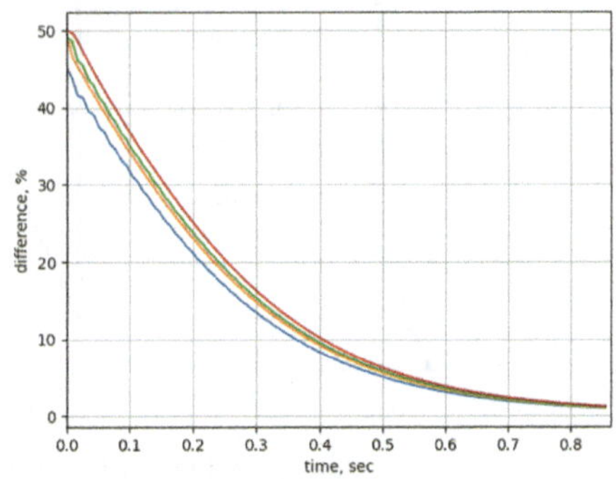

Fig. 6. Difference rate between point-source and "infinite permeability" models in percents

A.4 Appendix 4

(See Figs. 7 and 8)

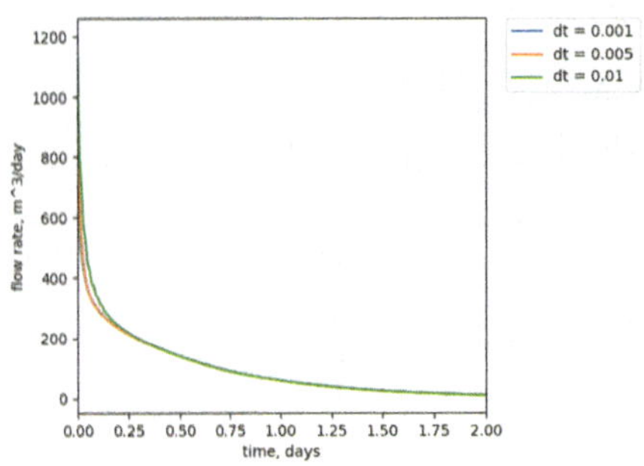

Fig. 7. Flow rate with different time steps

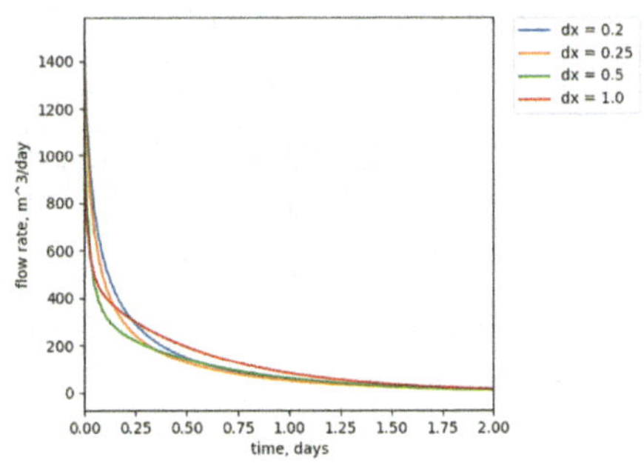

Fig. 8. Flow rate with different spatial steps

References

1. Basniev, K.S., Dmitriev, N.M., Rozenberg, G.D.: Neftegazovaya gidrome-hanika. M.-Izhevsk: Institut komp'yuternyh issledovanij (2005)

2. Turchak, L.I., Plotnikov, P.V.: Osnovy chislennyh metodov. – Obshche-stvo s ogranichennoj otvetstvennost'yu Izdatel'skaya firma "Fiziko-matematicheskaya literatura" (2002). S. 280
3. Simonov, M.V., Akhmetov, A.V., Roshchektaev, A.P.: Semi-analytical model of transient fluid flow to multilateral well. In: SPE Annual Caspian Technical Conference and Exhibition. – Society of Petroleum Engineers (2017)
4. Simonov, M.V., Roschektaev, A.P.: Multifractured horizontal well inflow model for shale gas and oil. PRONEFT". Professional'no o nefti, no. 2(4), pp. 25–30 (2017)
5. Khamidullin, M.R.: Numerical Simulation of One-Phase Flow to Multi-Stage Hydraulically Fractured Horizontal Well. Uchenye zapiski kazanskogo universiteta. seriya fiziko-matematicheskie nauki. T. 158. – No. 2 (2016)
6. Bogachev, K., Shelkov, V.: A New Approach in Modeling Hydraulic Fractures and Auto Fractures at Injectors in Full-field Models. SPE-138071-MS. SPE Russian Oil and Gas Conference and Exhibition, 26–28 October 2010, Moscow, Russia (2010)
7. Bogachev, K., Zhabitsky, Y., Eydinov, D., Shelkov, V., Robinson, T.: A New Approach to Numerical Simulation of Fluid Flow in Fractured Shale Gas Reservoirs. SPE-147021-MS. Canadian Unconventional Resources Conference, 15–17 November 2011, Alberta, Canada (2011)
8. Chen, Z., et al. A finite horizontal-well-conductivity model for pressure transient analysis in multiple fractured horizontal wells. SPE Latin American and Caribbean Petroleum Engineering Conference. – Society of Petroleum Engineers (2015)
9. Chen, Z., et al.: A semianalytical approach for obtaining type curves of multiple-fractured horizontal wells with secondary-fracture networks. SPE J. **21**(02), 538–549 (2016)
10. Douglas, J.: Alternating direction methods for three space variables. Numeische Mathematik **4**(1), 41–63 (1962)
11. MoradiDowlatabad, M., et al.: Novel approach for predicting multiple fractured horizontal wells performance in tight reservoirs. SPE Offshore Europe Conference and Exhibition. – Society of Petroleum Engineers (2015)
12. Wang, T.Y., Chen, C.C.P.: Thermal-ADI-a linear-time chip-level dynamic thermal-simulation algorithm based on alternating-direction-implicit (ADI) method. IEEE Trans. Very Large Scale Integr. (VLSI) Syst. **11**(4), 691–700 (2003)

Double Aging of Heat-Treated Aluminum Alloy of (7075) and (6061) to Increase the Hardness Number

Suleiman E. Al-Lubani[1](✉) and Ahmad I. Ateyat[2]

[1] Al-Balqa Applied University, Amman, Jordan
slubani@bau.edu.jo
[2] Philadelphia University, Jarash, Jordan

Abstract. The present study is an attempt to study the effect of double-aging process and its relevant parameters (like time, temperature and quenching media) on the mechanical properties of two types of aluminum alloys. AA7075 and AA6061 are selected in this study to examine their responses to the double-aging. Fabricated specimens are heat treated, and tested for hardness using micro-Vickers hardness test. They are then artificially aged at 150 °C for different aging periods. The specimens are then quenched in water and oil. Hardness test are then repeated to determine the optimum time and cooling rate that give the maximum (peak) hardness values. This procedure are repeated on other sets of specimens, but with double aging at 185 °C to find out which alloy has more significant response to double ageing. The outcomes of the study show that the 7075 aluminum alloy demonstrates better results than those obtained for the 6061 aluminum alloys. Increase in hardness from 135.66 HV to 150.61 HV is reported for the 7075 aluminum alloy.

Keywords: Heat treatment · AA7075 and AA6061 · Hardness · Double-aging

1 Introduction

Aluminum is known as one of the non-ferrous metals that turned into a practical rival in modern Engineering applications. This is because of the wide scope of physical and mechanical properties of this material that can be created from a high unadulterated aluminum to the most intricate alloys.

The most alluring properties of aluminum is its solidarity to weight proportion, low density; (one-third of that for steel and copper), its obstruction for erosion fabric-ability (formability), non-toxic, high heat and electrical conductivity, reflectivity for both heat and light. Unadulterated aluminum has likewise two noteworthy burdens (limitations), low melting point (about 660 °C) and its relatively low strength (about 10 MPa).

In order to improve the strength of the pure aluminum, the latter has to be mixed with other agents by the method of alloying. Aluminum alloys are usually classified into different series according to its alloying element, modification process, and the purity of the alloy (percentage of aluminum) [1, 2]. Aluminum alloys that are used cladding and for corrugated roofing are examples of non-heat-treatable alloys, which

D. A. Indeitsev and A. M. Krivtsov (Eds.): APM 2019, LNME, pp. 42–52, 2020.
https://doi.org/10.1007/978-3-030-49882-5_4

include 1–1.25% manganese and 2–7% magnesium, are of relatively high strength and are used for. While those used as fasteners and as light structural members are examples of heat-treatable alloys contain varying proportions of aluminum, magnesium, silicon, and sometimes copper. These have high strength and are, therefore, are widely used. Casted aluminum alloys usually contain silicon, silicon and copper, or silicon and magnesium [3]. A significant improvement in the alloy strength is achieved by solution heat treatment of the alloys for 8 h at 465 °C. An example for such an improvement, UTS is increased from about 250 MPa to 380 MPa [4]. Duplex aging enhances corrosion resistance since the grain boundary zone is removed. The 7075 alloy also demonstrates a high response to age hardening [4, 5].

Arturo Abúndez *et al.* [6] studied the relation between ageing and retrogression times and temperatures with Ultimate Tensile Strength (UTS) and hardness on Al6061 alloy. The outcomes of this study showed that a12% on UTS and 6% in hardness increment was achieved for combination of 200 °C ageing temperature and 250 °C retrogression temperature.

Rajaa *et al.* [7], have studied the effects of aging time on the mechanical properties of 6061 Aluminum alloys. It is found that the strength is increased through a process of artificial aging within two hours. An increase in the hardness value about 27.4% can be achieved through a 2-h artificial aging process. More than two hours of artificial aging time may cause a decrease in the hardness.

The research of Qing-Long Zhao *et al.* [8] has shown that the specimen of AA6061 preheated at a high temperature (540 °C) demonstrated a pronounced ageing hardening response to short-term ageing. Also, the rolling rolled-quenched alloy aged for 1 h exhibited high tensile strength (370 MPa), improved ductility and surface quality compared to T6 and T8 tempered 6061 alloys.

Masoud *et al.* have shown that the maximum hardness of 137 HV is obtained after aging the specimen at temperatures of 205 and 225 °C at 90 min and 120 min aging respectively [9]. The optimum aged could be achieved from 175 C to 195 C between 2 to 6 h of ageing time. It showed that 6 h of ageing could exhibit a maximum strength of 432.84 MPa, at a temperature of 185 °C [10]. The goal of this work was to evaluate the effects of double aging and parameters (i.e. time, temperature and quenching media) on the mechanical properties of two aluminum alloys (i.e. AA7075 and AA6061), to find out the differences in response to the double-aging.

2 Methodology

In the present study, two specimens of Aluminum Alloys (7075 and 6061) were selected. They are considered as the most common and the most used in applications among other types of Aluminum Alloys. Also, they both have the same alloying components, but having different amount percentages as described in Table 1. Therefore, a noticeable difference in their mechanical properties is reported. The 7075 Aluminum specimen has a circular cross section of a radius of 60 mm and a thickness of 20 mm. whereas; the 6061 Aluminum specimen has a squared cross section with a side length of 70 mm and a thickness of 20 mm. The details (shapes and dimensions) of the specimens are shown Table 2. Both specimens were heat treated at 470 °C for two hours and then they were cooled using water to the room temperature. Hardness tests were conducted on both specimens. The results were found to be 139.8 HV for the 7075 specimen, and 40.34 HV for the 6061 aluminum alloy specimen.

Table 1. chemical composition for 6061 and 7075 aluminum alloys

Alloy	Alloying elements percentages (wt%)									
	Si	Mn	Zn	Ni	Ti	Cu	Mg	Fe	Cr	Al
6061	0.73	0.08	0.01	0.015	0.03	0.31	1.51	0.18	0.08	97.0
7075	0.23	0.03	5.89	–	0.02	1.76	3.03	0.18	0.21	88.6

Table 2. Dimensions and shapes of aluminum alloys specimens

Alloy	7075		6061	
Shape	Disk		Square plate	
Dimensions	Radius	60 mm	Length * width	70 * 70 mm
	Thickness	20 mm	Thickness	20 mm

3 Heat Treatment of Aluminum Alloys

In many applications, alloying by itself is not sufficient to reach the needed strength of aluminum alloys; hence, other processes may be applied to the alloys to increase its strength and to produce more stable structure.

Quenching. This is a critical operation and must be carried out to precise limits if optimum results are to be obtained. The objective of the quenching is to ensure that the dissolved constituents remain in solution down to room temperature.

The usual quenching medium is water at room temperature. In some circumstances slow quenching is desirable as this improves the resistance to stress corrosion cracking of certain copper-free Al-Zn-Mg alloys.

Age Hardening. By the end of solution treatment and quenching processes, hardening is achieved either at room temperature (natural ageing) or with a precipitation heat treatment (artificial ageing). In some alloys sufficient precipitation occurs in a few days at room temperature to yield stable products with properties that are adequate for many applications. These alloys sometimes are precipitation heat treated to provide increased strength and hardness in wrought and cast alloys. Other alloys with slow precipitation reactions at room temperature are always precipitation heat treated before being used.

In some alloys, notably those of the 2xxx series, cold working of freshly quenched materials greatly increases its response to later precipitation treatment. Mills take advantage of this phenomenon by applying a controlled amount of rolling (sheet and plate) or stretching (extrusion, bar and plate) to produce higher mechanical properties. However, if the higher properties are used in design, reheat treatment must be avoided.

Re-aging (Multi-aging). The hardness achieved by age hardening (either natural or artificial) can be increased by re-aging; or in general; multi-aging process.

The multi-stage heat treatment known as: retrogression and re-ageing (RRA). This type of heat treatment is the process that can be used to enhance the mechanical and corrosion resistance properties of aluminum. RRA process was first developed by Cina and Gan [11] and their results showed that 7xxx series of aluminum alloys are known to respond to retrogression and re-aging thermal treatments [12].

RRA is accomplished in three major stages: Solutionization and Ageing, Retrogression, and Re-aging. Due to aging, a strength penalty of about 10%–20% may occur. To compensate this, retrogression is required.

Retrogression consists of two stages. The first stage is the partial dissolution of the hardening precipitates that result in a decrease in hardness/yield strength which is regained in subsequent re-aging. The second stage is the coarsening of precipitates that may occur, (for long time retrogression), which results in continuous decrease in hardness/strength.

4 Experimental Data

Specimen's Selection, Properties and Characteristics. Micro-Vickers testing machine (Fig. 1) used to test the specimens. As stated earlier, the specimens that were chosen for this study are 7075 (Fig. 2) and 6061 (Fig. 3) aluminum alloys because of their wide usage in engineering applications. Although they contain the same alloying elements (Si, Mg, Cr, Mn, Cu, Ti), however, there are differences in the weight percentage of each element, Therefore they may be considered as completely different alloys because of the differences in their mechanical properties.

The 7075 Aluminum Alloy. The main alloying element in the 7075 aluminum alloy is zinc (Zn) which gives the alloy its ability to resist stresses and its high strength compared with that of steel. The 7075 aluminum alloy or (aerospace/aircraft aluminum) as it called, has relatively high cost which limits its usage in that applications where the cheaper alloys can be used. The most common applications of 7075 aluminum alloys are in those where small weight to strength ratio is needed like aircrafts, marine and transportations industries, military applications (rifles reserves), boats, bikes, and climbing equipment [11].

The 6061 Aluminum Alloy. Silicon (Si) and magnesium (Mg) are the major alloying elements in the 6061 aluminum alloy. Unlike the other aluminum alloys, this alloy gives a superb wilding ability with a high machining and workability which make it sustain to wild verity of applications like construction structures, trucks frames, aerospace applications, vehicles and bikes frame and components. 6061 is also used in salt-sensitive environments such as small boats, fishing reels and its wildly used in SCUBA tanks (breathing air cylinders) [12].

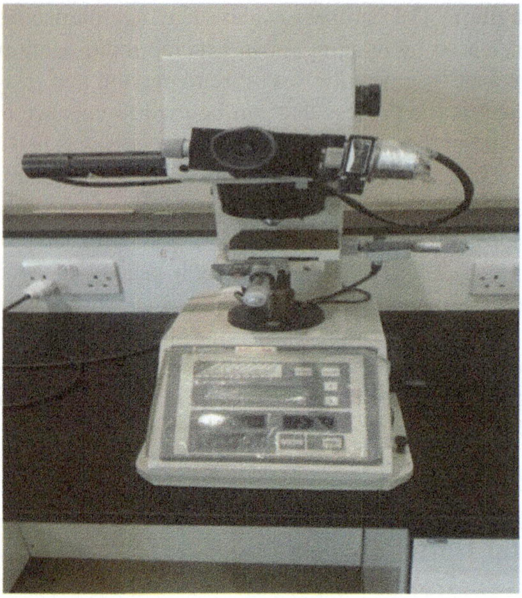

Fig. 1. Micro-Vickers testing machine

Fig. 2. AA 7075

Fig. 3. AA 6061

Chemical Compositions. The chemical compositions of the fabricated specimens were obtained by a specialized Jordanian company (JAMCO). Results are shown in Table 1 above.

Hardness Tests. The hardness of specimens was obtained using MITUTOYO apparatus. It is a computerized micro- Vickers testing machine. Hardness specimens were subjected to 100 g load for 15 s, and then the results were red from the device in HV units.

Furnace. The specimens were solution heat-treated and ageing using NOBERTHERM (L5/c6) laboratory muffle furnace. The unit is a property metallurgical laboratory of the faculty of Engineering Technology. It is capable of providing maximum temperature of 1100 °C.

Quenching. Two groups of specimens were prepared for quenching process. Each group contains three specimens of the 7075 aluminum alloy and three specimens of the 6061 aluminum alloy. The first group was water-quenched using tap water at room temperature, whereas the other group was quenched using petroleum oil. The type of cooling oil used in this study is similar to that usually used in power plants. The oil has a density of about 400 g/L. The quenching time and specimen's identification letters are shown in Table 3.

Table 3. Quenching time and identification letters of specimens

Time (h)	Water quenched		Oil quenched	
	7075	6061	6061	7075
2	A	B	A	B
4	C	C	D	D
6	F	E	F	E

- Solution heat-treatment (homogenizing): Both alloys (7075 and 6061) were soaked in the furnace for 2 h at 470 °C, then they were quenched in tap water at room temperature. It is known that the 6061 alloys are highly responsive to solution heat treatment; however, this step is important to achieve a homogenous crystal structure that is required for the precipitation hardening (ageing).
- Aging: All specimens were soaked at 150 °C for 2, 4 and 6 h, for the first aging, quenched, and tested. The same procedure were repeated at 185 °C for re-aging.

5 Experimental Procedure

Specimen's Preparation. The specimens were prepared with a high quality surface finishing and then they were stamped by an identification letters (A, B, C, D, E, and F) as listed in Table 4. The specimens are then tested for hardness.

Solution Heat-Treatment. The solution heat-treatment where performed for both alloys at the same condition which is 470 °C for two hours. The 6061 alloy has shown a lower response to solution heat treatment. This step, however, is important to achieve the desired results. All specimens were water quenched with a maximum delay of 40 s.

Age Hardening (First Ageing). The first ageing was performed for both alloys at constant temperature of 150 °C for 2, 4, and 6 h. The specimens were then immediately quenched in water and oil.

Re-aging. The same procedure of the first ageing (described above) was repeated at 185 °C and for the same time periods.

6 Experimental Results

Solution Heat-Treatment. The hardness of the two types of alloys was examined after treatment at 470 °C for two hours. The hardness was found to be 139.8 HV for the 7075 alloy, and 40.34 HV for the 6061 aluminum alloy.

Aging at 150 °C. Table 4, below, shows hardness results for both aluminum alloys that are obtained after the first aging at 150 °C for the different specified time periods. Figure 4, and Fig. 5, show the variation of hardness with time for both types of alloys and for the different types of quenching processes.

Table 4. Hardness of the specimens in (HV) aged at 150 °C for various aging times

Time (h)	6061 HV		7075 HV	
	Water quenched	Oil quenched	Water quenched	Oil quenched
0 (as quenched)	40.34	40.34	139.8	139.8
2	78	40.56	115.46	90.23
4	70.9	60.06	120.4	157
6	113.36	62.57	135.66	141.13

Re-ageing. The hardness achieved by the re-ageing of the specimens at 185 °C for the same time periods and the same quenching arrangement is shown in the Table 5. Figure 6 and Fig. 7, show the variation of hardness values of the two alloys with time for different quenching media (water and oil).

Table 5. Hardness of the specimens in (HV) aged at 185 °C for various aging times

Time (h)	6061 HV		7075 HV	
	Water quenched	Oil quenched	Water quenched	Oil quenched
0 (as quenched)	113.36	62.57	135.66	141.13
2	47.13	40.4	130.1	124.95
4	72.7	50.03	146.3	161.2
6	72.73	69.56	145.87	150.61

7 Discussion of Results

Solution Heat-Treatment. The hardness of 7075 alloy increases from 136.57 HV to 139.8 HV after homogenizing (solution heat-treatment) where the hardness of 6061 alloy has a 32 HV drop in hardness; from 70.32 to 40.34 after homogenizing.

Aging at 150 °C. Water quenched 7075 alloy the hardness of 7075 aluminum alloy decreases during the first two hours of treatment and then a considerable improvement is noted for the rest of time until it is removed from the furnace and quenched. Note that no peak values for hardness were reported.

Oil quenched 7075 Alloy. As in water quenching; the hardness was decreases during the first two hours to a minimum value of 90.23 HV, after that a peak value of 157 HV is recorded after four hours of treatment. The hardness then decreases slightly at the end of the six hours.

Water Quenched 6061 Alloy. The hardness of 6061 aluminum alloy decreases for the first two hours, a slight decrease in hardness for the period between two and four hours, and then the hardness increases to a peak value of 113.36 HV at the end of the six hours treatment.

Oil Quenched 6061 Alloy. For the first two hours of this process, no changes in hardness were reported. An increase in the hardness value of 20 HV was reported between two and four hours then the hardness keep constant until the end of the six hour time period.

8 Re-aging at 1985 °C

Water Quenched 7075 Alloy. The hardness of the specimen remains constant during the first two hours, and then it increases slightly to the value of 15 HV between two and four hours. No change in hardness was reported till the end of the four hour time period.

Oil Quenched 7075 Alloy. The hardness decreases during the first two hours then increases until it reaches a peak value of 161.2 HV at four hours then it decreases to a similar value to that of water quenched specimens discussed above.

Water Quenched 6061 Alloy. The hardness of this type of alloys decreases sharply during the first two hours. It then increases slightly for the time period between two and four hours and then remains constant for the rest of the time period.

9 Conclusions

From the present research work, following conclusions can be reported:

- 7075 aluminum alloy retract better for solution heat treatment (homogenizing) at 470 °C for 2 h.

- Oil quenched 7075 aluminum alloy, demonstrates peak values of hardness, while water quenched specimens show more stable hardness values for the same time period.
- Just for AA6061, water quenching can achieve higher hardness than oil quenching in both single-aging and double-aging processes.

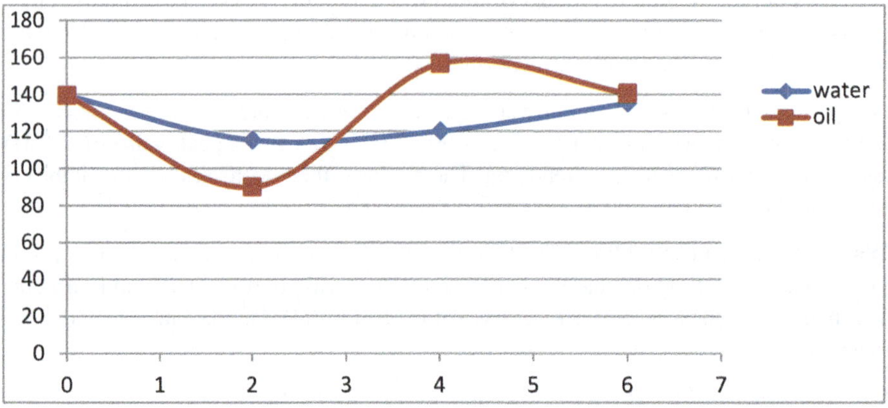

Fig. 4. Hardness verses time for the 7075 aluminum alloy aged at 150 °C

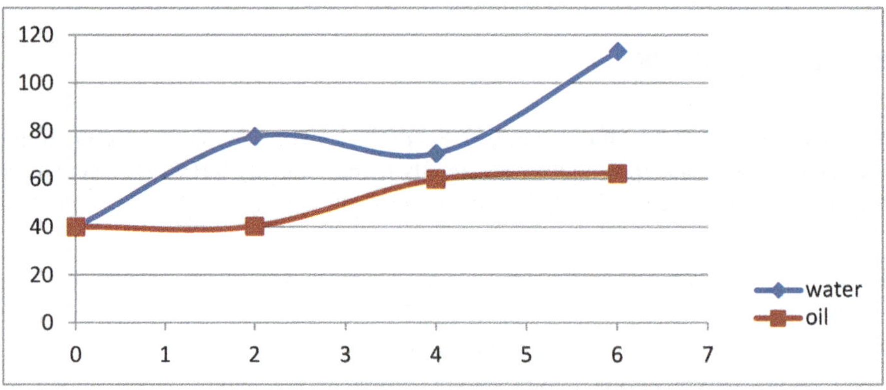

Fig. 5. Hardness verses time for 6061 aluminum alloy aged at 150 °C

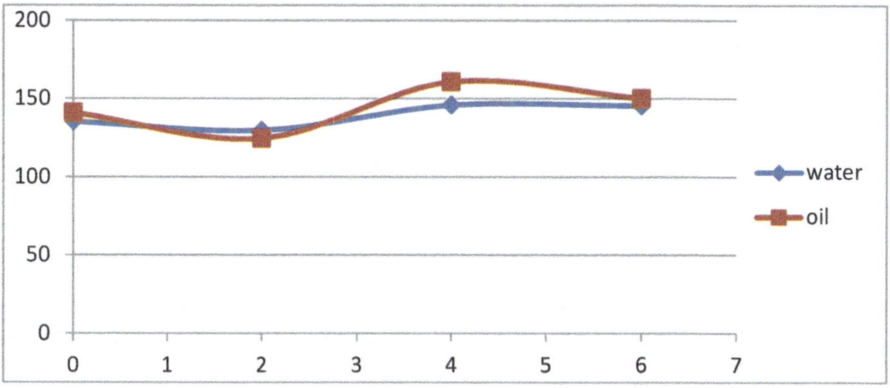

Fig. 6. Hardness verses time for 7075 aluminum alloy re-aged at 185 °C

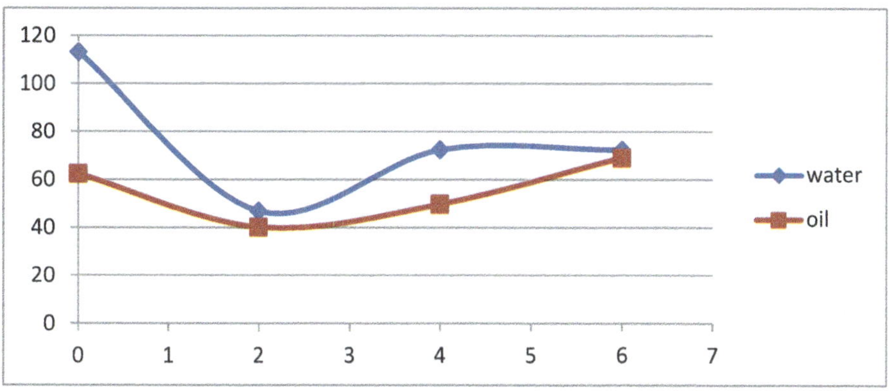

Fig. 7. Hardness verses time for 6061 aluminum alloy re-aged at 185 °C

References

1. ASM: Metals Handbook Desk Edition, 2nd edn., vol. 1 (1998)
2. Hatch, J.E.: Aluminum Properties and Physical Metallurgy. ASM International, Metals Park (1984)
3. GSA, Aluminum: Characteristics, Uses and Problems (2016). http://www.gsa.gov
4. Polmear, I.J.: Recent developments in light alloys. Mater. Trans. JIM **37**(1), 12–31 (1996)
5. Staley, J.T.: History of Wrought-Aluminium-Alloy Development. Aluminium Alloys: Contemporary Research and Applications, Treaties on Materials Science and Technology, Academic Press, **31** (1989). Table 1: chemical composition for 6061 and 7075 aluminum alloys
6. Abúndez, A., Pereyrab, I., Campillo, B.: Improvement of ultimate tensile strength by artificial ageing and retrogression treatment of aluminium alloy 6061. Mater. Sci. Eng. A **668,** 201–207 (2016)

7. Rajaa, S.M., Abdulhadi, H.A., Jabur, K.S., Mohammed, G.R.: Aging time effects on the mechanical properties of Al 6061-T6 alloy. Eng. Technol. Appl. Sci. Res. **8**(4), 3113–3115 (2018)
8. Zhao, Q.-L., Shan, T.-T., Geng, R., Zhang, Y.-Y., He, H.-Y., Qiu, F., Jiang, Q.-C.: Effect of preheating temperature on the microstructure and tensile properties of 6061 aluminum alloy processed by hot rolling-quenching. Metals **9**, 182 (2019). https://doi.org/10.3390/met9020182
9. Masoud, I.M., Abu Mansour, T., Al-Jarrah, J.A.: Effect of heat treatment on the microstructure and hardening properties of 6061 aluminum alloy. J. Appl. Sci. Res. **8**(10), 5106–5113 (2012). ISSN 1819-544X
10. Tan, C.F., Said, M.R.: Effect of hardness test on precipitation hardening aluminium alloy 6061-T6. Chiang Mai J. Sci. **36**(3), 276–286 (2009). www.science.cmu.ac.th/journal-science/josci.html
11. Cina, B.M., Gan, R.: Reducing the susceptibility of alloys particularly aluminum alloys to stress corrosion cracking. Israel Aircraft Industries Ltd., U.S. Patent 3856584, 1974-12-24
12. Shenoy, H.G., Chetty, S.V., Premkumar, S.: Evaluation of wear and hardness of al-si-mg based hybrid composite at different aging conditions (2012)

The Definition of Damage Parameter Changes from the Experimental High-Temperature Creep Curves

A. R. Arutyunyan(✉), R. A. Arutyunyan, and R. R. Saitova

St. Petersburg State University, Universitetsky pr., 28, St. Petersburg, Petrodvorets 198504, Russian Federation
a.arutyunyan@spbu.ru, rigastr@yandex.ru

Abstract. Under the action of relatively low stresses and high temperatures, metallic materials become brittle and fractured with a small value of residual deformations. This problem is known as a problem of thermal brittleness of metals. To solve this challenge in the mechanics of materials the damage conception was introduced. To describe the brittle region of the experimental long-term strength curve, the system of simple kinetic equations for the damage parameter and creep deformation was proposed, and the long-term strength criterion was formulated.

In this work, we propose to determine the damage parameter changes according to the experimental high-temperature creep curves. Only one kinetic equation for creep rate for compressible medium, recorded using the damage parameter is formulated. From this equation, the damage parameter is determined, depending on the creep rate and the creep deformation. Similarly, the value of the damage parameter is determined according to the Rabotnov solution. To describe the experimental creep curves various empirical dependences in the form of power, exponential, and mixed functions are used. Theoretical damage curves are plotted. The corresponding theoretical long-term strength curves are constructed.

Keywords: High-temperature creep · Damage parameter · Embrittlement · Kinetic equation for creep rate · Creep curves · Long-term strength · Rabotvov theory · Incompressible medium · Compressible medium

1 Introduction

The problem of thermal brittleness is considered, when under the action of relatively low stresses and high temperatures, metallic materials become brittle and fractured with a small value of residual deformation. The damage conception that was introduced in the mechanics of materials to describe long-term strength under conditions of high-temperature creep, have been developed in the fundamental works of Kachanov [1] and Rabotnov [2]. In their works, to describe the brittle region of the experimental long-term strength curve, the simple kinetic equation for the damage parameter was proposed, and the long-term strength criterion was formulated. The question of the relationship of creep deformation and damage in these works was not discussed.

D. A. Indeitsev and A. M. Krivtsov (Eds.): APM 2019, LNME, pp. 53–59, 2020.
https://doi.org/10.1007/978-3-030-49882-5_5

The next stage on the creep and damage problem solution relates to the work of Rabotnov [3], in which a system of two interrelated equations for creep deformation and damage parameter was proposed.

In the scientific literature on this problem, the following possible variants of the relationship of creep and damage are given. The processes of creep and damage developed in parallel and in the first approximation are not related to each other. Damage is the result of deformation, which creates fracture sources, leads to appearance of places with a high stress concentration and is a producer of point defects that are necessary for the development of slow fracture. Creep is a result of the micro-fracture processes in the material volume.

Low strain rates and high temperatures contribute to intergranular creep fracture. This fact let assume that damage and fracture could flow independent to plastic deformation. This is also referred by numerous cases of slow fracture with a very small value of the residual deformation. Studies of Ratcliffe and Greenwood [4], Betechtin [5] on the density changes in creep conditions have shown that the pores healing by a single and multiple application of hydrostatic pressure leads to a sharp creep deformation braking. So the time to fracture significant increase. At the same time, the creep rate practically does not changed. The processes of damage by density changes completely braking the development of fracture, which indicates the independence of the creep rate from damage. Apparently, all three possible variants of the relationship between creep and fracture are fair.

2 The Kachanov-Rabotnov Theory

Kachanov-Rabotnov criteria was developed to describe the brittle fracture region [1, 2]. In the Kachanov's brittle fracture model [1] the parameter of continuity ψ $(1 \geq \psi \geq 0)$ is introduced formally without giving to it a certain physical meaning. In the model of Rabotnov brittle fracture [2, 3] the damage parameter ω $(0 \leq \omega \leq 1)$ is introduced by the ratio $\omega = F_T/F_0$ (F_0 is initial, F_T is total pores area) and characterize the degree of reduction of cross-section area of the specimen. From the relation $F = F_0 - F_T$, it follows that $F = F_0(1 - \omega)$ (F is the current specimen cross section area).

In the Kachanov-Rabotnov model of brittle fracture, the rate of continuity parameter changes is given by the following equation

$$\frac{d\psi}{dt} = -A\left(\frac{\sigma_0}{\psi}\right)^n \tag{1}$$

where σ_0 is nominal stress, A, n are constants.

To take into account the deformation processes, Rabotnov introduced, in addition to Eq. (1), a kinetic equation for the creep rate $\dot{\varepsilon}$

$$\frac{d\varepsilon}{dt} = \dot{\varepsilon} = B\sigma_0^m \psi^{-\beta} e^{m\varepsilon} \tag{2}$$

where B, m are constants.

From (2) is follows

$$\psi = \left(\frac{\sigma_0^m B e^{m\varepsilon}}{\dot{\varepsilon}}\right)^{1/\beta} \quad (3)$$

3 The Theory for Compressible Medium

When the interrelated creep and damage equations was formulated, the physical content of the damage parameter should be given. In particular, irreversible changes in volume (loosening) [6] or density [7] were considered as a damage parameter. This parameter is the most representative characteristic of damage. In works [8, 9] the system of equations for creep rate and damage parameter for the compressible medium was considered.

In this work only one kinetic equation for creep rate for compressible medium, recorded using the damage parameter is formulated. From this equation, the damage parameter is determined, depending on the creep rate and the creep deformation.

A compressible medium with a continuity parameter $\psi = \rho/\rho_0$ (ρ_0 is initial, ρ is current specimen density) is introduced. Taking into account the mass conservation law, the kinetic equation for the creep rate is given as

$$\frac{d\varepsilon}{dt} = \dot{\varepsilon} = B\sigma_0^m \psi^{m-\beta} e^{m\varepsilon} \quad (4)$$

where β is constant.

From Eq. (4) we can receive

$$\psi = \left(\frac{\dot{\varepsilon} e^{-m\varepsilon}}{B\sigma_0^m}\right)^{1/m-\beta} \quad (5)$$

4 Deformation Creep Curves Description

Using formulas (3) and (5) it is possible to determine the time dependence of the parameter ψ according to the experimental creep deformation curves [10]. These curves are described by various empirical dependences in the form of power, exponential and mixed functions [10, 11]. In the paper [12], the case of power dependence was considered. Next, we will used the following dependence

$$\varepsilon = \frac{e^{kt}}{(f\,t + c)^n} + b \quad (6)$$

where f, c, k, b are constants.

On Fig. 1, the experimental creep deformation curves according [10] and the empirical dependence in the form of function (6) are presented. In the calculations the following coefficients are used: $c = 1 \cdot 10^5$, $b = 1 \cdot 10^{-1}$, $k = 2,6 \cdot 10^{-5}\ [h]^{-1}$, $n = 0,15$, $f = 8 \cdot 10^{-2}\ [h]^{-1}$.

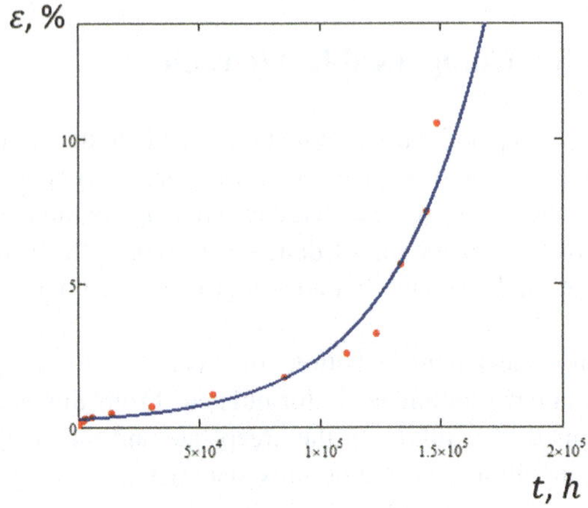

Fig. 1. Theoretical deformation creep curve according to relation (6) and experimental points [10].

5 Continuity Parameter Changes

Introduced the relation (6) into (3) we can receive the following equation for continuity parameter for Rabotnov solution

$$\psi = \left[\frac{B\sigma_0^m e^{m\left(\frac{e^{kt}}{(f\,t+c)^n}+b\right)}(f\,t+c)^n}{e^{kt}\left(k - \frac{nf}{f\,t+c}\right)} \right]^{\frac{1}{\beta}} \tag{7}$$

Taking into account the relations (6) and (5) the equation for continuity parameter for the case of compressible media can obtained

$$\psi = \left[\frac{e^{kt}\left(k - \frac{nf}{f\,t+c}\right) e^{-m\left(\frac{e^{kt}}{(f\,t+c)^n}+b\right)}}{(f\,t+c)^n B\sigma_0^m} \right]^{\frac{1}{m-\beta}} \tag{8}$$

The theoretical damage curves according to formulas (7) (curve 1) and (8) (curve 2) are shown on Fig. 2. In the calculations the following values of coefficients are used:

$c = 1 \cdot 10^5$, $b = 1 \cdot 10^{-1}$, $k = 2,6 \cdot 10^{-5}\ [h]^{-1}$, $n = 0,15$, $f = 8 \cdot 10^{-2}\ [h]^{-1}$, $m = 6$, $\beta = -2$, $\sigma_0 = 120\ MPa$, $B = 3 \cdot 10^{-19}\ [MPa]^{-6}$.

From Fig. 2 it can be seen that for the compressible medium (curve 2) the damage accumulation and, accordingly, the fracture processes are passed more intensive, compared with Rabotnov solution (curve 1 and formula (7)).

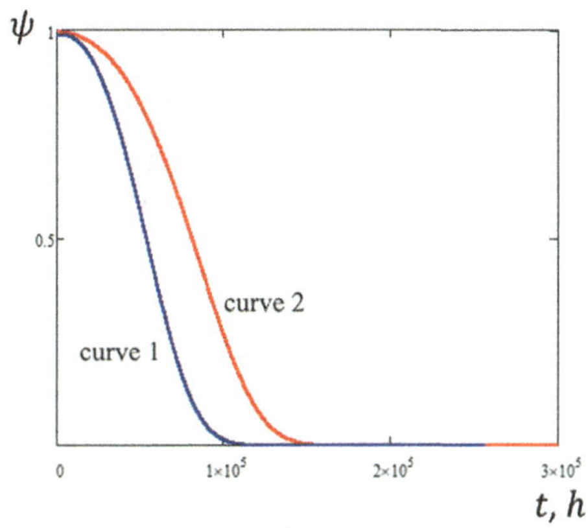

Fig. 2. Theoretical damage curves according to formulas (7) (curve 1) and (8) (curve 2).

6 Long-Term Strength Criterions

The long-term strength criterion can be obtained under the condition, when damage parameter is reached the critical value. Taking in (7) the fracture condition in the form $t = t_f$, $\psi = \psi_*$, we can obtain the following criterion of long-term strength

$$\sigma = \left[\frac{\psi_*^{\beta} e^{kt} \left(k - \frac{nf}{f\,t + c} \right)}{B e^{m\left[\frac{e^{kt}}{(f\,t+c)^n} + b \right]} (f\,t + c)^n} \right]^{\frac{1}{m}} \tag{9}$$

Taking in (8) the fracture condition in the form $t = t_f$, $\psi = \psi_*$, we can receive the long-term strength criterion for compressible medium

$$\sigma = \left[\frac{e^{kt} \left(k - \frac{nf}{f\,t + c} \right)}{(f\,t + c)^n e^{m\left[\frac{e^{kt}}{(f\,t+c)^n} + b \right]} B \psi_*^{m-\beta}} \right]^{\frac{1}{m}} \tag{10}$$

The long-term strength curves according the solutions (9) (curve 1) and (10) (curve 2) are shown on Fig. 3. In the calculations the following values of coefficients are used: $c = 1 \cdot 10^5$, $b = 1 \cdot 10^{-1}$, $k = 2,6 \cdot 10^{-5}\ [h]^{-1}$, $n = 0,15$, $f = 8 \cdot 10^{-2}\ [h]^{-1}$, $m = 6$, $\beta = -2$, $\sigma_0 = 120\ MPa$, $B = 3 \cdot 10^{-19}\ [MPa]^{-6}$, $\psi_* = 0,9$.

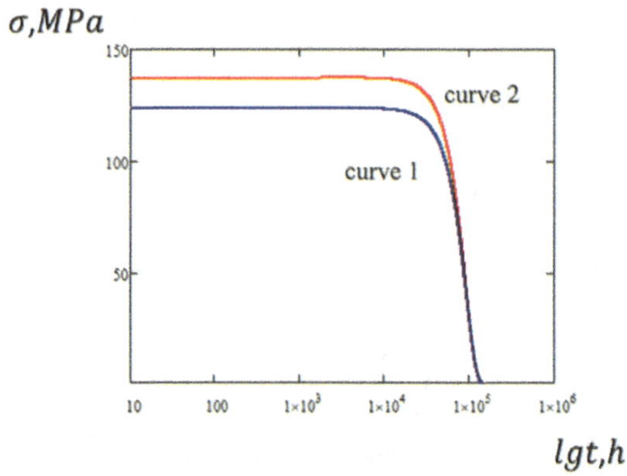

Fig. 3. The long-term strength curves according the solutions (9) (curve 1) and (10) (curve 2).

7 Conclusions

The compressibility of metal materials is taken into account and the relative density changes is considered as a continuity parameter. A method for determining the damage value using experimental creep curves according to the theory of Rabotnov and the theory for a compressible medium is proposed. To describe the experimental creep curves empirical dependence in the form of mixed power and exponential functions is used. Theoretical damage curves are plotted. The long-term strength criterion is obtained under the condition, when damage parameter is reached the critical value. The corresponding theoretical long-term strength curves are constructed. It is shown that for the case of a compressible medium, a more intensive damage accumulation and, accordingly, the fracture processes are observed, compared with the Rabotnov theory.

Acknowledgements. Financial support of the Russian Foundation for Basic Research is gratefully acknowledged (Grant N 18-01-00146).

References

1. Kachanov, L.M.: About Fracture Time Under Creep Conditions. Izv. USSR Academy of Sciences, OTN, vol. 8, pp. 26–31 (1958) (in Russian)
2. Rabotnov, Y.N.: On the mechanism of long-term fracture. In: Questions of Strength of Materials and Structures, Izd-vo USSR Academy of Sciences, pp. 5–7 (1959) (in Russian)
3. Rabotnov, Y.N.: Creep of Structural Elements. Science, Moscow (1966). (in Russian)

4. Ratcliffe, R.T., Greenwood, G.W.: Mechanism of cavitation in magnesium during creep. Phil. Mag. **12**, 59–69 (1965)
5. Betekhtin, V.I.: Porosity of solids. Trans. St.-Petersburg Acad. Sci. Strength Problems **1**, 202–210 (1997)
6. Novozhilov, V.V.: On plastic loosening. PMM **4**, 681–689 (1965). (in Russian)
7. Arutyunyan, R.A.: The Problem of Deformation Aging and Long-Term Fracture in the Mechanics of Materials. Publishing house of St.-Petersburg State University, St.-Petersburg (2004) (in Russian)
8. Arutyunyan, R.A.: High-temperature embrittlement and long-term strength of metallic materials. Mech. Solids **50**(2), 191–197 (2015)
9. Arutyunyan, A., Arutyunyan, R., Saitova, R.: The criterion of high-temperature creep of metals based on relative changes of density. WSEAS Trans. Appl. Theor. Mech. **14**, 140–144 (2019)
10. Aghajani, A., Somsen, Ch., Eggeler, G.: On the effect of long-term creep on the microstructure of a 12% chromium tempered martensite ferritic steel. Acta Mater. **57**(17), 5093–5106 (2009)
11. Agakhi, K.A., Basalov, Y.G., Kuznetsov, V.N., Fomin, L.V.: Creep process simulation considering pre-destruction stage and model identification. Vestnik of Samara State Tech. Univ. Ser. Phys. Mech. Sci. **2**(19), 243–247 (2009)
12. Arutyunyan, A.R., Arutyunyan, R.A., Saitova, R.R.: Evolution of the processes of damage to the elastic-plastic medium in high-temperature creep. In: Proceedings of the Scientific and Technical Conference on the Structural Mechanics of the Ship, Dedicated to the Memory of Professor V.A. Postnov and the 90th Anniversary of his Birth, St. Petersburg, pp. 92–93 (2017). (in Russian)

Evaluation Criteria for the Wear Resistance of High-Chromium Steels

S. A. Atroshenko(✉)

Institute for Problems of Mechanical Engineering Russian Academy of Sciences, Saint-Petersburg 199178, Russia
satroshe@mail.ru

Abstract. The paper is devoted to estimation on steels of the type of 12Cr and 8Cr in terms of efficiency. The criteria for wear resistance for these materials have been studied. The quality estimation on basic feature was carried out. The complex indicator of the quality level of investigated steels working under friction conditions has been determined. It was found that materials with a uniform distribution of carbides have the highest level of quality.

Keywords: 12Cr and 8Cr steels · Wear resistance · Quality level

1 Introduction

Currently, there is an increase in requirements for dimensional accuracy and quality of machined parts, which leads to the preferential use of machining processes in the final stages.

In modern machining production, more and more widespread use is found in expensive automated machine tools with microprocessor control. The operation of such equipment is characterized by a sharp increase in machine-minutes, toughening of the working conditions of the cutting tool, an increase in the consumption of tools per unit of output, which amounts to 5–10% of the total costs of cutting [1]. Thus, the role of the cutting tool, which largely determines the efficiency of machining, increases markedly.

The most important indicator of the operation of the cutting tool is the performance, which characterizes the state of the tool in which it is able to perform its functions, having a working surface wear that is less than the criteria value.

The following main consumer requirements for metal cutting tools can be distinguished: performance and reliability [2]. The performance depends on the processing modes (feed, speed and depth of cut), the geometry of the cutting part of the tool, the physicomechanical properties of the material of the workpiece and the tool material. Indicators of reliability of metal-cutting tools is its durability and reliability.

In this paper it is considered the issues of assessing the quality of tool material, as an important factor determining the performance of the cutting tool.

D. A. Indeitsev and A. M. Krivtsov (Eds.): APM 2019, LNME, pp. 60–65, 2020.
https://doi.org/10.1007/978-3-030-49882-5_6

2 Materials and Experiments

Tested samples of steel type X8 were [3]: steel 70H9F, 60H9M2VFTSCH and 105H8M2VFTSCH. Quenching of these steels was carried out in oil at a temperature: 1050 °C - for steel 70X9Ф, 1025 °C - for steel 60X9M2BФЦЧ and 1000 °C - for steel 105X8M2BФЦЧ. After quenching, the steel was subjected to low tempering at a temperature of 150 °C for 1 h.

For steels of type X12 (50H12FBCHTS, 70H12FBCHTS, 110H12FBCHTS) heat treatment consisted in quenching at a temperature of 1050 °C in oil and low tempering at 150 °C for 1 h.

For comparison, in the same conditions, cutting tools made of tool steels X12Ф1 and 9XC (GOST 5950-73) and high-speed steel P6M5 (GOST 19265-73), heat-treated according to the standard mode, were tested. For steel 9XC, heat treatment was carried out according to the following mode: quenching at a temperature of 870 °C in oil and tempering at 180 °C for 1 h. High-speed steel P6M5 was treated as follows: quenching at a temperature of 1220 °C and three times tempering at a temperature of 560 °C for 1 h.

The main indicator of the material that determines the performance of the tool is its wear resistance. Steel is called wear-resistant, if it well resists abrasion (wear) in various conditions of service: during sliding and rolling friction, with friction on hard and soft materials, with abrasive wear. Type X12 steels were tested on a heel friction machine in a pair with steel 9XC (friction velocity — 0.65 m/s; specific pressure — 2.5 Mn/m^2).

3 Results

As a rule, the wear resistance of steel is directly dependent on its hardness: the higher the hardness, the better the resistance to wear of the steel. However, in some cases, in addition to hardness, the structure of steel, for example, a large amount of hard and strong carbides, the presence of graphite precipitates, and the ability of steel to harden strongly (work hardening) during friction, have a great influence on wear resistance.

Such materials have the greatest wear resistance in which highly hard inclusions are evenly distributed in a sufficiently plastic material of the tool. Tool wear resistant also depends on the size of carbides. Carbides should be fine and evenly distributed in the structure. With coarse clusters, that is, with non-uniformity of decay at the interface of the carbide-matrix, cracks are possible. This impairs tool performance. In addition, with the clusters of carbides, there is a probability that the carbide system rather than the carbide – viscous matrix system, will come into contact with the metal being processed. Due to its brittleness, it is possible to chipping the tool, accelerating its wear. In addition, the surface of particles of the solid phase increases with the grinding of their sizes.

Certain conditions apply to the metal base. It should be as viscous as possible to prevent chipping of the working edge. Therefore, it is advisable to maintain a certain amount of residual austenite. A significant part of austenite, in a thin surface layer, under the action of microdeformations, arising during the operation of the tool, turns into martensite. This layer, with the formation of more hard areas, is located on a more viscous martensitic austenitic basis.

Fragile failure is one of the main reasons for the low performance of the tool. The brittle fracture of the tool is not due to the small initial strength and ductility of the steel, but to changes in its structure during operation, which leads to the formation of cracks at the interface of the carbide-matrix.

Wear resistance is the property of a material to resist wear under certain friction conditions, as measured by the reciprocal of the wear rate or wear intensity [4]. The wear rate is the ratio of the wear value to the time interval during which it occurred. There are instant (at a certain point in time) and average wear rate (for a certain time interval) [4, 5]. The wear intensity is the ratio of the wear value to the conditioned path on which wear occurred, or the volume of work performed. There are instant and average wear intensity [4, 5].

Wear is usually characterized by one of the following characteristics: linear wear intensity J_h, weight J_g, or energy ones J_w [6]. The linear wear characteristic is the height of the worn layer h, which falls on the friction path unit:

$$J_h = \frac{h}{L} \tag{1}$$

or

$$J_h = \frac{V}{LA_a}, \tag{2}$$

where L is the friction path, V is the worn material volume, A_a is the nominal contact area.

The weight characteristic of wear is the weight of a substance that is removed from a unit of nominal contact area per unit of friction path:

$$J_g = \frac{g}{A_a \cdot L}, \tag{3}$$

where g is the weight of the worn substance.

Energy intensity of wear, which determines the amount of worn material per unit of work of the friction force, is expressed by the following relationship:

$$J_w = \frac{V}{W_F}, \tag{4}$$

where W_F is the work of the force of friction.

Figures 1 and 2 show the graphs of changes in weight and linear wear for steels of type X12 when tested on a heel friction machine in pairs with steel 9XC (friction velocity is 0.65 m/s, specific pressure 2.5 Mn/m^2).

On the basis of the data obtained, it is possible to determine the quality level by the most important material indicator using the formula

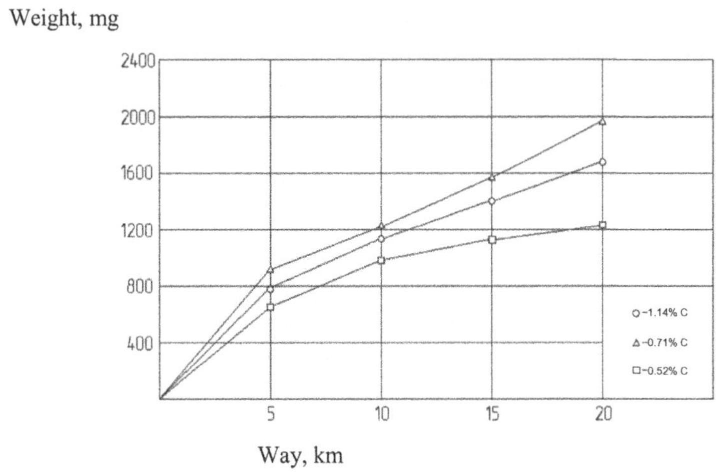

Fig. 1. Weight wear of steel type X12, depending on the path of friction and carbon content

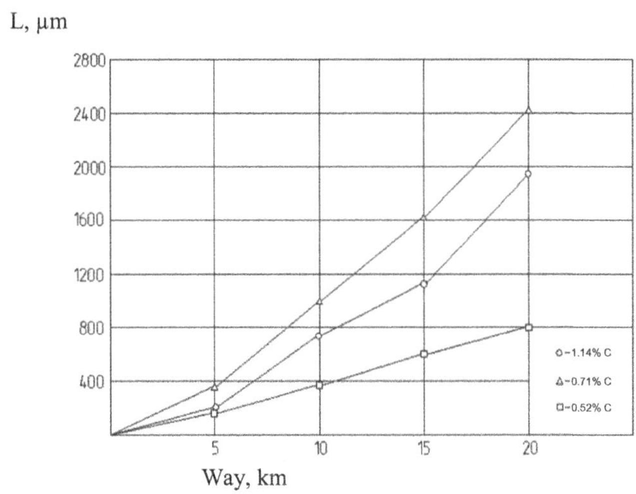

Fig. 2. Linear wear of steel type X12, depending on the path of friction and carbon content

$$Y_{\kappa} = \frac{P_{\det}}{P_{base}}, \tag{5}$$

where P_{det} is the value of the main (determining) indicator of the property of the product being evaluated, P_{base} is the base value of the same main (determining) indicator.

The quality level with a friction path of 20 km in terms of linear wear is equal to: $Y_{0.52} = 0.33$, $Y_{1.14} = 0.79$, $Y_{0.71} = 1$.

For a broader comprehensive assessment of the quality of tool materials, the following criteria can be distinguished: the average tool life (tool life in minutes corresponds to the width of the chamfer of wear on the rear face, 0.5 mm), the coefficient of variation of tool life, the time of trouble-free operation. The results of measurement of these quality indicators are shown in Table 1.

Table 1. Indicators of tool performance.

Steel	Average tool life, $\overline{T}$, min	The coefficient of variation, V	Trouble free time, T_p min
50H12FBCHTS	184	0,08	165,8
70H12FBCHZ	206	0,08	133,9
110H12FBCHZ	254	0,07	231,6
X12F1	83	0,62	16,96
P6M5	265	0,30	163,2
70H9F	90	0,38	46,5
60H9M2VFTSCH	368	0,26	245,4
105X8M2ВФЦЧ	46,8	0,13	38,8
9HS	189,5	0,34	107

However, in order to correctly assess the reliability of the instrument, it is necessary to proceed from the minimum coefficient of variation of durability and the maximum time of trouble-free operation, and not from the average tool life. Therefore, the expert method determined the coefficients of significance for these indicators. For the indicator of average tool life, the weight coefficient is equal to 0.25, for the variation of tool life - 0.35, for the time of no-failure operation - 0.4.

Complex indicator of the quality of the material for the weighted average arithmetic is determined by the formula.

$$Q = \sum_{i=1}^{n} aq, \tag{6}$$

where a is the weight coefficient, q is a relative indicator of the property level, n is the number of indicators of the properties.

Complex quality indicators for the evaluated materials are given in Table 2.

Table 2. Complex indicators of the quality of materials.

Steel	Complex indicator of quality
50H12FBCHTS	0,70
70H12FBCHZ	0,66
110H12FBCHZ	0,90
X12F1	0,12
P6M5	0,53
70H9F	0,20
60H9M2VFTSCH	0,74
105X8M2ВФЦЧ	0,28
9HS	0,38

4 Conclusions

According to the results, it can be concluded that steels with a reduced carbon content and a uniform distribution of carbides (50X12ФБЧЦ, 70X12ФБЧЦ, 110X12ФБЧЦ and 60X9M2ВФЦЧ with 1 and 2 carbide grade) have a higher level of quality. The rather high quality level of high-speed steel R6M5 also with a 3 grade carbide heterogeneity can be explained by higher heat resistance.

The reduced resistance of steel 70Kh9F can be explained by the fact that it contains less quantity of carbides than in steel 60Kh9M2VFTSCH and they are less heat resistant because steel 70KhF does not contain molybdenum and tungsten. The low resistance of steel 105X8M2ВФЦЧ is due to the fact that it contains a lot of carbides and they form clusters.

References

1. Vereshchaka, A.S.: Efficiency of cutting tools with wear-resistant coatings. - M.: Mashinostroenie (1993)
2. Sidulenko, O.A., Ostapenko, M.S., Vasilega, D.S.: Qualimetric assessment of the quality of metal-cutting tools. Bull. Tomsk Polytech. Univ. **318**(2) (2011)
3. Atroshenko, S.A., Naumova, N.S.: Evaluation of the quality of high-chromium steels in terms of performance. Herald Inzhekon. Series: Technical Sciences. **8**(51), p. 97 (2011)
4. GOST 27674 - 88 Friction, wear and lubrication. Terms and Definitions
5. GOST R 50740-95 Tribotechnical requirements and indicators. Principles of security. General provisions
6. Degtyarev, A.I.: Friction and wear of machine parts: Study Guide - Perm: Perm.Gos. tech. Univ – 2003

Moving Web and Dynamic Problem of Aerothermoelastic Vibrations and Instability

Nikolay Banichuk[1(✉)], Svetlana Ivanova[1], and Juha Jeronen[2]

[1] Ishlinsky Institute for Problems in Mechnics RAS, Prospect Vernadskogo 101, bld. 1, 119526 Moscow, Russia
banichuk@gmail.com

[2] Tampere University of Technology, Korkeakoulunkatu 6, 33720 Tampere, Finland

Abstract. The paper is devoted to the analysis of the axially travelling web supported by a system of fixed rollers and submerged in axially flowing gas medium. In order to accurately model the dynamics and stability of a lightweight moving web, the interaction between it and the surrounding air is taken into account. The light weight of the moving web leads to the inertial contribution of the surrounding air to the acceleration of the material becoming significant. In the context of this paper we apply a Galerkin method for dynamic stability analysis of the moving web based on developed added-mass model.

Keywords: Moving web · Aeroelastic vibrations · Instability

1 Basic Relations

The axially travelling thermoelastic web supported by a system of fixed rollers and submerged in axially flowing hot gas medium is considered. Only one span of the web is under consideration. To characterize the dynamics of the moving web (panel) performing in its axial movement additional transverse vibrations let us perform a numerical dynamic stability analysis. For numerical analysis a discrete approximation will be used for the partial differential equation [1,2]

$$\begin{aligned} \alpha^2 (1 + r_m) \frac{\partial^2 w}{\partial t^2} + 2\alpha c_0 (1 + r_v r_m) \frac{\partial^2 w}{\partial x \partial t} \\ + \left(c_0^2 - 1 + c_0^2 r_v^2 r_m\right) \frac{\partial^2 w}{\partial x^2} + \beta \frac{\partial^4 w}{\partial x^4} = g(x,t), \end{aligned} \tag{1}$$

with boundary conditions

$$(w)_{x=\pm 1} = 0, \quad \left(\frac{\partial^2 w}{\partial x^2}\right)_{x=\pm 1} = 0 \tag{2}$$

and initial conditions

$$(w)_{t=0} = g_1(x), \quad \left(\frac{\partial w}{\partial t}\right)_{t=0} = g_2(x), \tag{3}$$

D. A. Indeitsev and A. M. Krivtsov (Eds.): APM 2019, LNME, pp. 66–71, 2020.
https://doi.org/10.1007/978-3-030-49882-5_7

where g, g_1, g_2 are given functions, w is the transverse displacement. We consider in (1) - (3) the dimensionless variables $x' = x/l$, $t' = t/\tau$ (prime is omitted). We denote by l, h, D, E, ν, T, m respectively half-length, thickness, bending rigidity, Young's modulus, Poisson's ratio, tension, mass per unit area of the panel, while V_0, v_∞ are axial velocities of the web and gas, $D = Eh^3/12(1-\nu^2)$, $T = T_m - T_\theta$, $T_m = T_0$ is axial mechanical tension, T_θ is thermal compression due to fixed positions of supports. We define also the dimensionless quantities

$$\alpha = \frac{l}{\tau C}, \quad \beta = \frac{D}{l^2 T}, \quad r_m = \frac{m_a}{m}, \quad r_v = \frac{v_\infty}{V_0}, \quad c_0 = \frac{V_0}{C}, \quad \kappa = \frac{v_\infty}{C}, \quad \gamma = \frac{l}{m}\rho_f.$$

Here $C = \sqrt{T/m}$ is critical velocity of a traveling web, $m_a = \frac{\pi}{4} l \rho_f$ is added mass, ρ_f is gas density, $\tau = l/C$ is a scaling factor. The Eq. (1) is based on added-mass approximation and will be used to determine the dynamical behavior and loss of stability.

We will perform space-discretization using the finite element Galerkin method with C^2 - continuous Hermite elements. We set the load as $g \equiv 0$ and formulate the eigenvalue problem using the time-harmonic trial function

$$w(x,t) = \exp(st)W(x) \tag{4}$$

where s is the stability exponent (a complex number) and $W(x)$ is the vibration mode. We will solve the problem for eigenvalue-eigenfunction pairs (s, W). Loss of stability occurs at such values of the axial drive (transport) velocity V_0 where at least one eigenvalue s transits to the positive half-space (i.e., where the real part Re s becomes positive). The critical velocity is the smallest positive V_0 such the stability is lost.

2 Dynamic Stability Analysis

We will represent the displacement as a Galerkin series

$$W(x) = \sum_{n=1}^{\infty} c_n \Psi_n(x) \tag{5}$$

where Ψ_n are the global shape functions and c_n are the global degrees of freedom. The basis functions $\Psi_n(x)$ are defined piecewise.

Let us now develop the discrete eigenvalue problem. Inserting the time-harmonic trial function (4) into (1) and setting $g \equiv 0$ we have

$$\begin{aligned} & s^2\alpha^2 (1 + r_m) W + 2s\alpha c_0 (1 + r_v r_m) \frac{dW}{dx} \\ & + \left(c_0^2 - 1 + c_0^2 r_v^2 r_m\right) \frac{d^2W}{dx^2} + \beta \frac{d^4W}{dx^4} = 0. \end{aligned} \tag{6}$$

To obtain the weak form we multiply (6) by the test function $\Psi_j(x)$ and integrate over the dimensionless space domain $\Omega = \{x \in (-1,1)\}$:

$$s^2\alpha^2(1+r_m)\int_{-1}^{1} W\Psi_j dx + 2s\alpha c_0(1+r_v r_m)\int_{-1}^{1}\frac{dW}{dx}\Psi_j dx$$
$$+\left(c_0^2 - 1 + c_0^2 r_v^2 r_m\right)\int_{-1}^{1}\frac{d^2W}{dx^2}\Psi_j dx + \beta\int_{-1}^{1}\frac{d^4W}{dx^4}\Psi_j dx = 0.$$

Applying integration by parts (twice in the fourth order term) and the simply supported boundary conditions (2) we have the weak form

$$\begin{aligned} &s^2\alpha^2(1+r_m)\int_{-1}^{1} W\Psi_j dx + 2s\alpha c_0(1+r_v r_m)\int_{-1}^{1}\frac{dW}{dx}\Psi_j dx \\ &-\left(c_0^2 - 1 + c_0^2 r_v^2 r_m\right)\int_{-1}^{1}\frac{dW}{dx}\frac{d\Psi_j}{dx}dx + \beta\int_{-1}^{1}\frac{d^2W}{dx^2}\frac{d^2\Psi_j}{dx^2}dx = 0. \end{aligned} \tag{7}$$

Inserting the Galerkin series (5) into the weak form (7) we have the following system:

$$\begin{aligned} &\sum_{n=1}^{n_{max}}\left\{s^2\alpha^2(1+r_m)\mathrm{A}_{jn} + 2s\alpha c_0(1+r_v r_m)\mathrm{B}_{jn}\right. \\ &\left.+\left[-\left(c_0^2 - 1 + c_0^2 r_v^2 r_m\right)\mathrm{C}_{jn} + \beta\mathrm{D}_{jn}\right]\right\} c_n = 0. \end{aligned} \tag{8}$$

Here $j,\ n = 1, 2, 3, ...$, and the matrices A_{jn}, B_{jn}, C_{jn}, and D_{jn} are defined by

$$\begin{aligned} \mathrm{A}_{jn} &= \int_{-1}^{1}\Psi_n(x)\Psi_j(x)dx, \quad \mathrm{B}_{jn} = \int_{-1}^{1}\frac{d\Psi_n(x)}{dx}\Psi_j(x)dx, \\ \mathrm{C}_{jn} &= \int_{-1}^{1}\frac{d\Psi_n(x)}{dx}\frac{d\Psi_j(x)}{dx}dx, \quad \mathrm{D}_{jn} = \int_{-1}^{1}\frac{d^2\Psi_n(x)}{dx^2}\frac{d^2\Psi_j(x)}{dx^2}dx. \end{aligned} \tag{9}$$

Further, defining

$$\begin{aligned} &\mathrm{M}_2 = \alpha^2(1+r_m)\mathrm{A}_{jn}, \quad \mathrm{M}_1 = 2\alpha c_0(1+r_v r_m)\mathrm{B}_{jn}, \\ &\mathrm{M}_0 = -\left(c_0^2 - 1 + c_0^2 r_v^2 r_m\right)\mathrm{C}_{jn} + \beta\mathrm{D}_{jn} \end{aligned} \tag{10}$$

we obtain the (discrete) quadratic eigenvalue-eigenvector pairs $(s,\ c_0)$

$$\left\{\mathrm{M}_2 s^2 + \mathrm{M}_1 s + \mathrm{M}_0\right\} c_0 = 0. \tag{11}$$

Its companion form is the following twice large generalized linear eigenvalue problem

$$\begin{Vmatrix} -\mathrm{M}_1 & -\mathrm{M}_0 \\ \mathrm{I} & 0 \end{Vmatrix}\begin{Vmatrix} sc_0 \\ c_0 \end{Vmatrix} = \begin{Vmatrix} \mathrm{M}_2 & 0 \\ 0 & \mathrm{I} \end{Vmatrix} s \begin{Vmatrix} sc_0 \\ c_0 \end{Vmatrix} \tag{12}$$

Equation (12) can be solved using a standard solver. Problem parameters values used in the numerical examples corresponding to paper materials are the following: $\rho_f = 1.225\,[\mathrm{kg/m^3}]$, $v_\infty = 0\,[\mathrm{m/s}]$, $T_0 = 500\,[\mathrm{N/m}]$, $m = 0.08\,[\mathrm{kg/m^2}]$, $l = 1\,[\mathrm{m}]$, $h = 10^{-4}\,[\mathrm{m}]$, $E = 10^9\,[\mathrm{N/m^2}]$, $\nu = 0.3$, $\tau = l/\sqrt{T/m}\,[\mathrm{s}]$, $\alpha_\theta = 3\cdot 10^{-6}\,[1/\mathrm{K}]$. The value of the linear thermal expansion coefficient has been chosen from the range typical for paper materials according to [3]. Specifically, it is valid for the machine direction (i.e. x-coordinate).

Consider a setup where the temperature is kept constant within the material. In this case

$$T_\theta = \frac{Eh}{1-\nu}\varepsilon_\theta = \frac{Eh}{1-\nu}\alpha_\theta\theta \approx 0.428570.$$

The effect is rather small, but nevertheless if the system is already operating near its stability limit, thermal expansion may reduce the critical velocity just enough to make the system stability loss. As an example, let us consider a rather extreme temperature difference between 20 °C and 90 °C, which may occur when a paper web initially at room temperature enters the dryer section. The temperature difference is $\theta = 70\,°\mathrm{C}$, which yields $T_\theta = 30\,\mathrm{N/m}$. With a typical level of applied axial tension $T_0 = 500\,\mathrm{N/m}$, the decrease in tension due to thermal expansion from 500 N/m to 470 N/m is 6%.

In the reference setup, where $\theta = 0$ and $T = T_0$, the four pairs of s with the smallest magnitude solved from Eq. (12) are shown in Fig. 1.

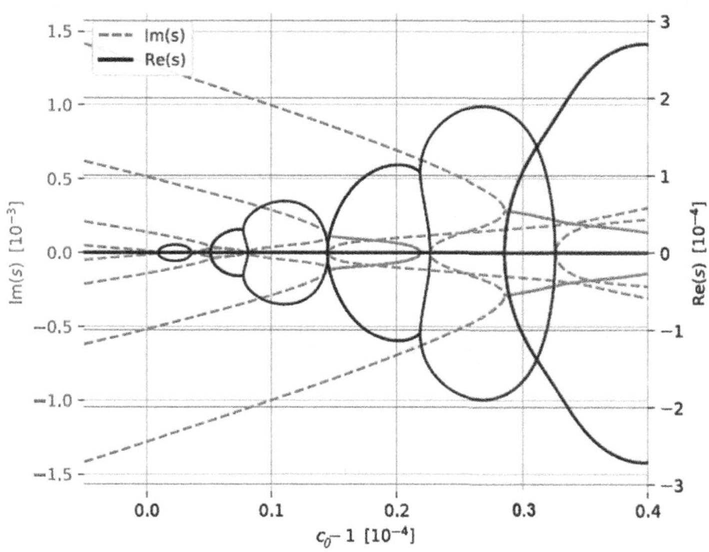

Fig. 1. Stability exponent s as a function of the dimensionless axial drive velocity c_0.

Critical values of the dimensionless axial drive velocity c_0 are listed in Table 1. Based on the table, the lowest critical velocity is

$$V_0^{cr} = c_0^{cr}C = (1 + 0.009 \cdot 10^{-4})\sqrt{T/m} \approx 79.05701\,\mathrm{m/s}.$$

Table 1. Critical points where the real part of at least one solution enters or exists the positive half-plane.

$c_0 - 1$ [10^{-4}] with $\theta = 0$	$c_0 - 1$ [10^{-4}] with $\theta = 70$K	Mode	Type	Note
0.009	0.009	1	Loss of stability	Classic divergence gap
0.036	0.038	1	Gain of stability	
0.051	0.054	1+2	Loss of stability	
0.081	0.086	1+2	Gain of stability	Other branch still unstable
0.145	0.153	1+2	Gain of stability	End of unstable branch
0.145	0.153	2+3	Loss of stability	
0.226	0.240	2+3	Gain of stability	Other branch still unstable
0.285	0.303	3+4	Loss of stability	
0.325	0.346	2+3	Gain of stability	End of unstable branch

In the thermally expanded case when $\theta = 70\,\mathrm{K}$, we have $T = 470\,\mathrm{N/m}$. Visually the solution is not significantly different from Fig. 1 so the graph is omitted. The critical points are listed in Table 1. The lowest critical velocity is

$$V_0^{cr} = c_0^{cr} C = (1 + 0.009 \cdot 10^{-4})\sqrt{T/m} \approx 76.64862\,\mathrm{m/s}$$

showing a decrease of 3% with regard to the reference setup. Note the different value of the normalization constant C, because the tension is now lower.

In both cases the presence of bending rigidity ($D > 0$) pushes the first critical point above $c_0 = 1$, but only by a small amount, because for paper materials, D is small. We have

$$D = \frac{Eh^3}{12(1-\nu^2)} \approx 9.1575 \cdot 10^{-5}\,\mathrm{Nm},$$
$$\beta = \frac{D}{l^2 T} \approx 1.8315 \cdot 10^{-7}.$$

Nevertheless, because β is the coefficient of the highest-order term in (1), the presence of finite bending rigidity, no matter how small, introduces a singular perturbation to the equation, changing its qualitative behavior (on singular perturbation, see [4]).

3 Conclusions

An effective partly-analytical and partly-numerical approach has been presented to model a thermoelastic web moving at a constant velocity subjected to potential flow. An aerothermoelastic model based on a homogeneous temperature distribution and an added-mass approximation for the fluid reaction pressure was developed and analyzed. This model was then applied to some particular cases of dynamic interaction. As a result, in the case of non-stationary behavior of the web, the analysis was performed using the Galerkin method. The presented approach has applications, for example, in elasticity, aeroelasticity, aerothermoelasticity and axially moving materials.

Acknowledgements. *The study was performed in Ishlinsky Institute for Problems in Mechanics RAS and supported by the Russian Science Foundation (project 17-19-01247).*

References

1. Banichuk, N.V., Ivanova, S.Y.: Mathematical modelling of the axially moving panels subjected to thermomechanical actions. Mech. Based Des. Struct. Mach. **46**(1), 101–109 (2018)
2. Banichuk, N., Jeronen, J., Neittaanmaki, P., Saksa, T.: Tero Tuovinen. Springer International Publishing Switzerland, Mechanics of Moving Materials (2014)
3. Kouko J. Effects of Heating, Drying and Straining on the Relaxation and Tensile Properties of Wet Paper. PhD thesis, Department of Physics, University of Jyväskylä. Research Report 12/2014
4. Bender, C.M., Orszag, S.A.: Advanced Mathematical Methods for Scientists and Engineers I: Asymptotic Methods and Perturbation Theory. Springer (1978). Reprint 1999

New Ansatzes for Solution of Nonlinear Nonautonomous Klein-Fock-Gordon Equation

A. N. Bulygin(✉) and Yu. V. Pavlov

Institute for Problems in Mechanical Engineering of Russian Academy of Sciences, 61 Bol'shoy, V.O., Saint Petersburg 199178, Russia
bulygin_an@mail.ru

Abstract. Methods of finding of exact analytical solutions of nonlinear nonautonomous Klein-Fock-Gordon (KFG) equation are proposed. Solutions $U(x, y, z, t)$ are sought in the form of compound function $U = f(W)$. The argument $W(x, y, z, t)$ is called ansatz. Function $f(W)$ and ansatz W are sought from the different equations. The function $f(W)$ satisfies to nonlinear ordinary differential equation. Its solution is received in the form of integral. Ansatz W is defined as a root of the special algebraic equation. Different ansatzes W are proposed and different solutions of the nonlinear nonautonomous KFG equation are found.

Keywords: Klein-Fock-Gordon equation · Nonautonomous equation · Exact solution · Ansatz

1 Introduction

The nonlinear nonautonomous Klein-Fock-Gordon equation

$$U_{xx} + U_{yy} + U_{zz} - \frac{U_{tt}}{v^2} = p(x, y, z, t)\, F(U) \tag{1}$$

is widely applied in modern fundamental science and its many applied sections. Here v is a constant, $p(x, y, z, t)$ and $F(U)$ are arbitrary functions, and the subscript means the derivative with respect to the corresponding variable.

Effective methods for solution of the autonomous equation ($p(x, y, z, t) = p_0 = \text{const}$) are developed and in literature there are practically no methods of solution of nonautonomous ($p(x, y, z, t) \neq \text{const}$) KFG equation. Ways of construction of exact analytical solutions of the Eq. (1) for wide, but special type functions $p(x, y, z, t)$ are proposed in works [1–5]. In these works the solution of the Eq. (1) is sought in the form of compound function

$$U = f(W). \tag{2}$$

Then the Eq. (1) will take a form

$$f_{WW}\left[W_x^2 + W_y^2 + W_z^2 - \frac{W_t^2}{v^2}\right] + f_W\left[W_{xx} + W_{yy} + W_{zz} - \frac{W_{tt}}{v^2}\right] = p\, F[f]. \tag{3}$$

D. A. Indeitsev and A. M. Krivtsov (Eds.): APM 2019, LNME, pp. 72–80, 2020.
https://doi.org/10.1007/978-3-030-49882-5_8

The Eq. (3) can be solved if to impose restrictions for functions $W(x, y, z, t)$ and $p(x, y, z, t)$. Let's accept that $W(x, y, z, t)$ and $p(x, y, z, t)$ satisfy to the following equations

$$\textbf{1.}\ W_x^2 + W_y^2 + W_z^2 - \frac{W_t^2}{v^2} = 0, \quad W_{xx} + W_{yy} + W_{zz} - \frac{W_{tt}}{v^2} = p(x, y, z, t), \tag{4}$$

$$\textbf{2.}\ W_x^2 + W_y^2 + W_z^2 - \frac{W_t^2}{v^2} = p(x, y, z, t), \quad W_{xx} + W_{yy} + W_{zz} - \frac{W_{tt}}{v^2} = 0, \tag{5}$$

$$\begin{aligned} \textbf{3.}\ & W_x^2 + W_y^2 + W_z^2 - \frac{W_t^2}{v^2} = q(x, y, z, t)P(W), \\ & W_{xx} + W_{yy} + W_{zz} - \frac{W_{tt}}{v^2} = q(x, y, z, t)Q(W). \end{aligned} \tag{6}$$

Here $q(x, y, z, t)$, $P(W)$, $Q(W)$ are arbitrary functions.

From (4)–(6) we can see that $f(W)$ will be the solution of the Eq. (3) if it satisfies the following nonlinear ordinary differential equations

$$\textbf{1.}\quad f_W = F(f), \quad \int \frac{df}{F(f)} = W, \tag{7}$$

$$\textbf{2.}\quad f_{WW} = F(f), \quad \int \frac{df}{\sqrt{E + V(f)}} = \pm\sqrt{2}W, \tag{8}$$

$$\textbf{3.}\quad P(W)f_{WW} + Q(W)f_W = F(f), \quad \int \frac{df}{\sqrt{E + V(f)}} = G(W), \tag{9}$$

$$G(W) = \int \frac{dW}{PQ}, \quad p(x, y, z, t) = \frac{q(x, y, z, t)}{2A} P(W) \exp\left(-2\int \frac{Q}{P}\, dW\right). \tag{10}$$

Here $V_f = F(f)$ and (E, A) are constants of integration.

As it appears from (7)–(10) the function $f(W)$ can be found by integration for arbitrary nonlinear function $F(f)$ if the corresponding integrals exist. The solution of $f(W)$ is found in an explicit form if the integrals allow inversion. Thus, the proposed ways lead the solution of the Eq. (1) to finding of the ansatzes $W(x, y, z, t)$ from solutions (4)–(6).

2 Ansatzes and Methods of Their Construction

Ansatzes are possible to find from the Eqs. (4)–(6) if to use methods which were developed for construction of functionally invariant solutions of the wave equation [6–8]. Let's assume that

$$W = \Psi(\tau), \tag{11}$$

$\Psi(\tau)$ are the arbitrary function and τ is a root of the algebraic equation

$$[x - \xi(\tau)]^2 + [y - \eta(\tau)]^2 + [z - \zeta(\tau)]^2 = v^2(t - \tau)^2. \tag{12}$$

The arbitrary functions $\xi(\tau), \eta(\tau), \zeta(\tau)$ define the root $\tau(x, y, z, t)$ of the Eq. (12). Let's consider simple special cases:

$$1.\ \ \xi = \eta = \zeta = 0, \quad \tau = t \pm \frac{R}{v}, \quad R = \sqrt{x^2 + y^2 + z^2}, \tag{13}$$

$$\mathbf{2.}\ \ \xi = v x_1 \tau, \quad \eta = v x_2 \tau, \quad \zeta = v x_3 \tau,$$

$$\tau = -\frac{X \pm \nu_1}{v R_0}, \quad \nu_1 = \sqrt{X^2 + R_0 s^2}, \quad (R_0 \neq 0), \tag{14}$$

$$\tau = -\frac{s^2}{2vX}, \quad (R_0 = 0), \tag{15}$$

$$X = x_1 x + x_2 y + x_3 z - vt, \quad s^2 = x^2 + y^2 + z^2 - v^2 t^2, \quad R_0 = 1 - (x_1^2 + x_2^2 + x_3^2). \tag{16}$$

Root τ given by the formula (14) contains a square root. Radicand is a square form of four variables x, y, z, t. Its Sylvester determinants are

$$\Delta_1 = 1 - x_2^2 - x^3, \quad \Delta_2 = (1 - x_3^2) R_0, \quad \Delta_3 = R_0^2, \quad \Delta_4 = 0,$$

and eigenvalue are

$$\lambda_1 = 0, \quad \lambda_2 = R_0, \quad \lambda_3 = R_0, \quad \lambda_4 = 2 - R_0.$$

Therefore, under a root a non-negative square form (if $x_1^2 + x_2^2 + x_3^2 \leq 1$), and root τ is real.

It is possible to prove by direct calculations that

$$\tau_x^2 + \tau_y^2 + \tau_z^2 - \frac{\tau_t^2}{v^2} = 0, \tag{17}$$

$$\tau_{xx} + \tau_{yy} + \tau_{zz} - \frac{\tau_{tt}}{v^2} = -\frac{2}{v \nu_1}, \tag{18}$$

$$\nu_{1,x}^2 + \nu_{1,y}^2 + \nu_{1,z}^2 - \frac{\nu_{1,t}^2}{v^2} = R_0, \tag{19}$$

$$\nu_{1,xx} + \nu_{1,yy} + \nu_{1,zz} - \frac{\nu_{1,tt}}{v^2} = \frac{2R_0}{\nu_1}. \tag{20}$$

The formulas

$$x_1 \nu_{1,x} + x_2 \nu_{1,y} + x_3 \nu_{1,z} + \frac{\nu_{1,t}}{v} = 0, \tag{21}$$

$$x \nu_{1,x} + y \nu_{1,y} + z \nu_{1,z} + t \nu_{1,t} = \nu_1. \tag{22}$$

were taking into account in receiving the Eqs. (17)–(20). From the Eqs. (4), (17) and (18) one can see that if to construct ansatz $W = \Psi(\tau)$ then it is possible to solve the Eq. (1) with

$$p(x, y, z, t) = -2\frac{\Psi_\tau}{v \nu_1}. \tag{23}$$

An ansatz can be chosen as

$$W = \Psi(\nu_1). \tag{24}$$

Taking into account (19), (20) the Eqs. (3) can be written in the form

$$\Psi_{\nu_1}^2 f_{WW} + \left(\Psi_{\nu_1\nu_1} + \frac{2}{\nu_1}\Psi_{\nu_1}\right) f_W = \frac{p}{R_0} F(f). \tag{25}$$

The Eq. (25) can be solved on the basis of the proposition (5) if we assume that

$$p(x, y, z, t) = \frac{1}{R_0 \Psi_{\nu_1}^2}, \qquad \Psi = a + \frac{b}{\nu_1}, \tag{26}$$

where a, b are the integration constants.

In addition to functions τ and ν_1 we will enter new function

$$\lambda(x, y, z, t) = vw(\tau)\nu_1. \tag{27}$$

Here $w(\tau)$ is an arbitrary function of the root $\tau(x, y, z, t)$. If to calculate private derivative of the functions λ of the first and second orders, then one can prove that

$$\lambda_x^2 + \lambda_y^2 + \lambda_z^2 - \frac{\lambda_t^2}{v^2} = \lambda\sigma, \qquad \sigma = 2w_\tau \left(\frac{vwR_0}{2\nu_1 w_\tau} - 1\right), \tag{28}$$

$$\lambda_{xx} + \lambda_{yy} + \lambda_{zz} - \frac{\lambda_{tt}}{v^2} = 2\sigma. \tag{29}$$

By direct calculations it is also possible to prove that function τ, ν_1, λ satisfy to the following equations

$$\tau_x \nu_{1,x} + \tau_y \nu_{1,y} + \tau_z \nu_{1,z} - \frac{\tau_t \nu_{1,t}}{v^2} = -\frac{1}{v}, \tag{30}$$

$$\tau_x \lambda_x + \tau_y \lambda_y + \tau_z \lambda_z - \frac{\tau_t \lambda_t}{v^2} = -w, \tag{31}$$

$$\lambda_x \nu_{1,x} + \lambda_y \nu_{1,y} + \lambda_z \nu_{1,z} - \frac{\lambda_t \nu_{1,t}}{v^2} = vR_0 w - \nu_1 w_\tau. \tag{32}$$

The formulas

$$x_1 \tau_x + x_2 \tau_y + x_3 \tau_z + \frac{\tau_t}{v} = \frac{1}{v}, \tag{33}$$

$$x\tau_x + y\tau_y + z\tau_z + t\tau_t = \tau \tag{34}$$

are taking into account for proof of Eqs. (30)–(32).

Introduction of function λ expands a set of ansatzes of W which can be constructed on the basis of functions τ, ν_1, λ. At first, ansatz

$$W = \Psi(\lambda) \tag{35}$$

allows to write the Eq. (3) in the following form

$$\lambda \Psi_\lambda^2 f_{WW} + (\lambda \Psi_{\lambda\lambda} + 2\Psi_\lambda) f_W = \frac{p}{\sigma} F(f). \tag{36}$$

It is also solved on the basis of the proposition (5) if

$$p(x,y,z,t) = \frac{P}{\lambda\sigma\Psi_\lambda^2}, \quad \Psi = a + \frac{b}{\lambda}, \quad (a,b) = \text{const}. \tag{37}$$

It is possible to prove that functions

$$W = \Psi(\theta) \tag{38}$$

with arguments which are the product of two functions $(\tau, \lambda, \nu_1, \sqrt{w(\tau)})$

$$\theta = \left\{\tau\lambda, \tau\nu_1, \nu_1\sqrt{w(\tau)}\right\} \tag{39}$$

will be ansatz W. For ansatzes (38), (39) the Eq. (3) coincides with (36)

$$\theta\Psi_\theta^2 f_{WW} + (\theta\Psi_{\theta\theta} + 2\Psi_\theta)\, f_W = \bar{p}F(f) \tag{40}$$

but with different $\bar{p}$

$$\bar{p} = \left\{\frac{p}{2w(\tau d - 1)}, \; \frac{p}{\frac{2}{v}\left(\frac{\tau}{c\nu_1} - 1\right)}, \; \frac{p}{\frac{1}{v}\left(2d + \frac{w_\tau}{w}\right)}\right\}, \quad c = \frac{2}{vR_0}, \quad d = \frac{\sigma}{2w}. \tag{41}$$

The Eq. (40) is solved also as (36), and allows to find solutions of the nonlinear nonautonomous KFG equation for different functions p.

On the basis of functions τ, λ, ν_1 it is possible to construct ansatz W which depends on two variables

$$W = \{\Psi(\tau,\lambda), \; \Psi(\tau,\nu_1), \; \Psi(\lambda,\nu_1)\}. \tag{42}$$

In the case of ansatzes (42)

$$W_x^2 + W_y^2 + W_z^2 - \frac{W_t^2}{v^2} = \begin{cases} 2w\Psi_\lambda\left[\lambda\Psi_\lambda d - \Psi_\tau\right], \\ R_0\Psi_{\nu_1}\left(\Psi_{\nu_1} - c\Psi_\tau\right), \\ \left[\frac{R_0}{\nu_1^2}\left(\nu_1\Psi_{\nu_1} + \lambda\Psi_\lambda\right) - 2w_\tau\Psi_\lambda\right]\left(\nu_1\Psi_{\nu_1} + \lambda\Psi_\lambda\right), \end{cases} \tag{43}$$

$$W_{xx} + W_{yy} + W_{zz} - \frac{W_{tt}}{v^2} = \begin{cases} \frac{2w}{\lambda}\left[d\lambda\frac{\partial}{\partial\lambda} - \frac{\partial}{\partial\tau}\right](\lambda\Psi_\lambda + \Psi), \\ \frac{R_0}{\nu_1}\left(\frac{\partial}{\partial\nu_1} - c\frac{\partial}{\partial\tau}\right)(\nu_1\Psi_{\nu_1} + \Psi), \\ \left[\frac{R_0}{\nu_1^2}\left(\lambda\frac{\partial}{\partial\lambda} + \nu_1\frac{\partial}{\partial\nu_1}\right) - 2w_\tau\Psi_\lambda\frac{\partial}{\partial\lambda}\right] \\ \times(\nu_1\Psi_{\nu_1} + \lambda\Psi_\lambda + \Psi). \end{cases} \tag{44}$$

From (43), (44) it is visible that ansatz W can be chosen as the root of different equations. If to accept that

$$d\lambda\Psi_\lambda - \Psi_\tau = 0, \qquad \Psi = \Phi(\ln\lambda - d\tau), \tag{45}$$

$$\Psi_{\nu_1} - c\Psi_\tau = 0, \qquad \Psi = \Phi(c\nu_1 + \tau), \tag{46}$$

$$\lambda\Psi_\lambda + \nu_1\Psi_{\nu_1} = 0, \qquad \Psi = \Phi\left(\frac{\lambda}{\nu_1}\right), \tag{47}$$

then, as we can see from (43), W will satisfy the equation

$$W_x^2 + W_y^2 + W_z^2 - \frac{W_t^2}{v^2} = 0 \tag{48}$$

and, therefore, the Eq. (3) can be solved on a basis (4). For the cases

$$\lambda\Psi_\lambda + \Psi = 0, \qquad \Psi = \frac{\Phi(\tau)}{\lambda}, \tag{49}$$

$$\nu_1\Psi_{\nu_1} + \Psi = 0, \qquad \Psi = \frac{\Phi(\tau)}{\nu_1}, \tag{50}$$

$$\lambda\Psi_\lambda + \nu_1\Psi_{\nu_1} + \Psi = 0, \qquad \Psi = \frac{1}{\lambda}\Phi\left(\frac{\lambda}{\nu_1}\right), \tag{51}$$

the ansatz W is the wave functions. Therefore, the Eq. (3) is solved on the basis of the proposition (5). In (45)–(51) $\Phi(u)$ is an arbitrary function of u. Besides, we used the simplest solution of the corresponding equation which defines the ansatz W.

The number of wave functions can be increased if to pass to new coordinates which allow to find new wave functions by using one wave function. So, if $f(x, y, z, t)$ is the wave function, then

$$\frac{1}{s^2} f\left(\frac{x}{s^2}, \frac{y}{s^2}, \frac{z}{s^2}, \frac{t}{s^2}\right) \tag{52}$$

and

$$\frac{1}{z - vt} f\left(\frac{x}{z - vt}, \frac{y}{z - vt}, \frac{s^2 - 1}{2(z - vt)}, \frac{s^2 + 1}{2v(z - vt)}\right) \tag{53}$$

are the wave functions also [7].

The Eq. (40) can be solved also on the basis of the proposition (6). If to take into account that

$$f' = \frac{f_\theta}{\Psi_\theta}, \qquad f'' = \frac{1}{\Psi_\theta^2}\left(f_{\theta\theta} - \frac{\Psi_{\theta\theta} f_\theta}{\Psi_\theta}\right), \tag{54}$$

then the Eq. (40) takes the form

$$\frac{d}{d\theta}\left(\theta'' \frac{f_\theta^2}{2}\right) = \bar{p}\,\theta^3 \frac{dV_f}{d\theta}, \qquad V_f = F(f). \tag{55}$$

If $\bar{p}\theta^3 = 1$ then one has from (55)

$$\int \frac{df}{\sqrt{E+V(f)}} = \pm\sqrt{2}\left(a - \frac{1}{\theta}\right). \tag{56}$$

Here E, A are integration constants.

3 Special Solutions

Below we consider as examples the cases when

$$F(U) = \{\sin U, \quad \sinh U, \quad \exp(mU)\}. \tag{57}$$

For these functions integrals (7) and (8) exist and can be inverted. Finally we find

$$F(U) = \left\{2\tan^{-1} e^W, \quad 2\tanh^{-1} e^W, \quad \frac{-1}{m}\log(E - mW)\right\}, \tag{58}$$

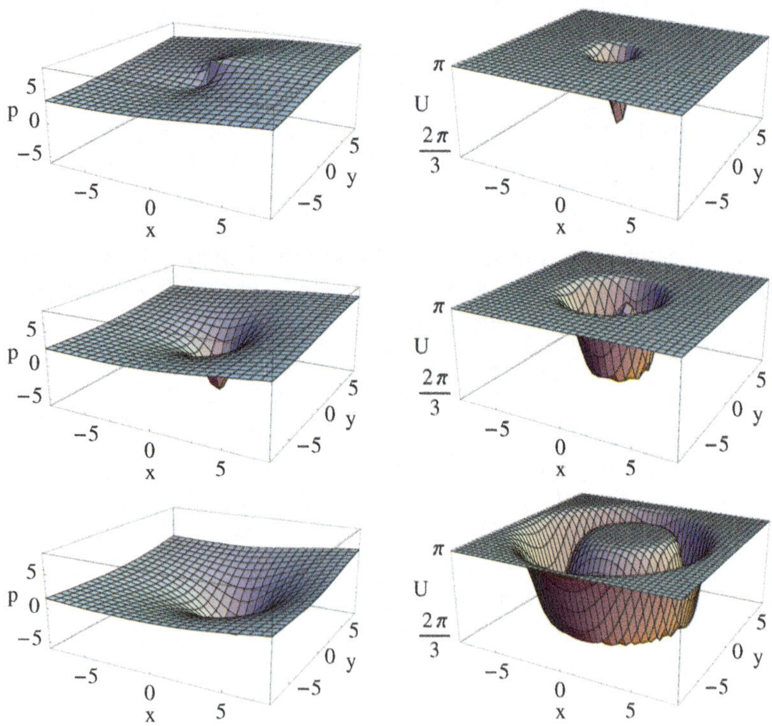

Fig. 1. Function p and solution U in the case (14) for $z = 0$, $t = 0$ (above), $t = 2$ (in the middle), $t = 5$ (below).

$$U = \begin{cases} 2\tan^{-1}\left[\sqrt{1-r^2}\,\mathrm{tn}\left(\frac{W}{r}, r\right)\right], \\ 2\tanh^{-1}\left[\mathrm{sn}\left(\frac{W}{\sqrt{1-r^2}}, r\right)\right], \\ \frac{2}{m}\log\left[\frac{\sqrt{Em}}{\sinh\left(W\frac{m\sqrt{E}}{\sqrt{2}}\right)}\right], \end{cases} . \tag{59}$$

Here $\mathrm{tn}(u,r) = \mathrm{sn}(u,r)/\mathrm{cn}(u,r)$, $\mathrm{sn}(u,r)$, $\mathrm{cn}(u,r)$ are elliptic sine and cosine, r is module of the corresponding elliptic functions ($0 \le r \le 1$), E is a constant of integration.

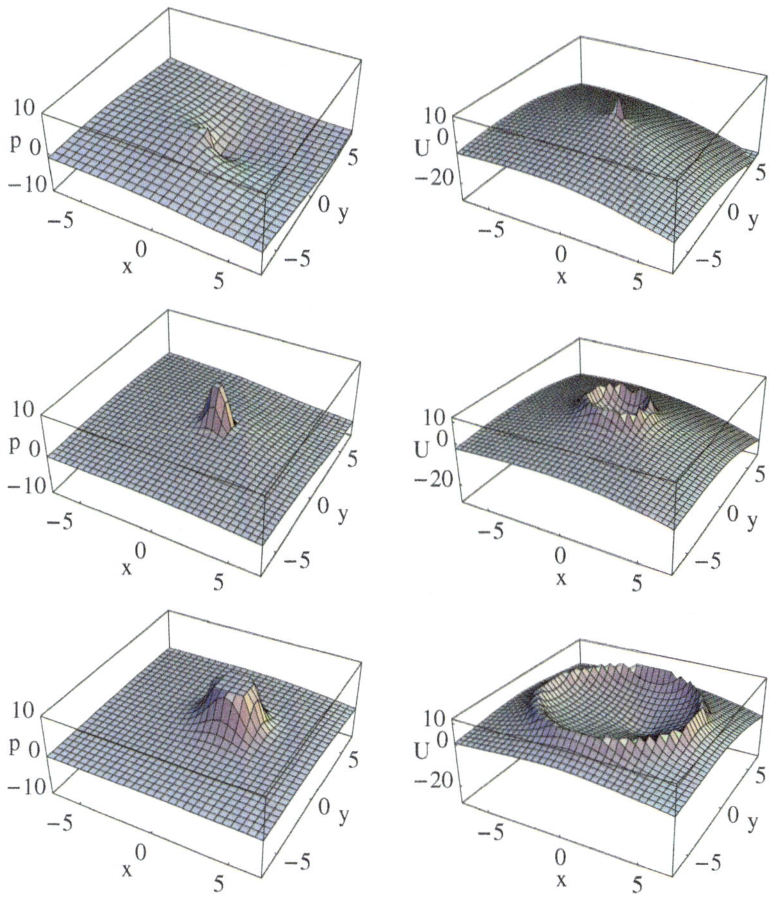

Fig. 2. Function p and solution U in the case (49) for $z = 0$, $t = 0$ (above), $t = 2$ (in the middle), $t = 5$ (below).

Shown in Fig. 1 is the spatial variations of the amplitude of amplitude of $p(x, y, z, t)$ and solution $U(x, y, z, t)$ are given for different times ($t = 0, 2, 5$) in the case $F(U) = \sin U$, $W = \Psi(\tau)$, $w(\tau) = \tau^2$, $\tau = -(X + \nu_1)/(vR_0)$, $R_0 \neq 0$ (see (14)). One can see that $p(x, y, z, t)$ and $U(x, y, z, t)$ represent the localized waves whose shape vary in time. The same form is represented by $p(x, y, z, t)$ and $U(x, y, z, t)$ on Fig. 2 for solution (49) with $F(U) = \exp U$, $W = \Phi(\tau)/\lambda$, $w(\tau) = \tau$, $\tau = -(X + \nu_1)/(vR_0)$, $R_0 \neq 0$.

4 Conclusion

New ansatzes are proposed. They allowed to construct new exact solutions of the nonlinear nonautonomous KFG equation. The nonautonomous KFG equation, unlike autonomous, more adequately describes physical processes as it allows to consider heterogeneity of real media in which the corresponding physical phenomena take place. It is possible to hope that the found solutions will be useful at the description and modeling of these phenomena and processes.

Acknowledgements. This work was supported by the Russian Foundation for Basic Researches, Grant Nos. 17-01-00230-a and 17-02-00365-a.

References

1. Aero, E.L., Bulygin, A.N., Pavlov, Yu.V.: Functionally invariant solutions of nonlinear Klein-Fock-Gordon equation. Appl. Math. Comput. **223**, 160–166 (2013). https://doi.org/10.1016/j.amc.2013.07.088
2. Aero, E.L., Bulygin, A.N., Pavlov, Yu.V.: Solutions of the sine-Gordon equation with a variable amplitude. Theor. Math. Phys. **184**, 961–972 (2015). https://doi.org/10.1007/s11232-015-0309-8
3. Aero, E.L., Bulygin, A.N., Pavlov, Yu.V.: Exact analytical solutions for nonautonomic nonlinear Klein-Fock-Gordon equation. In: dell'Isola, F., Eremeyev, V.A., Porubov, A. (eds.) Advances in Mechanics of Microstructured Media and Structures, pp. 21–33. Springer, Cham (2018). https://doi.org/10.1007/978-3-319-73694-5_2
4. Bulygin, A.N., Pavlov, Yu.V.: Solutions of nonlinear non-autonomous Klein-Fock-Gordon equation. In: Proceedings of the XLVI Summer School – Conference "Advanced Problems in Mechanics", APM 2018, St. Petersburg, pp. 33–43 (2018)
5. Bulygin, A.N., Pavlov, Y.V.: Methods of finding of exact analytical solutions of nonautonomous nonlinear Klein-Fock-Gordon equation. In: Altenbach, H., et al. (eds.) Dynamical Processes in Generalized Continua and Structures, pp. 147–161. Springer, Cham (2019)
6. Forsyth, A.R.: New solutions of some of the partial differential equations of mathematical physics. Messenger Math. **27**, 99–118 (1898)
7. Bateman, H.: The Mathematical Analysis of Electrical and Optical Wave-Motion on the Basis of Maxwell's Equations. Cambridge University Press, Cambridge (1915)
8. Smirnov, V., Sobolev, S.: Sur une méthode nouvelle dans le problème plan des vibrations élastiques. Tr. Seism. Inst. **20**, 1–37 (1932). (English transl.: On a new method in the plane problem on elastic vibrations. In: Selected Works of S.L. Sobolev, vol. I, Springer, New York, 2006, pp. 45–80). https://doi.org/10.1007/978-0-387-34149-1_2

Development of the Legs Fixation Mechanism for Lokomat Therapy Training Device

Andrei Dolgirev(✉) and Natalia Maltseva

Peter the Great St. Petersburg Polytechnic University (SPbPU), Polytechnicheskaya 29, 195251 St. Petersburg, Russia
andrewdolgirev@gmail.com

Abstract. This article describes the application of constructive solutions for the modernization and calculation of the effects of loads on structural parts that hold the lower limbs of a person who is undergoing treatment on a robotic walking training device located at the St. Petersburg Research Institute of Phthisiopulmonology of the Ministry of Health of the Russian Federation.

Considering the specifics of the training device that's main task is to recover walking function, it is important to pay particular attention to the convenience and safety of the structure.

At the end of the research and manufacturing described hereunder, the existing training device will be able to begin its work at the Central Research Institute of Phthisiopulmonology, in particular it will provide free rehabilitation for patients. At the moment, only a few simulators for locomotor therapy are installed in St. Petersburg. Most of the treatment sessions are paid.

Support for solving the problem is provided by specialists of the Research Institute of Phthisiopulmonology.

The project was fully implemented on the basis of the Center for Scientific and Technical Creativity of Youth Fablab Polytech. The work on the project described in the article also demonstrates the competencies of digital production, obtained by the authors at Fablab Polytech helping to solve a wide variety of technical problems. In the process of work, such production methods as computer modeling, generative design, computational operations with the influence of loads and CNC machines were used.

Keywords: Locomotor therapy · Rehabilitation · Fablab · Digital manufacturing · Walking simulator

1 Disadvantages of the Locomotor Therapy System

When using the locomotor simulator, specialists of the Central Research Institute of Phthisiopulmonology revealed a number of disadvantages that almost completely exclude its operation (Fig. 1).

D. A. Indeitsev and A. M. Krivtsov (Eds.): APM 2019, LNME, pp. 81–92, 2020.
https://doi.org/10.1007/978-3-030-49882-5_9

Fig. 1. Robotic walking training device

Identified deficiencies:

- During operation there is a significant load on the knee joint when exposed to the knee stop, in addition, there is no fixation of the knee by the straps provided by the design;
- The padlock is not equipped with a mechanism for fixing belts and lifting the toe of the foot, which causes discomfort and danger during operation;
- The aesthetic appearance of the simulator is inferior to world analogues;
- The simulator has a fairly large weight and steel structure (Fig. 2).

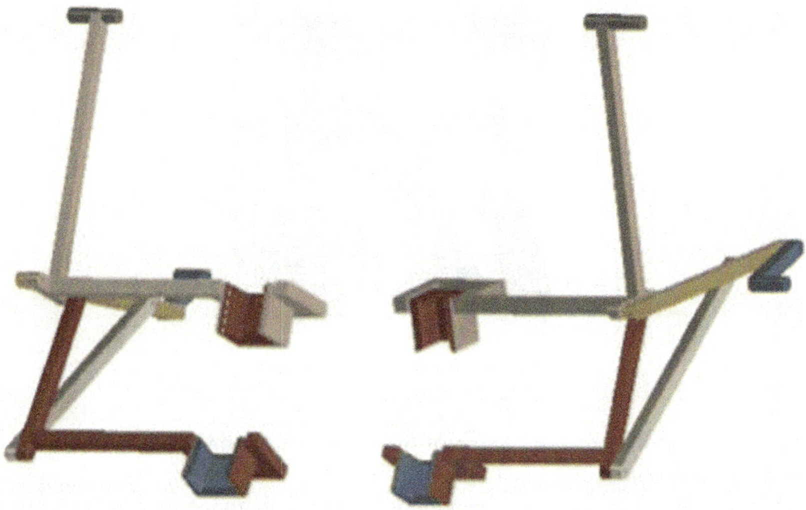

Fig. 2. Knee and ankle support

2 Modification of the Knee Support

For the safe and efficient operation of the locomotion therapy device, it was decided to carry out a number of improvements. First of all, it was necessary to unload the knee, replacing the existing imperfect parts to fix it. The new mechanism, which is essentially an orthosis, should eliminate the discomfort that occurs due to an incorrectly positioned support, which serves as part of the knee fixation device. The whole structure must be changed, the fulcrum should be moved to the thigh and ankle and should have a movable connection with the bearing. This construction will allow the patient for the free knee-bend.

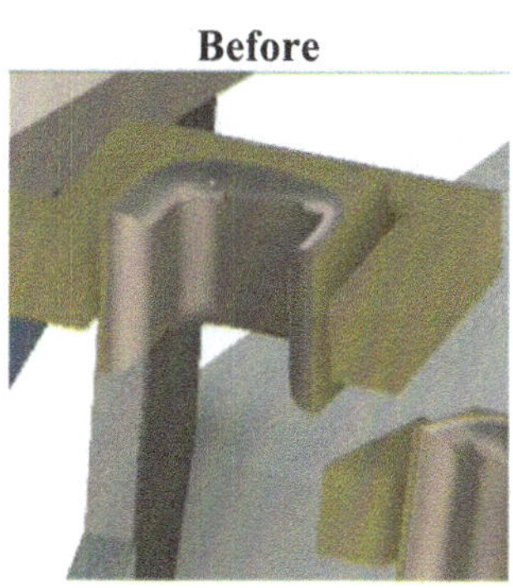

Fig. 3. Original knee support

After

Fig. 4. Modified knee support

Figures 3 and 4 show the outline of the knee fixation mechanism before and after modernization respectively.

After construction analysis, searching for ways to modernize it, preparing and compiling project documentation, a model was created in computer-aided engineering system (Fig. 5).

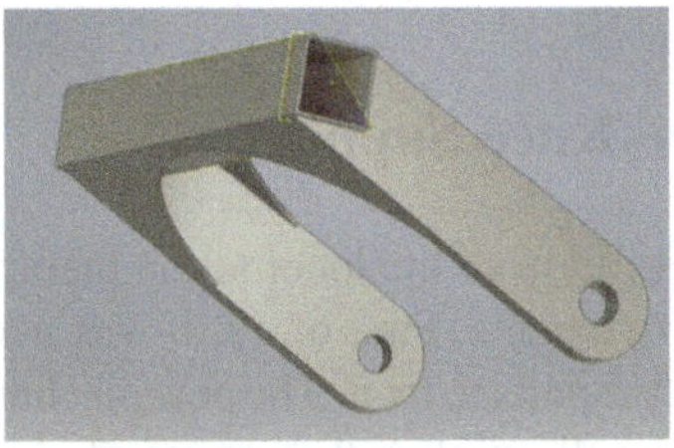

Fig. 5. Detail in CAE

Fig. 6. Assembly in CAE

Figure 6 illustrates the assembly of the details of the prototype of the modernized mechanism of fixation of the knee joint.

3 Topological Optimization

3.1 General Idea

Generative design is a workflow where an engineer can set conditions and restrictions and the system will automatically generate optimal parameters. Many engineering problems cannot be solved by analytical methods; therefore, numerical methods are used to obtain satisfactory results. Due to the fact that part of the design calculation is transferred to the software, the interaction between the engineer and the program is carried out as between team members.

Generally, optimization is the process of searching and choosing the best version of construction from the set of possible ones under given conditions and restrictions. The end result of such process will be the optimal solution. Thus, it is possible to optimize in any field that interests us, in compliance with two conditions:

- there is more than one compared option;
- There is one or more optimality criteria.

In particular, it is possible to optimize the geometric structure of an object - properties of construction that remain unchanged during deformation. Such process is important in the field of design.

In this chapter the basic concepts and methods of topological optimization will be considered using the example of the modernization of the walking training device. In subsequent work on the project, it will allow not only to use unique design solutions but also to reduce the weight of the structure using less materials with more and more popular methods for manufacturing parts. Example would be the manufacture of parts by 3D printing and selective laser sintering.

The main difference between generative design and topological optimization is the creation of an effective solution to an engineering problem that does not require further improvement. For example, in generative design it is possible to solve a problem without specifying an initial volume.

To solve the problems of topological optimization, a CAE system will be used in the work. At the beginning, a model that should describe the shape of the future product is created. Those structural elements that will not be modified are studied in detail (for example, holes for fasteners). It is necessary to set the initial conditions for the problem being solved. For example, for problems of linear-elastic statics, external loads, material properties, and restrictions on the degrees of freedom of the structure are specified. For structural problems, the finite element method is mostly often used.

3.2 Topological Optimization of Locomotor Therapy System Parts

Let's consider the finite element method on the example of parts for the modernization of the locomotor therapy system.

Analytically, such a situation can be considered as an infinite number of points, for each of them we try to find a solution. The analytical method is not suitable for complex structures because it requires a lot of rather cumbersome calculations. The solution of this problem can be discretization, in particular, replacing the problem with an infinite number of indeterminates with a similar one with finite number of elements.

Concerning example, we will divide our part into points; the distance from one point to another will be called the finite element.

Let's consider the disadvantages of the method, and how to solve them. For example, the calculation can be obtained only for the specified number of points and is not accurate, it will be just approximate to the real solution.

Therefore, the number of finite elements should be increased in order to obtain more accurate result. To represent the error of this method, let's recall the approximation of a circle by a polygon.

Our part is fixed in the upper part, because it is firmly fixed to another detail. In this case, we apply force of 500 N at the end to the lower part of the bearing bore. At this point the load will be localized, which can be created by the weight of a person in case of emergency. Remember that we must exclude any, even the most unlikely cause of harm to the patient.

As shown in Fig. 7, due to the applied force, the part extended for 0.07 mm.

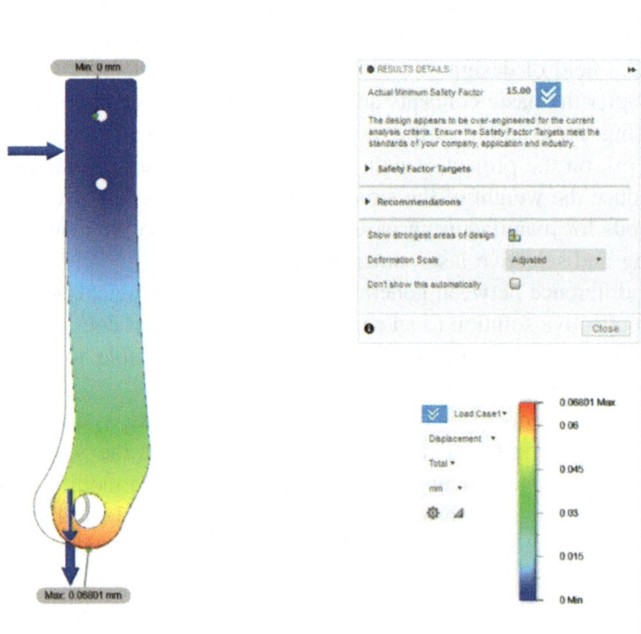

Fig. 7. Maximum deformation 0.7 mm

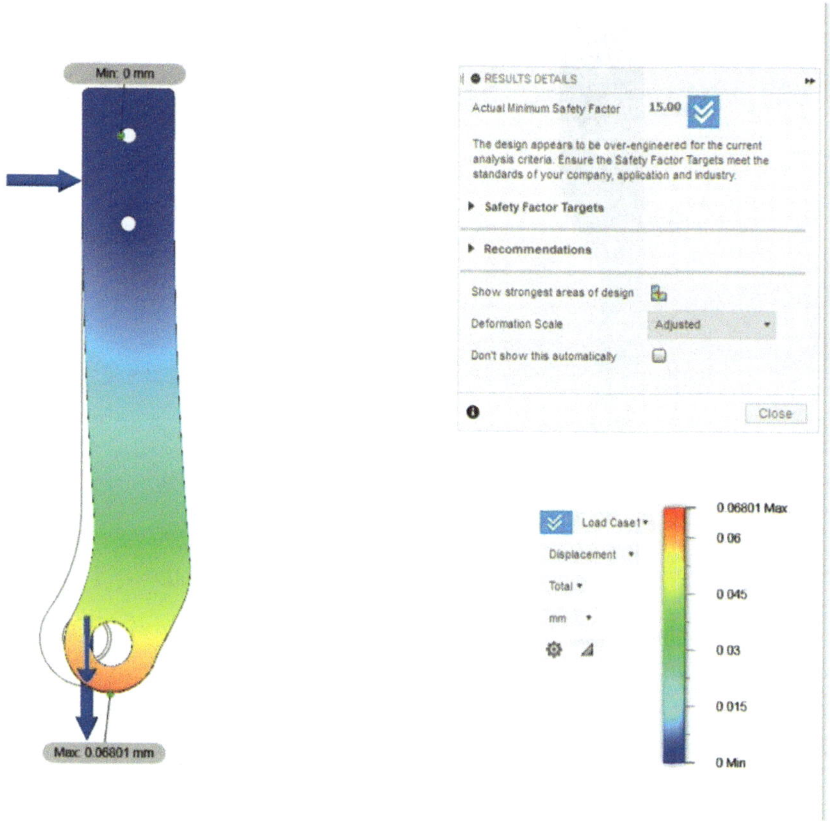

Fig. 8. Mises maximum stress 14.02 MPa

With advances in technologies, the solution of constitutive equation was automated using the finite element method which made it possible to achieve significant results in the field of structural optimization. There were developed such strategies for the description of the material as: homogenization approach, SIMP (Solid Isotropic Material with Penalisation) method, ESO (Evolutionary Structural Optimization) method. One of the modern strategies is the Level-sets approach to describing the structure topology. Also, the SIMP method will be analyzed in more detail, as it is the most studied and developed, it is used in most software to implement topological optimization (Fig. 8).

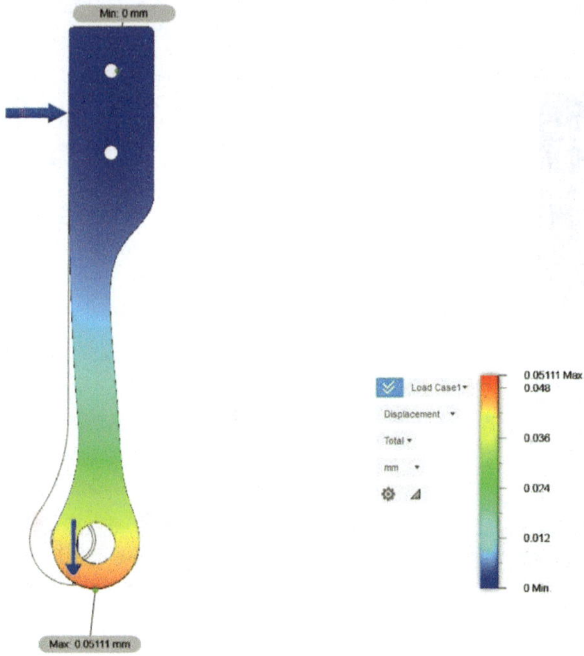

Fig. 9. Deformation of the optimized part

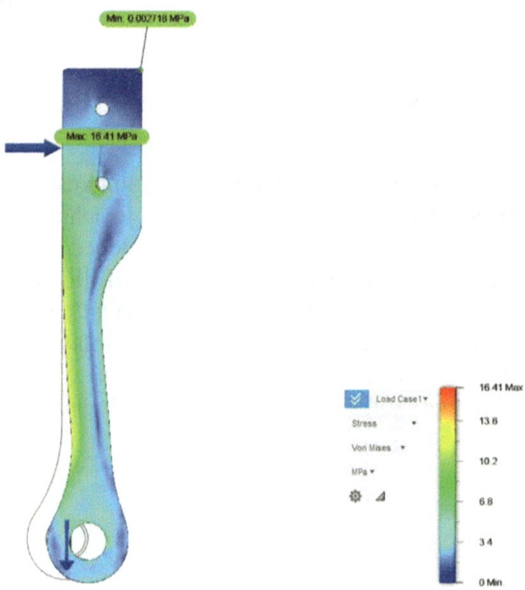

Fig. 10. Mises stress optimized details

Figures 9 and 10 show the result of calculations of the optimized part:

- Maximum deformation is 0.05 mm;
- Mises stress is 16.4 MPa.

In future topological optimization will make it possible to satisfy modern world requirements for the aesthetic appearance of such simulators, which will make Russian locomotor training device more competitive. Moreover, with the help of topological optimization, the mass of the entire structure is significantly reduced, and with certain manufacturing methods it is possible to reduce the cost of production by reducing the consumption of materials.

4 Weld Joints Calculations

Since it was decided to use welding technology to connect some parts, the calculation of the allowable load on these elements is required. In this calculation, as in the topological optimization, which is described in detail in the previous chapter, the finite element method is used.

5 Manufacturing

Technical support is fully provided by youth scientific and technical creativity center Fablab Polytech. The center also granted access to CNC machines, which are one of the main tools of digital production.

For milling parts from 2 mm sheet aluminum, the FlexiCAM S2 1525 milling and engraving machine was used with a working field of 1540 * 2580 mm.

Parts from sheet aluminum with a thickness of 8 mm were manufactured using the Ingro (First) MCV300 small-sized metalworking milling. The machine is designed for processing products in small-scale production.

After 3D modeling, milling settings were made in the SolidCAM program. SolidCAM is a program for creating control programs for milling and other CNC machines. SolidCAM created a ready-made program (G-Code) for the final version of the project. Further, the details were made on a metal-working CNC milling machine MCV-300.

Before making the final version of the project using the similar technology, the parts shown in Fig. 11 (left) were manufactured. After the tests construction was modernized and the parts shown in Figs. 11 (right) were manufactured in the same way.

Fig. 11. Details manufactured using MCV-300 first and final version

6 Futher Use

After the production of the first version, the prototype passed tests on locomotor training device under the supervision of employees of the Central Research Institute of Phthisiopulmonology (Fig. 12).

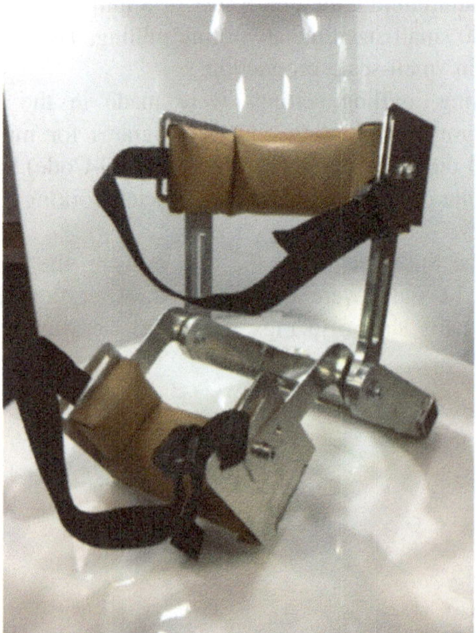

Fig. 12. The first version of the device

Next, a number of comments were collected, which was taken into account in the final version.

The final version has passed all the stages of manufacture and is ready for testing and further use (Fig. 13).

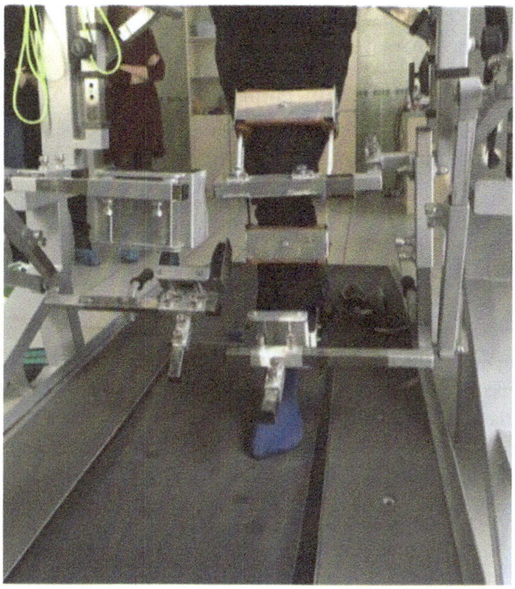

Fig. 13. Tests at the Central Research Institute

In this work, the full life cycle of the project on the modernization of the walking training device was completed: from the preparatory stage and implementation, to testing and providing the possibility of further use. The relevance of modernization was analyzed with the study of the disadvantages of the current version of the device and the advantages of world analogues. 3D models of parts in CAD were prepared, the method of topological optimization and generative design was studied and applied. The analysis of materials and calculation of the strength of the weld were made, various welding technologies were studied and taken into account for the subsequent safe operation of the device. The manufacture of the device parts was carried out according to the drawings made during this work, on a CNC metalworking milling machine.

Calculations on the strength of the weld showed that even with the maximum allowable loading mode, the displacement of parts and the strength coefficient are several times lower than critical values. Nevertheless, the fact that the modernized device will be further used in the treatment of people who have lost the walking function makes it possible to completely eliminate possible problems during its operation. Research of various aluminum welding technologies showed the relevance of using TIG welding to solve the problem.

In this paper, we studied and applied the method of topological optimization to the details of the training device, which will make it possible to further use such advanced manufacturing methods as 3D printing and selective laser sintering, while in mass production, significantly saving materials and reducing weight. Also, the generative design method applied to construction details can make decisions in the appearance of domestic medical rehabilitation units, making them more aesthetic and at the same time competitive in the market.

During the work, the project received support from:

- St. Petersburg Research Institute of Phthisiopulmonology of the Ministry of Health of the Russian Federation.
- Center for Scientific and Technical Creativity of Youth Fablab Polytech.

References

1. Autodesk: Introduction to CAD for Engineers (date of the application: 20.04.2019)
2. Autodesk: Introduction to Generative Design (date of the application: 01.05.2019)
3. Gravlee, J.R.:. Braces and Splints for Musculoskeletal Conditions. www.aafp.org/afp/2007/0201/p342.html (date of the application 25.04.2019). American Family Physician
4. Browne, P.A.: Topology Optimization of Linear Elastic Structures. University of Bath Department of Mathematical Sciences (2013)

Modeling of the Superelastic Behavior of CuAlNi - Single Crystals Accounting Anisotropy of Elastic Properties

Tatiana Y. Chernysheva(✉), Margarita E. Evard, Aleksandr E. Volkov, and Fedor S. Belyaev

St.Petersburg State University, St. Petersburg 199034, Russia
st040151@student.spbu.ru
https://spbu.ru/

Abstract. The lattice deformation tensor for $\beta_1 \leftrightarrow \beta_1'$ martensitic transformation was found from available crystallographic data. This tensor was used for modeling of the isothermal deformation of CuAlNi shape memory alloys in the frames of a microstructural model. The simulated stress-strain curves obtained for pseudoelastic austenitic and pseudoplastic martensitic CuAlNi are in a good qualitative agreement with literature experimental data for single crystals with different orientations.

Keywords: CuAlNi · Single crystal · Microstructural model · $\beta_1 \leftrightarrow \beta_1'$ · Martensitic transformation

1 Introduction

Mechanisms of shape memory and pseudoelasticity in Cu-based shape memory alloys (SMA) were scrutinized by the end of the 20th century (see [1] and [2]). At the same time, development of practical applications met a variety of difficulties, mainly brittle fracture, fatigue and aging of these alloys [3]. However, rapid development of materials science, new methods for grain size control [4], refinement of single-crystal growth technique allowing production of single crystals with preset parameters [5], design of complex quaternary alloying [6] aroused new interest to Cu-based SMA in the recent decade [7]. It stimulated interest to working out theoretical approaches for simulation of mechanical behavior of these alloys.

An existing microstructural model (see [8,9] and [10]) provides a good description of the functional properties of a TiNi SMA including the tension – compression asymmetry of the phase deformation [11]. This model also secured a good simulation of a vibration protection device comprising functional TiNi parts (see [9,12]). A use of an alternative SMA demands adjusting of this model by accounting of the specific features of the martensitic transformation. One of the basic material constants in the model is the tensor of Bain's deformation transforming the crystallographic lattice of the parent austenitic phase into that

D. A. Indeitsev and A. M. Krivtsov (Eds.): APM 2019, LNME, pp. 93–100, 2020.
https://doi.org/10.1007/978-3-030-49882-5_10

of the martensitic phase. This tensor for B2 - B19′ transformation in TiNi alloy was calculated in the work of K. Knowles and R. Smith [13]. The type of the martensitic transition in CuAlNi depends on temperature and stress conditions as well as on the composition (see [1,2] and [14]). The austenitic phase, which is usually referred as the β_1 structure, can be ordered by the $D0_3$ or $L2_1$ ways depending on the stage of the next-nearest neighbor ordering [14]. The object of interest in this work is the $\beta_1(D0_3) \leftrightarrow \beta_1'$ transformation occurring either on loading above Ms or on cooling and heating under a stress exceeding some critical level (see [15,16] and [17]). The crystallography of this transformation is described in the work [2]. An analytic solution of the invariant plane problem for cubic to 18R transformation was given in 1978 by De Vos, Aernoudt and Delaey [18]. Later the non-invariant plane of the austenite-martensite interface was experimentally identified (see [19] and [20]) and described by introduction of an isotropic dilatation of the habit plane [21]. Better correspondence with experimental data was obtained in the work [22] where the total $\beta_1 \leftrightarrow \beta_1'$ transition was considered as a sequence of the transformation from parent β_1 phase to twinned γ_1' phase and the transformation of the γ_1' - phase to detwinned β_1' - phase. This work also gives twelve deformation gradient matrices of the shear during $\beta_1 \leftrightarrow \beta_1'$ transformation.

In the present work the Bain's deformation tensor for $\beta_1 \leftrightarrow \beta_1'$ transformation is calculated following the work [2] and then it is used for modeling of the stress-strain diagrams and the transformation plasticity effect in CuAlNi single and poly crystals.

2 Model

All calculations were carried out in the frames of the microstructural model (see [9] and [10]). In this model two structural levels are considered. The representative volume of SMA consists of grains characterized by their crystallographic orientations. Total deformation tensor of any grain is resolved into elastic, thermal and phase (due to the phase transformation) components. Plastic deformation also can be taken into account. The deformation of the representative volume is calculated by averaging of the grain deformations. On the microscopic level (inside a grain) the austenite phase and N domains (crystallographically equivalent orientation variants) of the martensite phase are distinguished. The grain deformation tensor is obtained by averaging

$$\varepsilon^{\text{gr}} = (1 - \Phi^{\text{gr}})\,\varepsilon^{A} + (1/N)\sum_{N} \Phi_{N}\varepsilon^{(n)}, \qquad (1)$$

where $(1/\text{N})\Phi_n$, ε^A and $\varepsilon^{(n)}$ are the volume fraction and the deformation tensors of austenite and of the n-th variant of martensite, $\Phi^{\text{gr}} = (1/N)\Sigma_n\Phi_n$ the total volume fraction of martensite in the grain.

The phase deformation of a grain is supposed to be the average over the martensite variants:

$$\varepsilon^{\text{Ph}} = (1/N)\Sigma_n\Phi_n(\omega)D_n \qquad (2)$$

were D_n is the Bain's deformation of the n-th variant of martensite. The procedure for calculation of the evolution of internal variables Φ_n is expanded elsewhere [9].

For calculation of matrix D for one of the variants we use the sheme of $D0_3 \leftrightarrow 18R$ transformation described in [2]. The transformation is considered to consist of two steps. First step is a combination of tension and contraction in the plane $(110)_p$. On this step the austenitic $(110)_p$ plane transforms into the martensitic $(001)_m$ plane by contraction along $[001]_p$ and elongation along $[110]_p$ so that the indicated angle α changes from $70^\circ\ 32'$ to 60° typical for a close-packed structure Fig. 1a. On the second step a shear on the plane $(110)_p$ occurs by the vector $s = \frac{1}{18}[110]$. Figure 1b illustrates the shear and shuffle mechanism to bring a $D0_3$ lattice into 18R martensite.

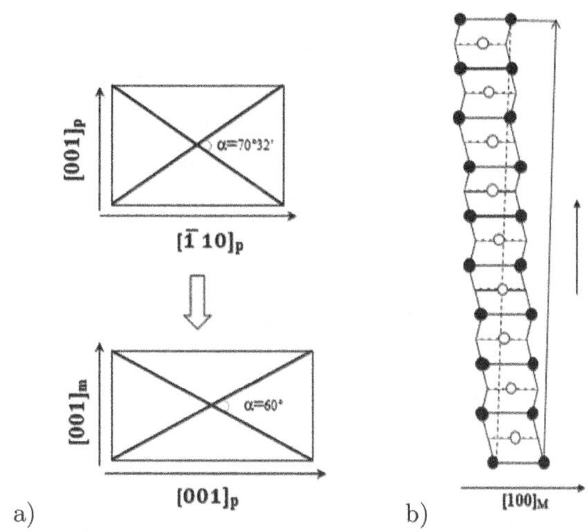

Fig. 1. The structural change from austenitic DO_3 phase into martensitic 18R: a - deformation in $(110)_p$; b - shear and shuffle (after [1] and [21]).

To find the deformation gradient $(\underline{\underline{F}}^{(1)})_f$ on the 1-st step denote the relative elongation along $[110]_p$ by ε_1 and the relative contraction along $[001]_p$ by ε_2 $(\varepsilon_1, \varepsilon_2 > 0)$. Then we can write the deformation gradient matrix in the normalized basis $f = \left\{ \frac{1}{\sqrt{2}}[110], \frac{1}{\sqrt{2}}[\bar{1}10], [001] \right\}$:

$$(\underline{\underline{F}}^{(1)})_f = \begin{pmatrix} 1 & 0 & 0 \\ 0 & 1+\varepsilon_1 & 0 \\ 0 & 0 & 1-\varepsilon_2 \end{pmatrix} \tag{3}$$

Obviously, the matrix $(A)_e^f$ realizing the transition from the austenitic basis $e = \{[100], [010], [001]\}$ to the basis f is:

$$A = (A)_e^f = \frac{1}{\sqrt{2}} \begin{pmatrix} 1 & -1 & 0 \\ 1 & 1 & 0 \\ 0 & 0 & \sqrt{2} \end{pmatrix} \tag{4}$$

So we can calculate the matrix of the deformation gradient $\left(\underline{\underline{F}}^{(1)}\right)_e = A \left(\underline{\underline{F}}^{(1)}\right)_f A^{-1}$ in the basis e:

$$\left(\underline{\underline{F}}^{(1)}\right)_e = \frac{1}{2} \begin{pmatrix} 2+\varepsilon_1 & -\varepsilon_1 & 0 \\ -\varepsilon_1 & 2+\varepsilon_1 & 0 \\ 0 & 0 & 2(1-\varepsilon_2) \end{pmatrix} \tag{5}$$

Deformation $\underline{\underline{F}}^{(1)}$ transforms the basis **e** into a new basis ***g***. In this basis the vectors s and $\tilde{\mathbf{n}}$ describing the shear have the coordinates $(s)_g = \frac{1}{18}[\bar{1}10]$, $(\tilde{\mathbf{n}})_g = (110)$. Since the deformation gradient of the shear is $\underline{\underline{F}}^{(2)} = \underline{\underline{I}} + \mathbf{s} \otimes \tilde{\mathbf{n}}$ find its matrix in the basis g ($\underline{\underline{I}}$ is the unity tensor):

$$\left(\underline{\underline{F}}^{(2)}\right)_g = \begin{pmatrix} \frac{17}{18} & \frac{1}{18} & 0 \\ -\frac{1}{18} & \frac{19}{18} & 0 \\ 0 & 0 & 1 \end{pmatrix} \tag{6}$$

Note that this matrix corresponds to the deformation gradient of the 5th variant of the transformation in the enumeration used by Y. Zhang and Q. Sun [22]. Finally, we find the deformation gradient matrix in the parent phase basis ***e*** for the $DO_3 \leftrightarrow 18R$ transformation:

$$\left(\underline{\underline{F}}\right)_e = \left(\underline{\underline{F}}^{(1)}\right)_e \left(\underline{\underline{F}}^{(2)}\right)_g = \frac{1}{18} \begin{pmatrix} 17+9\varepsilon_1 & 1-9\varepsilon_1 & 0 \\ -1-9\varepsilon_1 & 19+9\varepsilon_1 & 0 \\ 0 & 0 & 18(1-\varepsilon_2) \end{pmatrix} \tag{7}$$

To find unknown values ε_1 and ε_2 we take into account that the angle α changes from $70°32'$ to $60°$ and assume that the relative volume change $\theta = det\left(\underline{\underline{F}}\right) - 1 = (1-\varepsilon_2)(1+\varepsilon_1) - 1$ can be set equal to zero. These assumptions give: $\varepsilon_1 \approx 0.1067, \varepsilon_2 \approx 0.0964$. Thus, the deformation gradient matrix for the $\beta_1 \leftrightarrow \beta_1'$ transformation is

$$\left(\underline{\underline{F}}\right) = \begin{pmatrix} 0.9978 & 0.0022 & 0 \\ -0.1089 & 1.1089 & 0 \\ 0 & 0 & 0.902 \end{pmatrix} \tag{8}$$

and the matrix of Green-Lagrangian deformation tensor for the martensitic variant under consideration is

$$\underline{\underline{D}} = \begin{pmatrix} 0.0039 & -0.0594 & 0 \\ -0.0594 & 0.1149 & 0 \\ 0 & 0 & -0.0932 \end{pmatrix} \tag{9}$$

The deformation matrices for other crystallographic variants of the transformation can be obtained by rotation of the matrix (10) with rotations belonging to the cube symmetry group.

3 Simulation

It is well known that the β_1 austenitic structure in CuAlNi SMA demonstrates strong anisotropy of elastic moduli [23]. Based on the experimental results [24], the following matrix of the elastic moduli for austenite was calculated (all the values are in GPa):

$$C_A = \begin{pmatrix} 136 & 124.3 & 124.3 & 0 & 0 & 0 \\ 124.3 & 136 & 124.3 & 0 & 0 & 0 \\ 124.3 & 124.3 & 136 & 0 & 0 & 0 \\ 0 & 0 & 0 & 90 & 0 & 0 \\ 0 & 0 & 0 & 0 & 90 & 0 \\ 0 & 0 & 0 & 0 & 0 & 90 \end{pmatrix} \tag{10}$$

As direct measurements of the elastic moduli for a single variant of martensite is of great difficulties and in the literature there are no theoretical estimations of these characteristics, there was supposed that the martensitic phase is elastically isotropic with the Young's modulus E_m = 24 GPa and the Poisson's ratio ν_m = 0.33. This assumption can be justified by the fact that during the direct martensitic transformation several different variants of martensite appear, so that their elastic properties can be described in terms of the average values.

For simulation of stress-strain tensile diagrams of single crystal at temperature above A_f [23] the following values were used: the characteristic temperatures of the transformation M_f = 207 K, M_s = 208 K, A_s = 211 K, A_f = 216 K, the latent heat of the transformation $q_0 = -45\,\mathrm{MJ/m^3}$. Plastic deformation was not taken into account. The stress-strain curves are presented on Fig. 2. Tensile direction orientations relative to crystallographic axis of the β_1 phase are shown on the stereographic triangle (Fig. 2c). One can see that results of modeling are in a good qualitative agreement with experimental data.

Figure 3 demonstrates stress-strain tensile diagrams of a single crystal at different temperatures. For this material the characteristics temperatures M_f = 117 K, M_s = 136 K, A_s = 159 K, A_f = 178 K correspond to the experimental values [1]. Here the experimental data argue that along with $\beta_1 \leftrightarrow \beta_1'$, the material undergoes the $\beta_1 \leftrightarrow \gamma_1'$ martensitic transformation which is not directly considered by our model. Still, if it is possible to interpret the γ_1' phase as twinned β_1'-phase, one can expect that the calculated stress-strain diagram for this alloy in the martensitic state will reproduce the experimental one. This is the case depicted on Fig. 3a the critical maximum values of stress and strain are in a good correspondents with the experimental values.

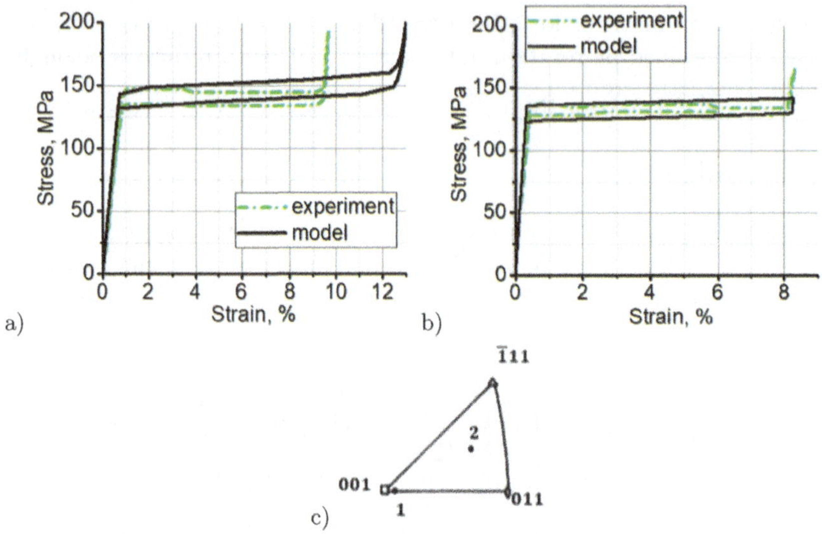

Fig. 2. Stress-strain diagrams of CuAlNi single crystals: a - orientation 1, b - orientation 2; c - tension direction orientations relative to crystallographic axis of the β_1 phase. $M_f = 208$K, $M_s = 209$ K, $A_s = 211$ K, $A_f = 216$ K, temperature 293 K.

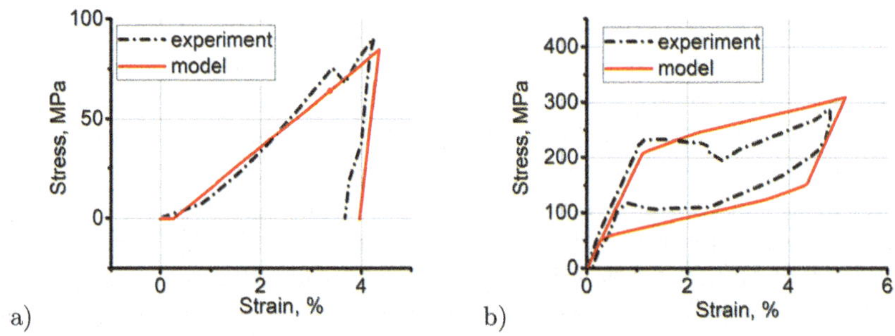

Fig. 3. Stress-strain diagrams of CuAlNi single crystals at temperature 132 K (a) and 192 K (b). $M_f = 117$ K, $M_s = 136$ K, $A_s = 159$ K, $A_f = 178$ K.

4 Conclusions

1. The considered microstructural model can correctly describe stress-strain diagrams of CuAlNi single crystals both in the austenitic and in the martensitic state.
2. A good correspondence between modeling and experimental results provides the evidence that the knowledge of the lattice deformation tensor (Bain's tensor) plays an important role for microstructural modeling of the mechanical behavior of SMA. Another example of the crucial role of using the correct Bain's deformation tensor is the description of the tension-compression asymmetry in TiNi alloy [11]. For CuAlNi one can also expect a correct

characterization of this asymmetry as well as of the mechanical behavior at combined stress loading.

3. Since the microstructural model is appropriate to describe functional properties of CuAlNi single crystals, one can expect reliable results at its application for simulation of the functional-mechanical behavior of devices and structures with CuAlNi SMA parts.

Acknowledgements. This research was supported by the grant of Russian Foundation of Basic Research 19-01-000658.

References

1. Funakubo, H. (ed.): Shape Memory Alloys, vol. 275. Gordon and Breach Science Publication, New York (1987)
2. Otsuka, K., Wayman, C.M. (eds.): Shape Memory Materials, vol. 282. Cambridge University, Cambridge (1998)
3. Miyazaki, S., Kawai, T., Sakamoto, H., Otsuka, K.: Study of fracture in Cu-Al-Ni shape memory bicrystals. Trans. Jpn. Inst. Met. **22**(4), 244–252 (1982). https://doi.org/10.2320/matertrans1960.22.244
4. Sugimoto, K., Kamei, K., Matsumoto, H., Komatsu, S., Akamatsu, K., Sugimoto, T.: Grain-refinement and the related phenomena in quaternary Cu-Al-Ni-Ti shape memory alloys. J. de Phys. **12**, 43 (1982)
5. Gastien, R., Corbellani, C.E., Araujo, A., Zelaya, E., Beroia, J.I., Sade, M., Lovey, F.C.: Changes of shape memory properties in CuAlNi single crystals subjected to isothermal treatment. Mat. Charateriz. **84**, 240 (2013). https://doi.org/10.1016/j.matchar.2013.08.008
6. Akash, K., Akash, K.J., Karmarkar, G., Jadhav, A., Narayane, D.C., Patra, N., Palani, I.A.: Investigations on actuation characteristics and life cycle behaviour of CuAlNiMn shape memory alloy bimorph towards flappers for aerial robots. J. Mater. Des. **144**, 64–71 (2018). https://doi.org/10.1016/j.matdes.2018.02.013
7. Kneissl, A.C., Mehrabi, K., Bruncko, M., McKay, B.J., Uhlenhaut, D.: Characterization and properties of NiTi(W) and CuAlNi shape memory alloys. Int. JMR **100**, 1038 (2009)
8. Evard, M.E., Volkov, A.E.: Modeling of martensite accommodation effect on mechanical behavior of shape memory alloys. JEMT **121**, 102 (1999)
9. Volkov, A.E., Evard, M.E., Red'kina, K.V., Vikulenlov, A.V., Makarov, V.P., et al.: Simulation of payload vibration protection by shape memory alloy parts. JMEP **23**, 2719 (2014)
10. Resnina, N., Rubanik, V. (eds.): Shape Memory Alloys: Properties, Technologies, Opportunities. Edition: Materials Science Foundations, vols. 81–82. Trans Tech Publications Ltd., Zurich (2015). Volkov, A.E., Evard, M.E., Belyaev, F.S.
11. Volkov, A.E., Emelyanova, E.V., Evard, M.E., Volkova, N.A.: An explanation of phase deformation tension-compression asymmetry of TiNi by means of microstructural modeling. J. Alloy. Compd. **577**, S127–S130 (2013). https://doi.org/10.1016/j.jallcom.2012.05.131
12. Volkov, A.E., Evard, M.E., Vikulenlov, A.V., Uspenskiy, E.S.: Simulation of vibration isolation by shape memory alloy springs using a microstructural model of shape memory alloy. Mater. Sci. Forum **738–739**, 150–154 (2013). https://doi.org/10.4028/www.scientific.net/MSF.738-739.150

13. Knowles, K.M., Smith, D.A.: The crystallography of the martensitic transformation in equiatomic nickel-titanium. J. Acta Met. **20**, 101 (1981). https://doi.org/10.1016/0001-6160(81)90091-2
14. Nakata, Y., Iizuka, Y., Ono, T.: The effects of aging on the degree of order in Cu-Al-Ni shape memory alloys. J. Mater. Trans. **57**, 257–262 (2016). https://doi.org/10.2320/matertrans.MB201511
15. Novak, V., Malimanek, J., Zarubova, N.: Martensitic transformations in single crystals of CuAlNi induced by tensile stress. Mat. Sci. Eng. A **191**, 193 (1995). https://doi.org/10.1016/0921-5093(94)09628-A
16. Novak, V., Sittner, P., Zarubova, N.: Anisotropy of transformation characteristics of Cu-base shape memory alloys. Mat. Sci. Eng. A **24–236**, 414 (1997). https://doi.org/10.1016/S0921-5093(97)00175-5
17. Sittner, P., Hashimoto, K., Kato, M., Tokuda, M.: Stress induced martensitic transformations in tension/torsion of CuAlNi single crystal tube. Scr. Mat. **48**, 1153 (2003). https://doi.org/10.1016/S1359-6462(02)00583-3
18. De Vos, J., Aernoudt, E., Delaey, L.: Remarks on ordering in ternary β CuZnAl alloys. J. Metallkunde **69**, 438 (1978)
19. Sun, Q.P., Zhang, X.Y., Terry, X.T.: Some recent advances in experimental study of shape memory alloys. Solid Mechanics and its Applications, vol. 62 (1997). https://doi.org/10.1007/0-306-46936-7_39
20. Sun, Q.P., Zhang, X.Y., Terry, X.T.: On deformation of A-M interface in single crystal shape memory alloys and some related issues. JEMT **121**, 38 (1999). https://doi.org/10.1115/1.2815997
21. Kato, H.: Habit plane analysis of the Cubic/18R(9R) martensite transformation in copper-based shape memory alloys. Scr. Mat. **28**, 1125 (1998). https://doi.org/10.1016/S1359-6462(97)00579-4
22. Zhang, X., Sun, Q., Shouwen, Yu.: A non-invariant plane model for the interface in CuAlNi single crystal shape memory alloys. JMPS **48**, 2163 (2000). https://doi.org/10.1016/S0022-5096(99)00102-7
23. Horikawa, H., Ichinose, S., Morri, K., Miyazaki, S., Otsuka, K.: Orientation dependence of stress-induced martensitic transformation in CuAlNi alloy. Metall. Trans. A **20**, 779–780 (1989). https://doi.org/10.1007/BF02628376
24. Sedlak, P., Seiner, H., Landa, M., Novak, V., Sittner, P., Mañosa, Ll.: Elastic constants of BCC austenite and 2H orthorhombic martensite in CuAlNi shape memory alloy. Acta Materialia **53**, 13 (2005). https://doi.org/10.1016/j.actamat.2005.04.013

Experimental Investigations of 3D-Deformations in Additively Manufactured Pantographic Structures

Gregor Ganzosch[1(✉)], Emilio Barchiesi[2], Rafal Drobnicki[3], Aron Pfaff[4], and Wolfgang H. Müller[1]

[1] Department of Mechanics, Chair of Continuum Mechanics and Constitutive Theory, Berlin University of Technology, Einsteinufer 5, 10587 Berlin, Germany
ganzosch@tu-berlin.de

[2] Dipartimento di Ingegneria e Scienze dell'Informazione e Matematica, Università degli Studi dell'Aquila, Via Vetoio 1, Coppito, 67100 L'Aquila, Italy

[3] Institute of Mechanics and Printing, Warsaw University of Technology, 85 Narbutta Street, 02-524 Warsaw, Poland

[4] Additive Design and Manufacturing, Fraunhofer Ernst Mach Institute, Ernst-Zermelo-Straße 4, 79104 Freiburg, Germany

Abstract. In the recent past new experimental techniques have been developed with the objective of linking generalized continuum theories with technology. So-called pantographic structures, which can be characterized as a meta-material, will be presented and investigated experimentally: Samples of different materials and dimensions are subjected to large deformation loading tests (tensile, shearing, and torsion) up to rupture, while their response to loading is recorded by an optical measurement system. 3D-digital image correlation is used to quantify the deformation.

Results show that the deformation behavior is strongly non-linear and that the structures are capable of performing large (elastic) deformations without complete failure. This extraordinary behavior makes pantographic structures very attractive as engineering material in technical applications for lightweight applications and in the medical industry.

Keywords: Experiment · Metamaterial · Digital image correlation · Additive manufacturing

1 Introduction

With the help of additive manufacturing fabrication of complex designed structures became cheaper over the last decades. Developments and implementations of customized substructures in combination with specially tailored materials became possible resulting in extraordinary macroscopic deformation behavior. Such a man-made structure is also referred to as a meta-material. Pantographic structures, which can be described as meta-materials with a substructure composed of two orthogonal arrays of beams, connected by internal cylinders or

D. A. Indeitsev and A. M. Krivtsov (Eds.): APM 2019, LNME, pp. 101–114, 2020.
https://doi.org/10.1007/978-3-030-49882-5_11

joints (see Fig. 1), were manufactured by using additive manufacturing techniques. Such meta-materials enable the fabrication of lightweight structures with high degrees of complexity in combination with high stiffnesses. Effective properties are carefully designed by tailoring the so-called microscopic constituents [3,17,18] in the substructure in order to achieve specially desired properties, where the unit cells of the substructures are repeated periodically. Therefore mechanical performance of meta-materials depends not only on the global structure, but also on the morphology of their subunits.

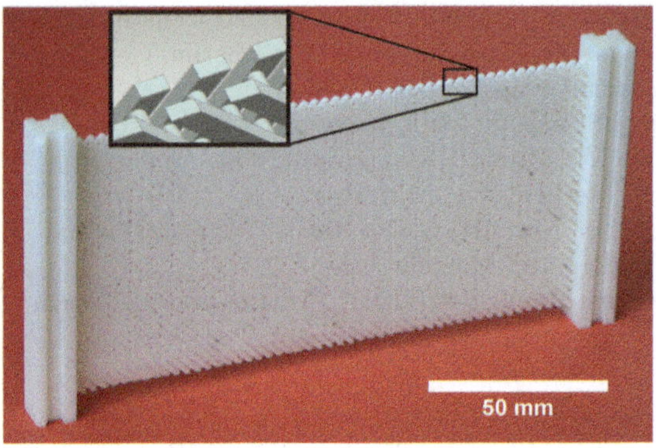

Fig. 1. Pantographic structure developed by [8] made out of polylactide, manufactured at the Institute of Mechanics at Technische Universität Berlin by means of fused deposition modeling technique.

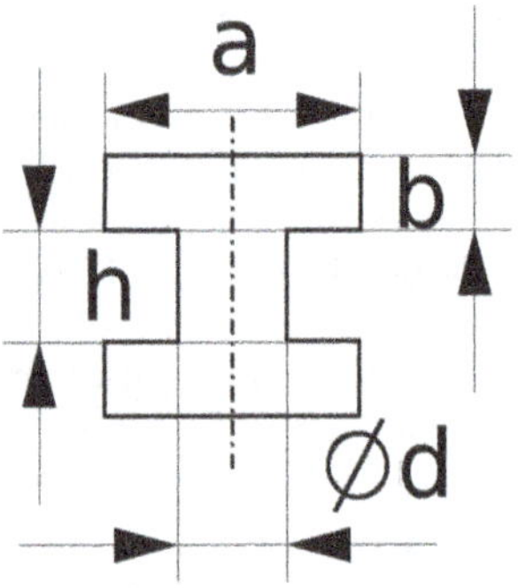

Fig. 2. Two-dimensional cross-section of an unit cell of a pantographic substructure.

The design and manufacture of meta-materials for specific engineering applications requires us to predict their performance. This can be achieved by the finite element method [5,35,39]. Normally the modeling of technical structures is based on the equations of the traditional CAUCHY-BOLTZMANN continuum. This requires a very detailed mesh if details need to be analyzed, which leads to higher computing time and associated higher costs. Alternatively, by using equations developed from generalized continuum theories, a low-detail mesh can be used to get the same results and, correspondingly, the computational costs are much less. The classical first gradient Cauchy theories need to be improved by introducing either additional degrees of freedom, for example those of the COSSERAT medium [6,7], micropolar parameters [13,14,29,30], or additional higher order gradients [1,2,9,24,25,27,31,32,36].

It is known that pantographic structures can be represented well by higher gradient theory. In fact they were treated as a second gradient continuum [3,10]. Here, the bending stiffness of the fibers is described by a second-gradient dependence (in displacement). However, new material parameters are introduced into the constitutive laws and experiments have to be designed to calibrate and determine these parameters [4,16]. By choosing a specially adjusted experimental setup in combination with an explicit theory it is possible to localize and to determine the unknown parameters [8,20–24,38]. On the one hand side one may say that newly identified parameters gained from theory have to be measured and determined in experiments [23,24,26], and on the other hand side the parameters obtained from theory or numerical analysis have to been validated by experiments [8,25,37].

In this paper three types of experiments applied to meta-materials made out of three different materials consisting of pantographic substructure will be performed. In addition a non-invasive optical measurement technique, the so-called Digital Image Correlation (DIC), will be used to detect and to measure the deformation on the specimens' surface. Pantographic metamaterials show a highly resilient and non-linear elastic material behavior resulting in large deformations.

2 Materials and Methods

Different materials and different experimental setups have been taken into account in this work. Three different additive manufacturing procedures were used to manufacture three differently sized samples, which will be described in Subsect. 2.1. Standardized test-setups of three different experimental methods will be presented in detail in Subsect. 2.2.

2.1 Manufacturing

Pantographic structures, consisting of rectangular beams and cylindrical pivots/joints, were 3D-printed using three different additive manufacturing methods:

i) Fused Deposition Modeling (FDM) by means of an Ultimaker 3 Extended (Ultimaker B.V., Geldermalsen, Netherlands) available at Technische Universität Berlin, Chair of Continuum Mechanics and Constitutive Theory, Germany,

ii) Selective Laser Sintering (SLS) by means of EOS Formagia P 100 (EOS GmbH, Munich, Germany) located at University of Technology Warsaw, Institute of Mechanics and Printing, Poland,

iii) Direct Metal Laser Sintering (DMLS) by means of EOS M 400 (EOS GmbH, Munich, Germany) located at Fraunhofer Ernst-Mach-Institute Freiburg, Germany.

Polylactic Acid (PLA - Ultimaker B.V., Geldermalsen, Netherlands) was used as raw material for FDM (specimen **PLA**). In order to increase the printing quality of the specimen, water-soluble Polyvinyl Acetate (PVA - Ultimaker B.V., Geldermalsen, Netherlands) was used additionally as support-structure during the

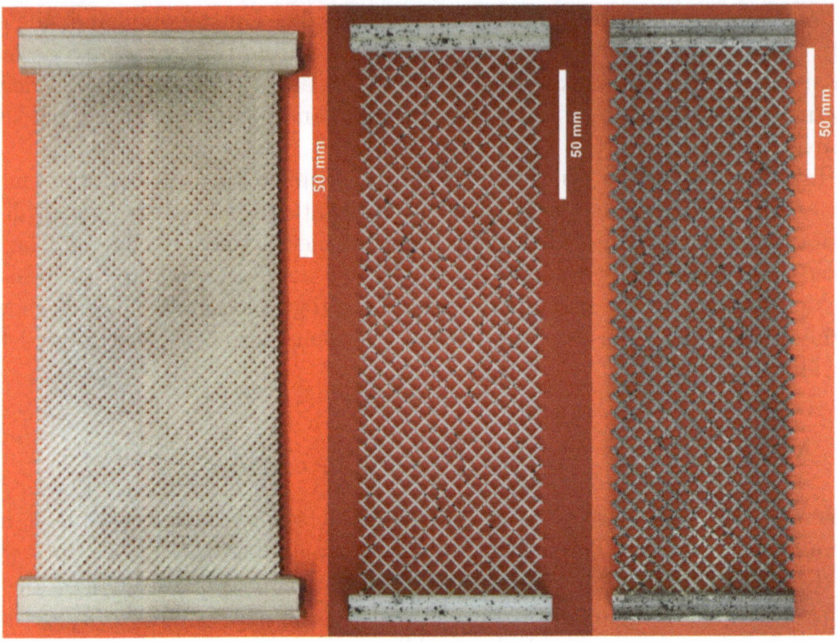

Fig. 3. Speckled pantographic structures made out of PLA (left), PA (middle), and AlSi10Mg (right).

printing process and was washed off afterwards (for further informations see [19]). Polyamide (PA2200 - EOS GmbH, Munich, Germany) was used as raw material in SLS (specimen **PA**). For further informations see [4,15]. Aluminium alloy (AlSi10Mg - EOS GmbH, Munich, Germany) was used as raw material using DMLS (specimens **ALU**, **ALU-H**). A special support structure and a complicated elaborated laser exposure strategy was employed in order to avoid thermal distortions due to the higher laser powers and energy input. Furthermore, a heat treatment was performed on sample **ALU-H** in order to reduce internal stresses (for further informations see [16]). Since the microscopic substructures influence the macroscopic deformation behaviors significantly [4,33], variations of specimens with different geometries were investigated as well. Figure 3 shows three types of specimens (sample **PLA**, **PA**, and **ALU**). In total four different specimens were investigated:

1) sample **PLA** in extension,
2) sample **PA** under shear,
3) sample **ALU** torsion-test,
4) sample **ALU-H** torsion-test.

Details of the inner and outer dimensions of all samples can be found in Table 1.

Table 1. Outer dimensions in mm (L = width, l = height, t = depth) of all samples as well as inner dimensions of the substructure of all samples corresponding to the schematic in Fig. 2 (a = width of beam, b = height of beam, $\o d$ = diameter of pivot, h = height of pivot).

Sample	L	l	t	a	b	$\o d$	h
PLA (extension)	140	70	3	1	1	1	1
PA (shear)	210	70	3	1	1	0.9	1
ALU (torsion)	210	70	5	1	1	0.9	3
ALU-H (torsion)	210	70	5	1	1	0.9	3

2.2 Experimental Setup

The experiments can be classified as extension, torsion and shearing tests. Extension tests have been performed on a MTS Tytron 250 testing device at Technische Universität Berlin, Chair of Continuum Mechanics and Constitutive Theoriy (CMCT). Torsion- and shearing-tests have been performed on a Zwick Z010 at Charité Berlin, Julius Wolff Institute (JWI) respectively. Quasi-static standard tests have been taken into account. The experimental setup of extension tests performed on sample **PLA** at CMCT is shown in Fig. 6A; the experimental setup of shearing-test performed at JWI is presented in Fig. 4. In this work we will also focus on torsion tests applied to **ALU** and **ALU-H** specimens. The schematic setup of torsion-test performed at JWI is shown in Fig. 5.

The MTS Tytron 250 testing-device controlled by the software Stationsmanager V 3.14 was used during extension tests on sample **PLA**. The applied force was measured by a load cell attached to the device, which is able to record axial forces in a range of $F = \pm 250\,\mathrm{N}$. The displacement, x, was imposed horizontally on the right of the specimen with a loading rate of $v = 15\,\mathrm{mm/min}$ (displacement-controlled). It was measured and monitored by the device's own encoder unit. Almost frictionless movement was achieved by using an air-film-bearing. External vibration was avoided by using a massive substructure and by arranging the system horizontally. Additionally to force-displacement (stress-strain) recording, pictures were taken (0.25 pictures/second) by means of a commercial Canon EOS 1000D camera with a resolution of 4272×2848 pixels. Triggered pictures and force/displacement were synchronized with the help of a TTL signal. 2D-DIC evaluation was performed in GOM Correlate 2017 software (GOM GmbH, Braunschweig, Germany).

For investigating specimens made out of aluminum (samples **ALU** and **ALU-H**), which need higher loads to generate measurable results, the Zwick Z010 testing-device, controlled by the software TestExpert was used. The shearing-tests on the **PA** sample were also performed with this device. The resultant

Fig. 4. Set-up of shear experiment of sample **PA** performed on the Zwick Z010 device at JWI (for further informations of experimental setup see [16]).

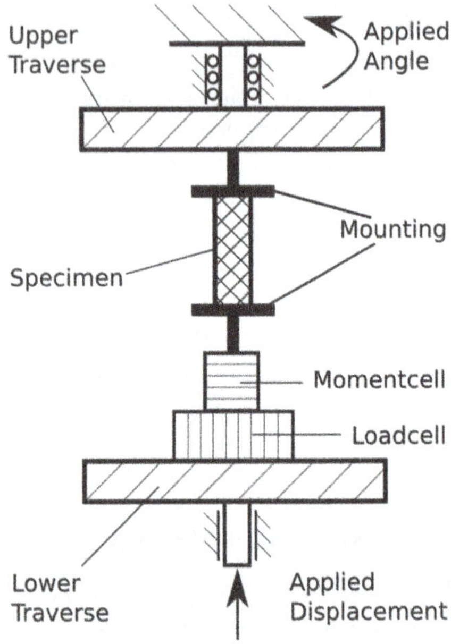

Fig. 5. Schematic set-up of extension and torsion experiments performed on the Zwick Z010 device.

applied axial force was measured by a device-own load cell (Zwick-Serie Xforce). The force transducer is able to record axial forces in the range of about $F = \pm 10000\,\mathrm{N}$, where the accuracy at $20\,\mathrm{N}$ is about 0.1%. The displacement x was controlled vertically. The upper traverse-part of the tensile-to-shear adaption device is fixed horizontally and vertically, while the lower part can be linearly moved in the vertical direction (see Fig. 5). The velocity of the shearing-test was set to $v = 15\,\mathrm{mm/min}$, which is quite slow for such tests (displacement-controlled and quasi-static). The displacement itself was recorded and monitored by a device-own encoder unit with an accuracy of $\pm 0.002\,\mathrm{mm}$. For the torsion tests a device-own torque sensor (Zwick-Serie M) was applied on the very fixed bottom of the lower traverse, while the torsion was induced on the top of the mounting with 1 degree/min on the upper traverse-part of the torque adaption-device (see Fig. 5). The torsion transducer is able to record moments up to $M = 20\,\mathrm{Nm}$ and resists maximal axial forces up to $\pm 5\,\mathrm{kN}$. Furthermore a non-invasive optical measurement device Q-400 (Dantec Dynamics GmbH, Ulm, Germany) was installed to record the state of three dimensional deformation of the surface of a sheet by the help of two cameras. A more-than-one camera system is able to recognize the 3D-motion within overlapping regions of the image sections. For enable the software of image correlation to separate small surface areas (so-called facets) and due to lack of contrast, the surfaces of all specimens had to be sprayed with a speckled pattern (see Fig. 3). During the deformation process, pictures were taken via direct TTL-signal every 2 s by means of the afore mentioned commercial camera system with a resolution of about 1600×1200 pixels. This way we were able to synchronize each picture to the related force-value in real time. By means of a calibration procedure of the camera setup, the commercial software Istra4D is able to re-calculate a three dimensional surface deformation.

3 Results

In order to obtain scalar results for an out-of-plane displacement of a sheet in DIC, a reference point in a single facet (a sub-area of image correlation) was selected for each sample. This point is located in the place where maximal out of-plane movement could be assumed. Due to the large deformations some facets moved out of the optical focus, which caused the image correlation to abort. Furthermore, image correlation may also be aborted when a sudden rupture occurs in between the shutter releases of the camera, so that the facets to be correlated are displaced too much. For further informations see [16,19]. Torsion tests applied to the aluminum specimens were performed on the same experimental setup with identical loading conditions.

3.1 Extension

The extension test with the **PLA** sample was performed on a MTS Tytron 250 device located at Institute of Mechanics at Technische Universität Berlin, Germany. Figure 7 shows the stress-strain relation of sample **PLA** in a biaxial tension test. The marked points **A**, **B**, and **C** correspond to the sequence pictures in Fig. 6, respectively. After linear elastic deformation (between points **A** - **B**), plastic deformation occurs resulting in a first rupture shortly after an elongation of about $\varepsilon = 6\%$, which is a quite low elongation for pantographics (see [16,34]). The pivot in the lower right corner breaks (Fig. 7C), but surprisingly total failure can be avoided. The structure is even able to resist higher loads after a second rupture. Necking, calculated by means of 2D-DIC in vertical direction, results in a shortening in vertical direction of about $\Delta y = 5$ mm and is shown in the right lower corner in Fig. 6. Obviously, high elastic performance as seen in previously investigated experimental studies [4,8,11,15,16,34] has not been experienced with this specimen because of the small height of the cylindric pivot/joint (see [34]).

3.2 Shearing

Shearing tests with the **PA** sample were performed on a Zwick Z010 testing device at the Julius Wolff Institute at Charité Berlin, Germany (see Fig. 4).

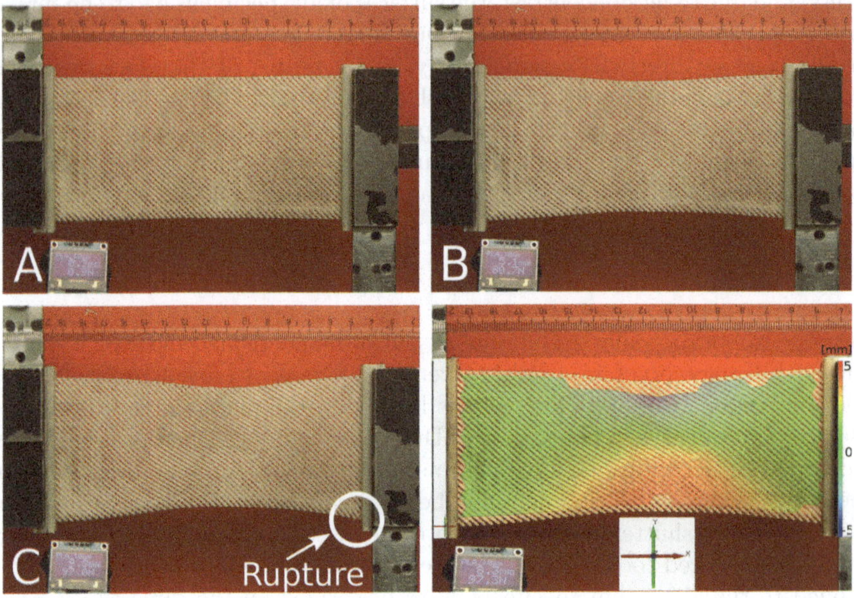

Fig. 6. Image sequence of specimen **PLA** during extension load. Picture **A** - **C** correspond to the marked points in Fig. 7. The picture in the lower right corner shows exemplary the calculated necking in y-direction by means of 2D-DIC shortly before first rupture occurs.

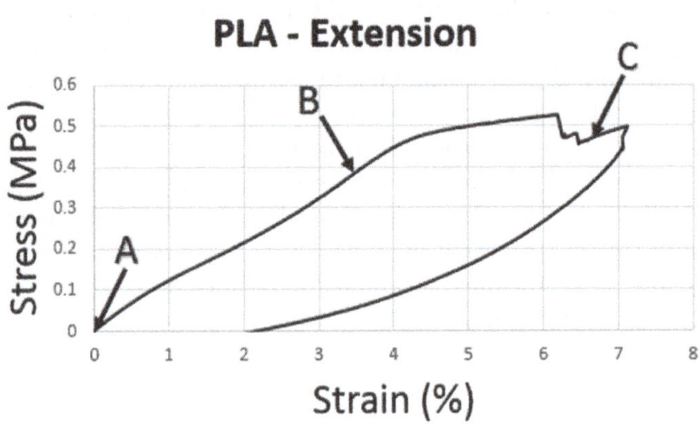

Fig. 7. Stress-strain curve of biaxial extension test of specimen **PLA** performed on MTS Tytron 250 device at CMCT. Arrows pointing to points A, B, and C correspond to pictures **A**, **B**, and **C** in Fig. 6, respectively.

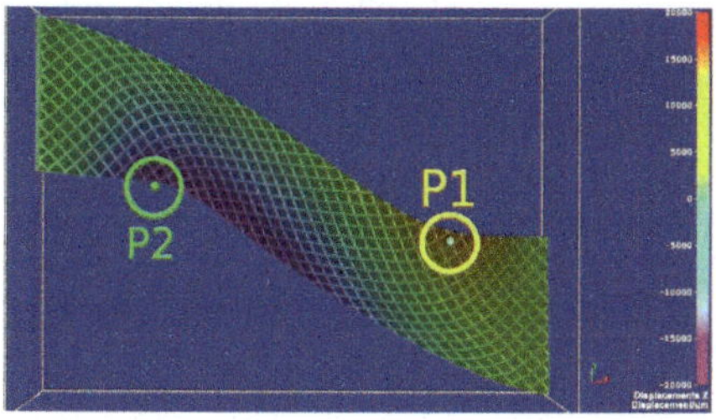

Fig. 8. 3D-DIC evaluation of specimen **PA** during shearing test with positive (red, point P1) and negative (blue, point P2) out-of-plane-movements.

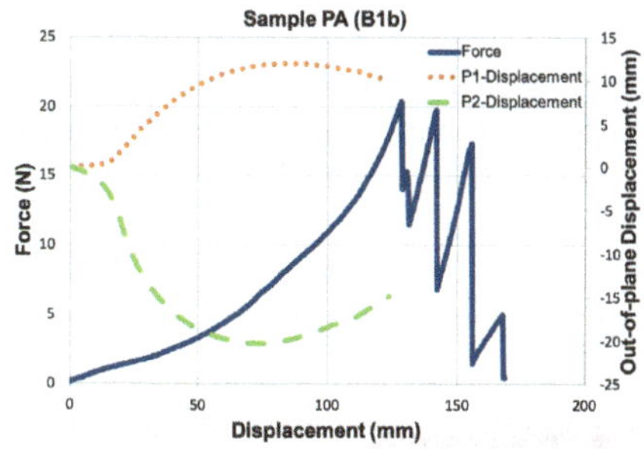

Fig. 9. Shear-force-displacement relation of specimen **PA** including asymmetric out-of-plane-movements of points P1 (positive) and P2 (negative), respectively.

Figure 8 shows the out-of-plane movement of the surface of sample **PA** during the shearing test. Because of the stiffness of pivots/cylinders (see [4,34]), which connect the arrays of beams of the different planes, out-of-plane buckling occurs, being in good agreement with previous investigations reported in [4,15,16]. In Fig. 9 out-of-plane movements of the points P1 and P2 corresponding to Fig. 8 and the force-displacement curve are shown. An almost linear elastic deformation behavior can be observed until about $y = 50\,\text{mm}$ (shear elongation of $\Delta e_{yy} = 71\%$). Surprisingly, strongly non-linear and non-symmetric out-of-plane movements start at about $y = 10\,\text{mm}$ while being still in the linear elastic deformation range ($\Delta y < 50\,\text{mm}$). After reaching a maximal shear-displacement at about $y_{max} = 70\,\text{mm}$ the out-of-plane buckling decreases. This is because of the high plastic deformation in the structure. One may say that the "shearing test becomes an elongation test" at this very point.

Furthermore, the first rupture occurs at about $y = 127\,\text{mm}$ of shear-deformation (shear elongation of $\Delta e_{yy} = 181\%$) at almost 21 N. The whole metamaterial is able to resist even further loading without leading to total failure resulting in a high resilient-kind deformation behavior. Even after the fourth rupture at about $y = 155\,\text{mm}$ shear-displacement (shear elongation of $\Delta e_{yy} = 221\%$), the whole structure does not fail and is able to resist higher loads up to 5 N before total failure.

3.3 Torsion

Torsion tests applied to sample **ALU** and sample **ALU-H** were performed on a Zwick Z010 testing device at Julius Wolff Institute at Charité Berlin, Germany. Figure 10 shows the heat-treated specimen sample **ALU-H** after one total rotation (360°). Figure 11 shows the out-of-plane buckling of the same specimen right after first torque was applied at the upper mounting part. The points P1 and P2 show about ±14 mm out-of-plane movement before the facets got lost and DIC-evaluation was aborted. In Fig. 12 moment-, angle-, and out-of-plane movement-time relations are shown. Linear elastic as well as linear plastic deformation behavior was observed. In order to investigate the plastic deformation of both specimens, rotations with load cycles of 10° steps were imposed up to 60° (0° – 10° – 0° – 20° – 0° – ... – 60° – 0° – 2000°). By reaching the negative area ($F < 0\,\mathrm{Nm}$) after the 20° step plastic deformation clearly occurs. Furthermore, non-linear deformation behavior can be recognized after 60° (at about 480 s).

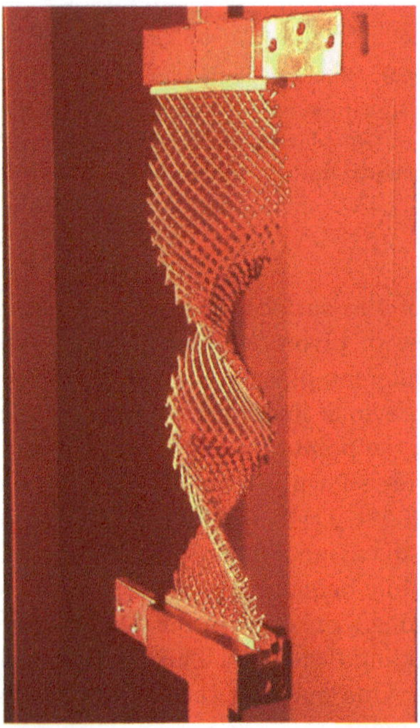

Fig. 10. Sample **ALU-H** after one full rotation during torsion test performed on Zwick Z010 at JWI, Charité in Berlin, Germany.

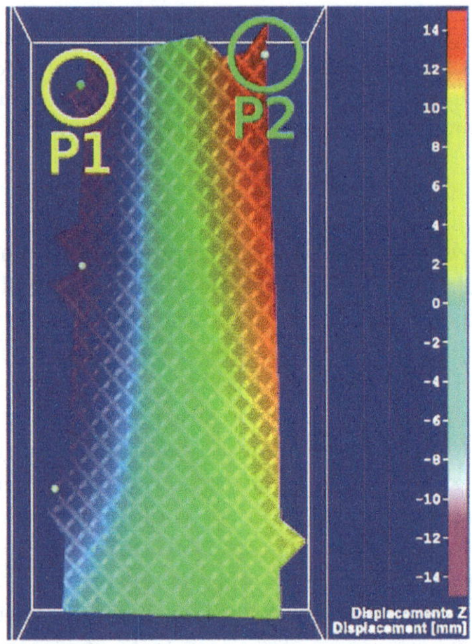

Fig. 11. Example of 3D-DIC evaluation of specimen **ALU-H** during torsion test with negative out-of-plane-movement (blue, point P1) and positive out-of-plane-movement (red, point P2).

By comparing the moment-time dependencies of the heat-treated specimen, **ALU-H**, with the untreated one, **ALU**, in Fig. 13, considerable differences in the deformation behavior can be observed. The specimen **ALU** reaches about 16% higher loads than the heat-threated one, **ALU-H**. But sample **ALU-H** is able to resist further loads without leading to a single failure because of its ductility. This is in contrast to sample **ALU**, in which local ruptures of beams and pivots result in total failure. But still, **ALU** was able to resist even higher torque-loads before total failure of the whole structure occurred. This resilient deformation behavior was also observed in the aforementioned extension test applied to specimen **PLA** and in the shearing test applied to specimen **PA**.

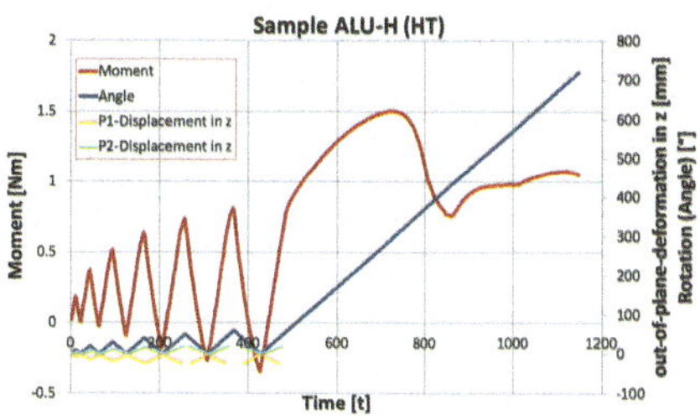

Fig. 12. Diagram of moment-, angle-, and out-of-plane movement over time of torqued specimen **ALU-H**.

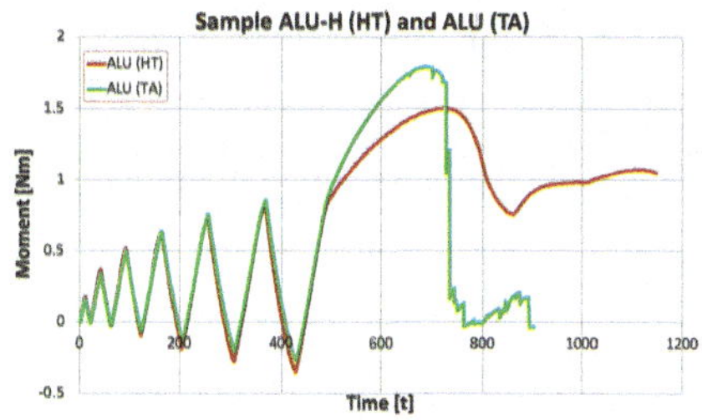

Fig. 13. Comparison of moment-time dependences of specimens **ALU-H** and **ALU** during a torsion test.

A reverse POYNTING effect was observed and is further discussed in [28] and [12]. It should also be mentioned that total failure of the heat-threated specimen **ALU-H** did not occur during the whole experiment ending after 5.7 total rotations (about 2000°). This extraordinary deformation behavior is comparable to investigations with cylindrical structures reported in [12].

4 Conclusion and Outlook

Pantographic structures made out of three different materials were additively manufactured using three differently types of additive manufacturing techniques. Polylactide was used as a raw material for 3D-printing of specimen **PLA** by means of fused deposition modeling at Technical University Berlin, Germany. Polyamide was used as a raw material for 3D-printing of specimen **PA** by means of selective laser sintering at Polytechnica Warsaw, Poland. An Aluminum alloy was used as raw material for 3D-printing of specimens **ALU** and **ALU-H** by means of direct metal laser sintering at Fraunhofer Ernst Mach Institute in Freiburg, Germany.

An extension test was performed on specimen **PLA**, a shearing test on specimen **PA**, and torsion tests with specimens **ALU** and **ALU-H**. In parallel, an image correlation procedure was performed. We were able to observe and to measure out-of-plane movements for specimens **PA**, **ALU**, and **ALU-H** qualitatively as well as quantitatively.

Specimen **PLA** deformed in a linear elastic way, until a first local plastic rupture occurred without resulting in total failure. This resilient deformation behavior may be used in failure-safe systems in the future to secure stability of mechanical systems, *e.g.*, in aircraft or automobile applications. In contrast to investigations that can be found in [34], out-of-plane buckling was observed during shearing tests applied to specimen **PLA**, due to the non-perfect pivots/joints. Independently, the aforementioned resilient deformation behavior was also observed for this shearing test. The untreated specimen, **ALU**, was able to carry 16% higher rotational loads during torsion tests then the heat-treated one, **ALU-H**. Nevertheless, specimen **ALU** failed after about one total rotation in contrast to specimen **ALU-H**, which was able to resist total failure during the whole experiment ending after 5.7 total rotations. Furthermore, a reverse POYNTING effect and extraordinary nonlinear deformation behavior was observed and is further discussed in [28] and [12].

One may say that because this structure stays in the elastic range, even at large deformations, it will play a crucial role in the future in different kind of industrial applications. Its light weight in combination with its high resilient deformation behavior makes this kind of meta-material very attractive to applications in industry (*e.g.*, for protection shields in civil or defense industries, damping or mounting device for aircraft or automobile industries).

Acknowledgement. We want to thank Dag Wulsten from the Julius Wolff Institute at Charité in Berlin, Germany, and Paul Zaslansky from the Zahnklinik at Charité in Berlin, Germany, for their support and help in the lab.

References

1. Abali, B.E., Müller, W.H., Eremeyev, V.A.: Strain gradient elasticity with geometric nonlinearities and its computational evaluation. Mech. Adv. Mater. Mod. Process. **1**, 1–11 (2015)
2. Abali, B.E., Müller, W.H., dell'Isola, F.: Theory and computation of higher gradient elasticity theories based on action principles. Arch. Appl. Mech. **87**, 1495–1510 (2017)
3. Barchiesi, E., Spagnuolo, M., Placidi, L.: Mechanical metamaterials: a state of the art. Math. Mech. Solids **24**, 212–234 (2018)
4. Barchiesi, E., Ganzosch, G., Liebold, C., Placidi, L., Grygoruk, R., Müller, W.H.: Out-of-plane buckling of pantographic fabrics in displacement-controlled shear tests: experimental results and model validation. Continuum Mech. Thermodyn. (2018)
5. Chen, C., Fleck, N.A.: Size effects in the constrained deformation of metallic foams. J. Mech. Phys. Solids **50**, 955–977 (2002)
6. Cosserat, E., Cosserat, F.: Sur la theorie de l'elasticite. Mathematiques, Premier memoire. Annales de la Faculte des sciences de Toulouse (1896)
7. Cosserat, E., Cosserat, F.: Théorie des corps déformables (1909)
8. dell'Isola, F., Lekszycki, T., Pawlikowski, M., Grygoruk, R., Greco, L.: Designing a light fabric metamaterial being highly macroscopically tough under directional extension: first experimental evidence. Zeitschrift für angewandte Mathematik und Physik **66**, 3473–3498 (2015)
9. dell'Isola, F., Sciarra, G., Vidoli, S.: Generalized Hooke's law for isotropic second gradient materials. In: Proceedings of the Royal Society of London A: Mathematical, Physical and Engineering Sciences (2009)
10. dell'Isola, F., Cuomo, M., Greco, L., Della Corte, A.: Bias extension test for pantographic sheets: numerical simulations based on second gradient shear energies. J. Eng. Math. **103**, 127–157 (2017)
11. dell'Isola, F., Turco, E., Misra, A., Vangelatos, Z., Grigoropoulos, C., Melissinaki, V., Farsari, M.: Force-displacement relationship in micro-metric pantographs: experiments and numerical simulations. C. R. Mec. **347**, 397–405 (2019)

12. Bachurikhin, V.P., Keller, I., Merzlyakov, A.F., Yurlov, M.A.: Experimental study of nonlinear effects under torsion of the uniform cylinder with initially circular cross section. In: Solid State Phenomena (2015)
13. Eringen, A.C., Suhubi, E.S.: Nonlinear theory of simple micro-elastic solids–I. Int. J. Eng. Sci. **2**, 189–203 (1964)
14. Eringen, A.C.: Mechanics of micromorphic continua. In: Mechanics of Generalized Continua. Springer (1968)
15. Ganzosch, G., dell'Isola, F., Turco, E., Lekszycki, T., Müller, W.H.: Shearing tests applied to pantographic structures. Acta Polytech. CTU Proc. **7**, 1–6 (2017)
16. Ganzosch, G., Hoschke, K., Lekszycki, T., Giorgio, I., Turco, E., Müller, W.H.: 3D-measurements of 3D-deformations of pantographic structures. Tech. Mechanik **38**, 233–245 (2018)
17. Gibson, L.J., Ashby, M.F.: Cellular Solids: Structure and Properties. Cambridge University Press (1999)
18. Gibson, L.J.: Biomechanics of cellular solids. J. Biomech. **38**, 377–399 (2005)
19. Juritza, A., Yang, H., Ganzosch, G.: Qualitative investigations of experiments performed on 3D-FDM-printed pantographic structures made out of PLA. In: New Achievements in Continuum Mechanics and Thermodynamics. Springer (2019)
20. Kong, S., Zhou, S., Nie, Z., Wang, K.: Static and dynamic analysis of micro beams based on strain gradient elasticity theory. Int. J. Eng. Sci. **47**, 487–498 (2009)
21. Lakes, R.S.: Experimental microelasticity of two porous solids. Int. J. Solids Struct. **22**, 55–63 (1986)
22. Lakes, R.S., Drugan, W.J.: Bending of a cosserat elastic bar of square cross section: theory and experiment. J. Appl. Mech (2015)
23. Lam, D.C.C., Yang, F., Chong, A.C.M., Wang, J., Tong, P.: Experiments and theory in strain gradient elasticity. J. Mech. Phys. Solids **51**, 1477–1508 (2003)
24. Liebold, C., Müller, W.H.: Measuring material coefficients of higher gradient elasticity by using AFM techniques and Raman-Spectroscopy. In: Generalized Continua as Models for Materials. Springer (2013)
25. Liebold, C., Müller, W.H.: Applications of higher-order continua to size effects in bending: theory and recent experimental results. In: Generalized Continua as Models for Classical and Advanced Materials. Springer (2016)
26. Liebold, C.: Größeneffekt in der Elastizität. Dissertation Technische Universität. Berlin (2015)
27. Mindlin, R.D., Eshel, N.N.: On first strain-gradient theories in linear elasticity. Int. J. Solids Struct. **4**, 109–124 (1968)
28. Misra, A., Lekszycki, T., Giorgio, I., Ganzosch, G., Müller, W.H., dell'Isola, F.: Pantographic metamaterials show atypical Poynting effect reversal. Mech. Res. Commun. **89**, 6–10 (2018)
29. Müller, W.H., Vilchevskaya, E.N.: Micropolar theory with production of rotational inertia: a rational mechanics approach. In: Advanced Structured Materials, Generalized Models and Non-classical Approaches in Complex Materials (2018)
30. Müller, W.H., Vilchevskaya, E.N.: Micropolar theory from the viewpoint of mesoscopic and mixture theories. Phys. Mesomech. **20**, 263–279 (2017)
31. Placidi, L., Barchiesi, E., Misra, A.: A strain gradient variational approach to damage: a comparison with damage gradient models and numerical results. Math. Mech. Complex Syst. **6**, 77–100 (2018)
32. Placidi, L., Misra, A., Barchiesi, E.: Two-dimensional strain gradient damage modeling: a variational approach. Z. Angew. Math. Phys. **69**, 56 (2018)

33. Rahali, Y., Giorgio, I., Ganghoffer, J.F., Dell'Isola, F.: Homogenization a la Piola produces second gradient continuum models for linear pantographic lattices. Int. J. Eng. Sci. **97**, 148–172 (2015)
34. Spagnuolo, M., Barcz, K., Pfaff, A., dell'Isola, F., Franciosi, P.: Qualitative pivot damage analysis in aluminum printed pantographic sheets: numerics and experiments. Mech. Res. Commun. **83**, 47–52 (2017)
35. Tekoğlu, C., Onck, P.R.: Size effects in two-dimensional Voronoi foams: a comparison between generalized continua and discrete models. J. Mecha. Phys. Solids **56**, 3541–3564 (2008)
36. Toupin, R.A.: Elastic materials with couple-stresses. Arch. Rational Mech. Anal. **11**, 385–414 (1962)
37. Turco, E., dell'Isola, F., Rizzi, N.L., Grygoruk, R., Müller, W.H., Liebold, C.: Fiber rupture in sheared planar pantographic sheets: numerical and experimental evidence. Mech. Res. Commun. **76**, 86–90 (2016)
38. Wei, Y., Wang, X., Wu, X., Bai, Y.: Theoretical and experimental researches of size effect in micro-indentation test. Sci. China Ser. A: Math. **44**, 74 (2001)
39. Yang, H., Ganzosch, G., Giorgio, I., Abali, B.E.: Material characterization and computations of a polymeric metamaterial with a pantographic substructure. Z. Angew. Math. Phys. **69**, 105 (2018)

Fatigue Reliability of Structures: Methodology of Assessment and Problems

Ruslan V. Guchinsky[1(✉)] and Sergei V. Petinov[2]

[1] TMH Engineering Ltd, Petrogradskaya nab., 34, St. Petersburg 197046, Russia
ruslan239@mail.ru

[2] Institute for Problems of Mechanical Engineering, Bolshoy pr., 61, V.O., St. Petersburg 199178, Russia
spetinov@mail.ru

Abstract. Fatigue reliability assessment of metallic structures in various applications according current design codes is based mostly on *S-N* criteria with uncertain characterization of fatigue properties of a particular material and the assumed damage. In case the crack is detected residual service life as recommended may be estimated by applying the Linear fracture mechanics techniques, again, with incomplete defining conditions of the crack growth and exhaustion of life.

Proposed earlier procedure of simulation the fatigue process based on the due finite element modeling of the affected area of a structure, application of the damage summation technique and appropriate criterion for fatigue failure of material allowed assessment of fatigue life from the onset of service loading up to development of a critical state, e.g. of the through crack in a structural component. Also, it was shown that the simulation scheme was capable of account the heterogeneity of the material structure fatigue resistance, the crack closure effects and elastic-plastic material response when the Strain-life criterion for fatigue failure was applied.

However, selection of the criterion for fatigue failure of material is but a straightforward decision: it is shown that *S-N* criterion even attributed to the same structural steel class as the Strain-life one does not provide in analysis of the fatigue process even an approximate convergence. This is mostly due to fairly indirect considering the inelastic properties of fatigue damage in *S-N* criteria and the methodology of fatigue testing specimens aimed at evaluation of *S-N* and $\Delta\varepsilon$-N criteria.

Further, the approach would need in more comparisons of simulated and test data in different structural applications.

1 Introduction

Proposed earlier procedure of simulation the fatigue process based on the due finite element modeling of the affected area of a structure, application of the damage summation technique and appropriate criterion for fatigue failure of material allowed assessment of fatigue life from the onset of service loading up to development of a critical state, e.g. of the through crack growth in a structural

D. A. Indeitsev and A. M. Krivtsov (Eds.): APM 2019, LNME, pp. 115–123, 2020.
https://doi.org/10.1007/978-3-030-49882-5_12

component [1,2], etc. Also, it was shown that the simulation scheme was capable of account the heterogeneity of the material structure fatigue resistance, the crack closure effects and elastic-plastic material response when the Strain-life criterion for fatigue failure was applied (e.g., [3]). Meanwhile, fatigue reliability assessment of metallic structures subjected to intensive alternating service loading in various applications according current design codes (e.g., [4]) is based mostly on the Stress-life (S-N) criteria which formally indicate the crack initiation at critical locations. In case the crack is detected in a component the residual service life as recommended may be estimated by applying the Linear fracture mechanics techniques, however, with certain limitations in defining conditions of the crack growth and exhaustion of life.

It was shown [5,6], that for the fatigue design purposes, in particular, application of the S-N curves collected by testing ≪classed≫ specimens (including typified welded joints [7]) terminated at almost complete failure could not provide identity of fatigue damage in structural details necessarily related to purposes of the design. Apart from that, application of different models of fatigue process (crack initiation defined by S-N criteria and crack growth assessed by the Linear fracture mechanics model) mechanically continuous is leading to uncertainties in defining fatigue life of structural components, partly, due to the uncertain gap between the states of the fatigue process at the ≪crack initiation≫ and ≪growth≫ assessed by different models.

However, the mentioned in above modeling of the fatigue process as a continuous one which allowed assessment of fatigue process from initiation of service loading up to development of a critical state may be applied for fatigue analysis purposes using the unique S-N criteria to ≪close≫ the mentioned gap. At the same time, application of the modeling principles in conjunction with the Stress-life criteria offering a certain consistency would necessarily need in comparing with proven, although far but exhausting, modeling based on using the Strain-life criteria.

The mentioned in above approaches and comparison of simulation results are commented in more details in the below focused on problems of practical application.

2 Description of the Approach

According the approach, the fatigue process in a structural component assessment is based on application of the finite element model of the component designed with the due attention to the mesh fineness in the area of expected crack initiation and propagation. The finite elements with respect to the polycrystalline material structure are assumed the grain clusters with approximately close (or, alternatively, random) slip resistance.

In the first procedure step the stress field at the expected crack extensions is analyzed in the ascending and descending half-cycles of loading and the stress range $S_i = 2\sigma_{ai}$ is obtained in every i-th element.

The cyclic elastic-plastic properties of material are characterized by Ramberg-Osgood approximation; the kinematic hardening and von Mises' criterion for plasticity are applied. The strain amplitude in finite elements (considered structural elements) is defined, respectively by:

$$\varepsilon_a = \sigma_a/E + (\sigma_a/K')^{1/n'}, \tag{1}$$

where E - elasticity modulus, n' - cyclic strain hardening exponent and K' - cyclic strength coefficient.

The number of load cycles prior to failure of elements at the initial step of the procedure is obtained by applying the Strain-life criterion:

$$\varepsilon_a = (\sigma'_f/E)(2N)^b + \varepsilon'_f(2N)^c, \tag{2}$$

where σ'_f - fatigue strength coefficient, ε'_f - fatigue ductility coefficient, b - fatigue strength exponent, c - fatigue ductility exponent.

Further, when the Stress-life criterion is applied, the *S*-*N* curve is formulated as:

$$\sigma_a = \begin{cases} \sigma'_{f1}(2N)^{b_1}, & 2N \le 2N_s \\ \sigma'_{f2}(2N)^{b_2}, & 2N > 2N_s \end{cases}, \tag{3}$$

where b_1, b_2 - fatigue strength exponent values in the bi-linear *S*-*N* curve, $\sigma'_{f1}, \sigma'_{f2}$ - fatigue strength coefficients in the curve model; $2N_s$ - is the "kink" number of cycles of the two-slope curve.

In further analyses the criterion (2) with description of cyclic properties of material (1) is presented in the form of Stress-life, $N(S)$, as the criterion (3). The damage accumulated at random loading estimated by the Palmgren-Miner rule is assumed uniformly distributed within elements. The damage accumulation in finite (structural) elements in the approach is provided by the loading history at the crack initiation site and in elements located at the crack path; it results also from the loading pre-history evolution in the course of failure (crack) extension through the preceding elements (from the initiation location). The linear damage accumulation rule, Bolotin's version [8]):

$$D = \sum_i n_i(S)/N_i(S) = N^* \int_{S_{min}}^{S_{max}} (p(S)/N(S))dS, \tag{4}$$

where $p(S) = (S/\sigma_s^2)\exp(-S^2/2\sigma_s^2)$ - probability density function of the Rayleigh distribution of the stress ranges, σ_s - stress scale parameter.

Number of cycles to failure of each element at the first step can be expressed by equating the accumulated damage to unity:

$$N_i^1(S_i, \sigma_{si}) = 1/\int_{S_{min}}^{S_{max}} (p(S, \sigma_{si})/N(S))dS, \tag{5}$$

where $S_{min} = 2 \cdot 0.55\sigma_{-1}$ - minimum stress range in distribution, corresponding to reduced fatigue limit σ_{-1} due to irregular loading; $S_{max} = S_i$ – maximum stress range in the distribution, corresponding to i-th element. Stress ranges below S_{min} are assumed non-damaging. As mentioned in above, the Raleigh stress range probability exceedance is defined by: $Q(S_i) = \exp(-S_i^2/2\sigma_{si}^2)$. Respectively, the scale parameter of stress in the stress probability distribution for each element is: $\sigma_{si} = S_i/\sqrt{-2\ln(Q)}$. Correspondingly, the number of cycles in the form (5) takes into account the irregular loading of each element at the first step of the procedure. Then the minimum number of cycles to failure of first element at the first step of the fatigue simulation procedure is calculated: $N_{min}^1 = \min\{N_i^1(S_i, \sigma_{si})\}$.

According the damage linear summation rule when the damage in an element reaches the critical unity, its compliance is artificially increased by several decimal orders, and nodal forces are step-wise redistributed in the surrounding elements. Damage of each element at the first step is defined as $d_i^1 = N_{min}^1/N_i$.

The damage calculation is repeated considering for the renewing stress (stress-strain) state ahead the crack tip. Respectively, an element damage criterion related to the critical load step number j_{cr}:

$$d_i^{j_{cr}} = \sum_{j=1}^{j_{cr}-1} d_i^j + N_{min}^{j_{cr}}/N_i^{j_{cr}} = 1. \tag{6}$$

When this condition is attained at an element, its stiffness is decreased by several decimal orders and «killed» by this procedure elements form the crack front progress.

This idea was first suggested in pioneering studies of Glinka and Ellyin focused on analysis of crack extensions in thin plates at the plane stress [9, 10]; further analyses had shown that the procedure may be applied in assessment of fatigue process including the crack initiation and growth of plane cracks in arbitrary bodies [1–3], etc.

For the purposes of the present study the fatigue failure criteria parameters, *S-N* and Strain-Life, should be in principle obtained by testing specimens prepared from the same material, say, from the same rolled plate, the same location in it (edge or the mid-part) to consider the peculiarities of the material microstructure. What is essential for the purposes of the comparative study, the means of evaluation of the failure may be different in experimental definition of the criteria parameters: *S-N* testing of standard specimens (e.g. IIW recommended [7]) is terminated typically prior to failure of specimens controlled by the rapid acceleration of the crack growth while in the strain-controlled testing the stress-strain diagram ascending part (in the tensile phase) distortion indicates the crack initiation and is used to complete the test.

3 Results and Discussion

Implementation of the approach was illustrated by simulation of failure of a thin plate with central elliptic hole fabricated from AlMg4.5Mn alloy (σ_y= 298 MPa; (σ_u = 363 MPa). The quarter part of the plate with dimensions $2h$ = 800 mm;

$2W = 400$ mm is shown in Fig. 1, hole dimensions: $2b = 50$ mm; $2R = 100$ mm. Plate was loaded by pulsating tension with maximum stress σ in load cycle. Two maximum nominal stress ranges in the random loading successions in the illustrative example were assumed as 100 MPa and 110 MPa. Maximum stress range probability exceedance was accepted $Q = 0.0001$ in both cases.

Material parameters for (1): $E = 0.71 \cdot 10^5$ MPa, $K' = 693$ MPa, $n' = 0.125$, $\sigma'_u = 654$ MPa. Strain-Life criterion (2) constants: $\varepsilon'_f = 0.45$, $b = -0.089$, $c = -0.755$ [11]. Criterion (2) was approximated by two curves with minimal interpolation error of 5%:

$$N = \begin{cases} 0.5(0.1/(\varepsilon_a - 2.852 \cdot 10^{-3}))^{1/0.45}, & \varepsilon_a > 3.53 \cdot 10^{-3} \\ 0.5(1.052 \cdot 10^{-2}/(\varepsilon_a - 3.92 \cdot 10^{-4}))^{1/0.11}, & \varepsilon_a \le 3.53 \cdot 10^{-3} \end{cases}, \qquad (7)$$

where the Stress-life criterion (3) constants are: $\sigma'_{f1} = 526$ MPa, $\sigma'_{f2} = 719$ MPa, $b_1 = -0.06$, $b_2 = -0.096$, $2N_s = 5897$ [11]. Fatigue limit of the alloy is: $\sigma_{-1} = 103.4$ MPa.

The crack nucleates in stress concentration area and propagates in horizontal direction following up the maximum principal stress plane. Finite-element mesh was refined in the crack growth area; the minimal element size (crack propagation increment) was designed equal to 0.3 mm (Fig. 1). In this illustration of the approach application the plane strain problem is considered and the course of ≪failure≫ of material elements the nodes of finite-element mesh are uncoupled to simulate the crack propagation.

The results of crack growth assesment are given in Figs. 2, 3 and 4. As seen, the crack growth rate increases rapidly after the crack becomes 2 mm long in both loading schemes characterized by (maximum) stress ranges. Also, the crack nucleation stage duration is practically identical in case of loading with maximum stress range 100 MPa when the Stress-Life or the Strain-Life criteria are applied (approximately $3.9 \cdot 10^7$ cycles).

In the case of more intensive loading (maximum stress range 110 MPa) the crack initiation stage estimated by the Strain-life criterion occurs shorter than in previous example. In both cases the crack propagation rate and stage duration predicted by this criterion, as seen, are substantially different from those assessed by the Stress-life scheme.

In Fig. 3 the crack growth is presented as related to the non-dimensional number of load cycles. The number of load cycles is divided by the fatigue life N_0 corresponding to failure of the first mesh node, meaning the crack nucleation. It is seen, that crack propapation stage in both examples is relatively short and extends over 10–15% of the crack nucleation stage. Using crack propagation extrapolation you can see that crack growth phase is no more than 25% of the total fatigue life.

Figure 4 illustrates results of the crack propagation simulation from 0.3 mm-deep initial crack at the hole. It is seen that the crack growth stage assessed by applying the Strain-Life criterion occures two times shorter than that obtained when the Stress-Life criterion was used.

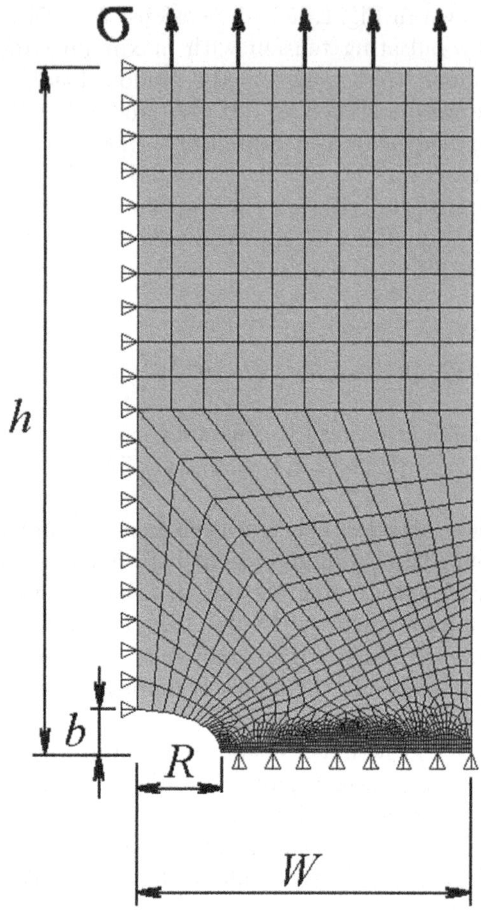

Fig. 1. The FE-model of the quarter part of the plate

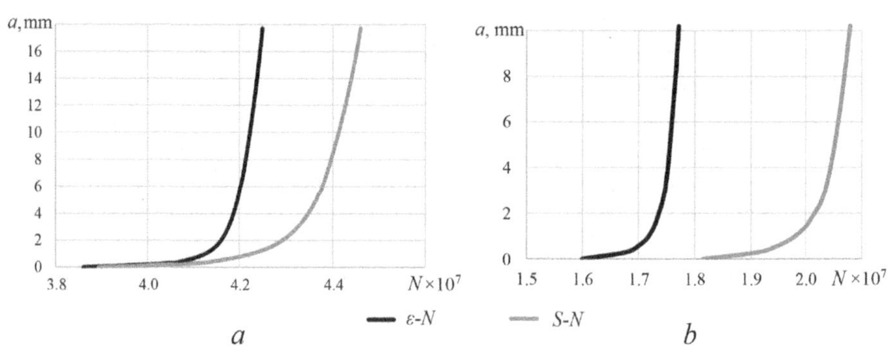

Fig. 2. Crack growth (initiation and propagation) vs the loading cycles number dependence on the maximum nominal stress range in the Raleigh's distribution: 100 MPa (*a*) and 110 MPa (*b*)

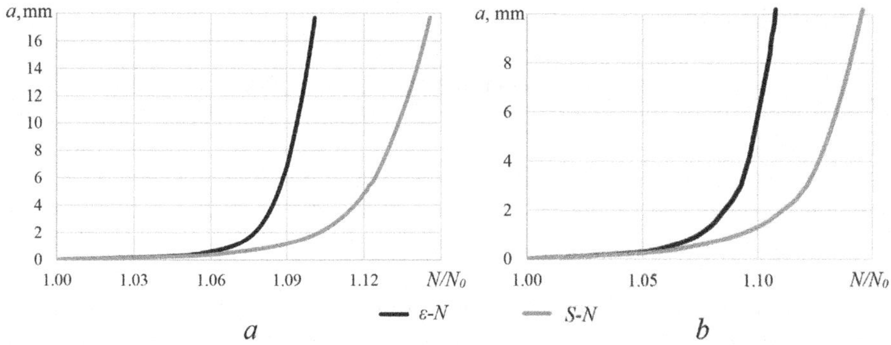

Fig. 3. Crack extension (initiation and growth) depending on the non-dimensional number of cycles and the maximum nominal stress range in the Raleigh's distribution: 100 MPa (*a*) and 110 MPa (*b*)

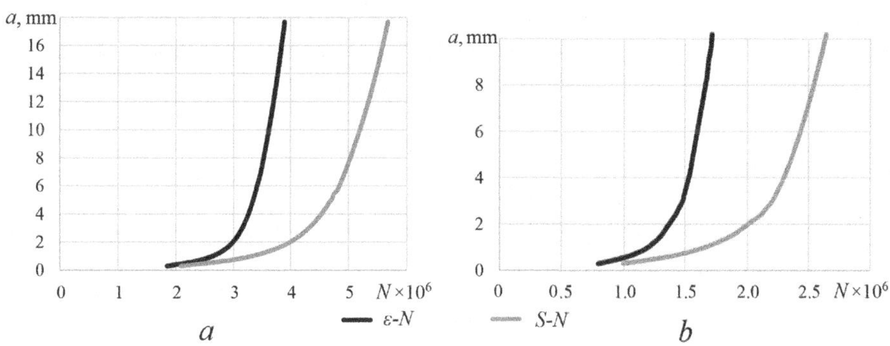

Fig. 4. Crack extension (from initial 0.3 mm size) depending on the number of load cycles and the maximum nominal stress range in the Raleigh's distribution: 100 MPa (*a*) and 110 MPa (*b*)

The discrepancies in the crack initiation and growth stages may be explained by the difference of typical testing specimens procedures, as mentioned in above: the *S-N* curve addressed testing is terminated when the compliance of the cracked specimen and acceleration of the loading frame is rapidly increasing, whereas when the parameters of Strain-Life criterion are obtained the test is terminated when the ascending part of the cyclic stress-strain hysterezis loop is distorted due to nucleation of macroscopic crack.

The further development of the FE modeling-based procedure of modeling fatigue process in structural components would need in comparison of simulated and observed fatigue damage in structural details with known loading histories similarly to attempted in [11].

4 Conclusions

The previously developed simulation of fatigue damage and crack growth based on respective FE modeling of a structural component, Strain-life criterion for fatigue and the damage accumulation procedure was extended into the same scheme where the Stress-life criterion typically used in current design and fatigue assessment codes was applied. The scheme of simulation fatigue process where the loading was assumed a cyclic one was completed with the means of considering the random loading histories.

The both versions of the procedure were tested in example of the randomly loaded plate with elliptic hole as the stress raiser and the fatigue damage and crack growth initiator. The crack initiation and growth stages predictions occurred different when the Strain-life and Strain-life criteria were applied. The discrepancies of results may be explained partly by specifics of experimental evaluation of the Strain-life and Stress-life criteria parameters, considering mechanics of material deformation and further studies are necessary in the prospect of development the unified approach to model the fatigue process in actual metallic structures.

References

1. Petinov, S.V., Guchinsky, R.V. Finite-element modeling of the semi-elliptical fatigue crack growth by the damage accumulation approach. In: Proceedings of the XLII International Summer School - Conference Advanced Problems in Mechanics, SPb-Repino, IPME RAS, pp. 296–306 (2014)
2. Guchinsky, R.V., Petinov, S.V.: Two-dimensional surface fatigue crack propagation and its finite element simulation. In: Applied Mechanics and Materials, Ch. 5, vol. 725–726, pp. 654–661 (2015)
3. Imran, M., Siddique, S., Guchinsky, R.V., Petinov, S.V., Walther, F.: Comparison of fatigue life assessment by analytical, experimental and damage accumulation modeling approach for steel SAE 1045. Fatigue Fract. Eng. Mater. Struct. **39**(9), 1138–1149 (2016)
4. Veritas, D.N.: Fatigue Assessment of Ship Structures. Classification Notes 30.7, Hovik, Norway (2014)
5. Petinov, S.V., Thayamballi, A.K.: The application of S-N Curves considering mismatch of stress concentration between test specimen and structure. J. Ship Res. **42**(1), 68 (1998)
6. Petinov, S.V., Reemsnyder, H.S. and Thayamballi, A.K.: The similitude of fatigue damage principle: application in S-N curves-based fatigue design. In: Marquis, G., Solin, J. (ed.) Fatigue Design and Reliability, pp. 219–228. ESIS Publication 23. Elsevier Science Ltd (1999)
7. EUROCODE 3: Design of Steel Structures. Part 1-9: Fatigue. BS EN 1993-1-9: 2009, p. 36
8. Bolotin, V.V.: Statistical Methods in Structural Mechanics. Holden-Day, San-Francisco (1969)
9. Glinka, G.A.: Cumulative model of fatigue crack growth. Int. J. Fatigue **4**, 59–67 (1982)

10. Ellyin, F., Fakinlede, C.O.: Probabilistic simulation of fatigue crack growth by damage accumulation. Eng. Fract. Mech. **22**(4), 697–712 (1985)
11. Fatemi, A., Plaseid, A., Khosrovaneh, A.K., Tanner, D.: Application of bi-linear log-log S-N model to strain-controlled fatigue data of aluminum alloys and its effect on life predictions. Int. J. Fatigue **27**, 1040–1050 (2005)

Inhomogeneous Distribution of Thermal Characteristics in Harmonic Crystal

Mikhail A. Guzev[1(✉)], Vladimir M. Sadovskii[2], and Chengzhi Qi[3]

[1] Institute for Applied Mathematics FEB RAS, Vladivostok, Russia
guzev@iam.dvo.r

[2] Institute of Computational Modeling SB RAS, Krasnoyarsk, Russia
sadov@icm.krasn.ru

[3] Beijing University of Civil Engineering and Architecture, Beijing, China
qichengzhi65@163.com

Abstract. In this paper, we consider a model of a uniform harmonic chain of particles for the analysis of non-stationary thermal effects in an ideal crystal system. The exact solution for the particle system is presented and the temperature is calculated as a measure of the average kinetic energy of the particles. The corresponding energy averaging is performed over the initial distribution of the displacements and velocities of the particles, provided that they obey the Boltzmann principle. Simple analytical formulae are presented for all energy derivatives with respect to time at the initial time and for the first derivative with respect to the number of particles. Over a small time interval, the temperature was shown to depend monotonically on the number of particles. This means that the non-uniformity of thermal characteristics distribution, i.e. dependence on the number of particles, occurs in the system without additional assumptions about the structure of the initial conditions on a macroscopic scale. The obtained formula for the distribution of kinetic energy is presented through Bessel functions. The functional dependence on the number of particles was shown to appear in the index of Bessel functions, and the parity of the number of particles affects the temperature distribution. The distribution of the kinetic energy for a large time was asymptotically analyzed as well.

Keywords: Harmonic crystal · Temperature · Boltzmann principle

1 Introduction

The results of numerical modeling of high-frequency oscillations of the particle kinetic and potential energies in an ideal crystal lattice are well known [1]: if at the initial time the particles are ordered in the ideal crystal lattice and their velocities are given randomly, then the transition of one type of energy to another one is accompanied by a high-frequency oscillatory process with damped amplitude. An approach to the theoretical analysis of this phenomenon was proposed in the model of a one-dimensional harmonic crystal [2], for which the heat

D. A. Indeitsev and A. M. Krivtsov (Eds.): APM 2019, LNME, pp. 124–138, 2020.
https://doi.org/10.1007/978-3-030-49882-5_13

transfer process was simulated. In the method used to analyze the thermal characteristics, the number of particles in the model is assumed to be $n \gg 1$. The results obtained in [2] show that the spatial non-uniformity of the distribution of characteristics in the crystal is determined by setting inhomogeneous initial conditions on a macroscopic scale, in which limits the additional averaging is performed over microscopic initial conditions.

In this paper, we remove the restriction $n \gg 1$ [2]. The result presented below shows that the non-uniformity of thermal characteristics distribution, i.e. dependence on the number of particles arises without additional assumptions about the structure of the initial conditions on a macroscopic scale.

2 Dynamic Relations for the Model

A one-dimensional crystal is some ideal model that has a characteristic internal spatial scale, for example, an equilibrium distance a between adjacent particles, and an external scale L determined by the length of the particles chain. In the general case, the ratio between the parameters L and a can be different, in particular, the condition $L \gg a$ is equivalent to a large number of particles $n \gg 1$ in the system. A harmonic crystal is considered as a chain of particles with the equal mass m, connected by the identical linear stiffness springs k, the interaction between which is set with the nearest neighbors., The coordinate x_j for a j-particle is representable in the form $x_j = ja + u_j$, where $0 \le x_1 \le ... \le x_n \le L$ and the function u_j sets the displacement of the particle from the equilibrium position.

Formulate the force conditions for the first and last particles: we will assume that they are connected by linear stiffness springs k with fixed points. Then the Hamiltonian function of the particle system under consideration is

$$H = m\frac{\dot{u}_i^2}{2} + k\frac{u_1^2}{2} + k\sum_{i=1}^{n-1}\frac{(u_{j+1} - u_j)^2}{2} + k\frac{u_n^2}{2}. \tag{1}$$

The equations of motion of the particles system, corresponding to the Hamilton function (1), are given by the formulas:

$$m\dot{u}_i = \frac{\partial H}{\partial \dot{u}_i},\ m\ddot{u}_i = -\frac{\partial H}{\partial u_i},$$

$$m\ddot{u}_i = k(u_{i+1} - 2u_i + u_{i-1}),\ i = 2, ..., n-1, \tag{2}$$

$$m\ddot{u}_1 = k(u_2 - 2u_1),\ m\ddot{u}_n = k(u_{n-1} - 2u_n).$$

Let us pass in (2) to dimensionless variables, choosing the distance a as the scale of length and assume the time scale to be equal $\sqrt{m/c}$. Then (2) is reduced to the system

$$\ddot{u}_1 = (u_2 - 2u_1),\ \ddot{u}_i = (u_{i+1} - 2u_i + u_{i-1}),\ i = 1, ..., n-1,\ \ddot{u}_n = (u_{n-1} - 2u_n) \tag{3}$$

with initial conditions: $u_i|_{t=0} = u_{i0}$, $\dot{u}_i|_{t=0} = \dot{u}_{i0}$.

The statement is true: the components u_i are determined from the relations

$$u_i = \sum_{j=1}^{n} \alpha_{ij} \left(q_{0j} \cos \omega_j t + \frac{p_{0j}}{\omega_j} \sin \omega_j t \right), \tag{4}$$

$$z_j = \frac{\pi j}{2(n+1)}, \omega_j = 2 \sin z_j, \; \alpha_{ij} = \sqrt{\frac{2}{n+1}} \sin 2 i z_j,$$

$$q_{0i} = \sum_{j=1}^{n} \alpha_{ij} \; u_j|_{t=0} \, , \; p_{0i} = \sum_{j=1}^{n} \alpha_{ij} \; \dot{u}_j|_{t=0} \, . \tag{5}$$

Indeed, differentiating (4) by time, we obtain

$$\ddot{u}_i = - \sum_{j=1}^{n} \alpha_{ij} \omega_j^2 \left(q_{0j} \cos \omega_j t + \frac{p_{0j}}{\omega_j} \sin \omega_j t \right). \tag{6}$$

Take into consideration that

$$\alpha_{(i-1)j} - 2\alpha_{ij} \; + \alpha_{(i+1)j} = \sqrt{\frac{2}{n+1}} [\sin 2(i-1) z_j - 2 \sin 2 i z_j + \sin 2(i+1) z_j]$$
$$= -4 \sin^2 z_j \cdot \sqrt{\frac{2}{n+1}} \sin 2 i z_j = -\omega_j^2 \alpha_{ij}.$$

From here and from (6), the equations follow for particles with numbers $i = 2, ..., n-1$ in (3):

$$u_{i+1} - 2u_i + u_{i-1} = \sum_{j=1}^{n} (\alpha_{(i-1)j} - 2\alpha_{ij} + \alpha_{(i+1)j}) \left(q_{0j} \cos \omega_j t + \frac{p_{0j}}{\omega_j} \sin \omega_j t \right)$$
$$= - \sum_{j=1}^{n} \omega_j^2 \alpha_{ij} \left(q_{0j} \cos \omega_j t + \frac{p_{0j}}{\omega_j} \sin \omega_j t \right) = \ddot{u}_i, \; i = 2, ..., n-1.$$

The validity of the first equation of the system (3) after substitution into it the solution (4) is easy to verify by using the relations

$$\ddot{u}_1 = - \sum_{j=1}^{n} \alpha_{1j} \omega_j^2 \left(q_{0j} \cos \omega_j t + \frac{p_{0j}}{\omega_j} \sin \omega_j t \right),$$

$$\alpha_{2j} - 2\alpha_{1j} = \sqrt{\frac{2}{n+1}} (\sin 4 i z_j - 2 \sin 2 z_j) = 2(\cos 2 i z_j - 1) \sqrt{\frac{2}{n+1}} \sin 2 z_j$$
$$= -4 \sin^2 z_j \cdot \sqrt{\frac{2}{n+1}} \sin 2 z_j = -\omega_j^2 \alpha_{1j}.$$

Substituting solution (4) into the last equation of system (3), we get

$$\ddot{u}_n = -\sum_{j=1}^{n} \alpha_{1j}\omega_j^2(-1)^j \left(q_{0j}\cos\omega_j t + \frac{p_{0j}}{\omega_j}\sin\omega_j t\right). \tag{7}$$

On the other hand,

$$\alpha_{(n-1)j} - 2\alpha_{nj} = (-1)^j\sqrt{\frac{2}{n+1}}(\sin 2iz_j - 2\sin z_j) = (-1)^j(\alpha_{2j} - 2\alpha_{1j})$$
$$= (-1)^{j+1}\omega_j^2\alpha_{1j}.$$

Then

$$u_{n-1} - 2u_n = \sum_{j=1}^{n}(\alpha_{(n-1)j} - 2\alpha_{nj})(q_{0j}\cos\omega_j t + \frac{p_{0j}}{\omega_j}\sin\omega_j t)$$
$$-\sum_{j=1}^{n}(-1)^j\alpha_{1j}\omega_j^2(q_{0j}\cos\omega_j t + \frac{p_{0j}}{\omega_j}\sin\omega_j t).$$

From here and from (7) it can be seen that the constructed solution u_n satisfies the last equation of system (3).

The fulfillment of the initial conditions (5) is ensured due to orthogonality property of the matrix $\|\alpha_{ij}\|$: $\sum_{j=1}^{n}\alpha_{ij}\alpha_{jk} = \delta_{ik}$, where δ_{ik} is the Kronecker symbol.

Relations (4) and (5) implement the general idea of constructing a solution to the system of equations (3) in the basis formed by the eigenvectors of the matrix of the difference operator of this system (see, for example, [3]). The physical meaning of the definition given above is assumed that Hamiltonian (1) is diagonalized by passing to the normal coordinates [4].

3 Kinetic Energy in a Small Time Interval

The kinetic energy of a particle is a measure of temperature. Local definition of the kinetic energy is

$$\bar{T}_j = \left\langle \frac{v_j^2}{2} \right\rangle_P. \tag{8}$$

The brackets $\langle\rangle_P$ denote averaging over the distribution $P = P(u_0, v_0)$ of microscopic initial displacements and velocities $(u_0, v_0) = (u_{01}, ..., u_{0n}, v_{01}, ..., v_{0n})$. In accordance with Boltzmann's principle [5], it is assumed that

$$P = C^{-1}\exp\left[-\frac{H(u_0, v_0)}{\beta}\right], \tag{9}$$

$$H(u_0, v_0) = \frac{|v_0|^2}{2} + W(u_0),\ C = \int_{R^n} du_0 dv_0 \exp\left[-\frac{H(u_0, v_0)}{\beta}\right].$$

Here, the dimensionless parameter β is equal to the ratio of the particle energy $ka^2/2$ to the value $k_B T_0$, where k_B is the Boltzmann's constant, T_0 has the meaning of the average crystal temperature. We take into account the conditions under which the model of an ideal crystal is being studied [2]: the initial displacements of the particles are zero. Then the solution (4) is reduced to the form:

$$u_i = \sum_{j=1}^{n} \alpha_{ij} \frac{p_{0j}}{\omega_j} \sin \omega_j t,\ \dot{u}_j = v_i = \sum_{j=1}^{n} \alpha_{ij} p_{0j} \cos \omega_j t.$$

From here and from (8), (9) it follows that

$$\bar{T}_j = \frac{\beta}{2(n+1)} \sum_{k=1}^{n} (\sin 2jz_k)^2 (\cos \omega_k t)^2$$

$$= \frac{\beta}{8(n+1)} \sum_{k=1}^{n} (1 - \cos 4jz_k + \cos 2\omega_k t - \cos 4jz_k \cos 2\omega_k t)\,. \tag{10}$$

Since

$$\sum_{k=1}^{n} \cos 2pz_k = \cos \frac{\pi p}{2} \frac{\sin \frac{\pi pn}{2(n+1)}}{\sin \frac{\pi p}{2(n+1)}} = \begin{cases} -1, \text{ if } p - \text{ even}, \\ 0, \text{ if } p - \text{ odd}, \end{cases} \tag{11}$$

then

$$\bar{T}_j = \frac{\beta}{8(n+1)} \left(n + 1 + \sum_{k=1}^{n} (\cos 2\omega_k t - \cos 4jz_k \cos 2\omega_k t) \right). \tag{12}$$

Taking into account (11), we obtain that $\bar{T}_j(0) = \beta/4$, i.e., at the initial time, the average kinetic energy of a particle does not depend on its number. Consider the change of the average kinetic energy of a particle. For this purpose we calculate the first and the second derivatives of the energy:

$$\frac{\partial \bar{T}_j}{\partial t} = \frac{\beta}{4(n+1)} \sum_{k=1}^{n} (-\omega_k \sin 2\omega_k t + \omega_k \cos 4jz_k \sin 2\omega_k t)\,,$$

$$\frac{\partial^2 \bar{T}_j}{\partial t^2} = \frac{\beta}{2(n+1)} \sum_{k=1}^{n} (-\omega_k^2 \cos 2\omega_k t + \omega_k^2 \cos 4jz_k \cos 2\omega_k t)\,.$$

From this follows that at the initial moment of time the rate of the energy change is zero: $\left.\frac{\partial \bar{T}_j}{\partial t}\right|_{t=0} = 0$. Taking into account the identity $\sin^2 z_k = (1 - \cos 2z_k)/2$, we can write the second derivative at the initial moment of time as follows

$$\left.\frac{\partial^2 \bar{T}_j}{\partial t^2}\right|_{t=0} = -\frac{\beta}{n+1}\sum_{k=1}^{n}(1-\cos 4jz_k - \cos 2z_k) - \frac{\beta}{n+1}\sum_{k=1}^{n}\cos 4jz_k \cos 2z_k.$$

The first sum is equal to $n+1$ by virtue of (11), and the second one is reduced to

$$\sum_{k=1}^{n}\cos 4jz_k \cos 2z_k = \frac{1}{2}\sum_{k=1}^{n}\left(\cos 2(2j-1)z_k + \cos 2(2j+1)z_k\right).$$

Using relation (11), it is easy to verify that each of the sums vanishes. Thus, local behavior $\bar{T}_j(t)$ is given by

$$\bar{T}_j(t) = \frac{\beta}{4} - \frac{\beta t^2}{2} + \ldots$$

i.e. $\bar{T}_j(t)$ is not dependent of j.

It is of interest to answer the question: how does the derivative of any order of the function $\bar{T}_j(t)$ depend on the number j at the initial moment of time? From (12) it follows that only even derivatives of the energy are non-zero at the initial moment of time:

$$\left.\frac{\partial^{2p} \bar{T}_j}{\partial t^{2p}}\right|_{t=0} = \beta\frac{2^{4p-3}}{n+1}\left[-\sum_{k=1}^{n}\sin^{2p} z_k + \sum_{k=1}^{n}\sin^{2p} z_k \cos 4\pi j z_k\right]. \tag{13}$$

For further calculations, the equation

$$\sin^{2p} z_k = \frac{1}{2^{2p}}\left\{2\sum_{r=0}^{p-1}(-1)^{p-r}\binom{2p}{r}\cos 2(p-r)z_k + \binom{2p}{p}\right\}. \tag{14}$$

is useful. Using (11), from (14) we get

$$\sum_{k=1}^{n}\sin^{2p} z_k = \frac{1}{2^{2p}}\left\{2\sum_{r=0}^{p-1}(-1)^{p-r}\binom{2p}{r}\sum_{k=1}^{n}\cos 2(p-r)z_k + n\binom{2p}{p}\right\}$$
$$= \frac{1}{2^{2p}}\left\{2\sum_{r=0}^{p-1}(-1)^{p-r}\binom{2p}{r}\cos[(n+1)(p-r)z_1]\frac{\sin n(p-r)z_1}{\sin(p-r)z_1} + n\binom{2p}{p}\right\}.$$

Since

$$\sin nz_1(p-r)\cos[(n+1)(p-r)z_1] = \frac{1}{2}\left[-\sin z_1(p-r) + \sin\left(\pi(p-r) - z_1(p-r)\right)\right]$$
$$= -\frac{1}{2}\sin z_1(p-r)[1+(-1)^{p-r}],$$

the previous ratio may be rewritten as

$$\sum_{k=1}^{n}\sin^{2p} z_k = \frac{1}{2^{2p}}\left\{-\sum_{r=1}^{p-1}(1+(-1)^{p-r})\binom{2p}{r} + n\binom{2p}{p}\right\}. \tag{15}$$

When calculating the second sum in (13), we will use the identity

$$\sum_{k=1}^{n} \cos 4jz_k \cos 2(p-r)z_k = \frac{1}{2}\left[\sum_{k=1}^{n} \cos 2z_k(2j-(p-r)) + \cos 2z_k(2j+(p-r)\right].$$

Applying (11) to this relation, we obtain

$$\sum_{k=1}^{n} \cos 2z_k(2j-(p-r)) = \cos[(n+1)(2j-(p-r))z_1]\frac{\sin n(2j-(p-r))z_1}{\sin(2j-(p-r))z_1}.$$

Usage of equality

$$\begin{aligned}
&\sin nz_1(2j-(p-r))\cos[(n+1)z_1(2j-(p-r))] \\
&= \frac{1}{2}\{-\sin z_1(2j-(p-r)) + \sin[(2n+1)z_1(2j-(p-r))]\} \\
&= -\frac{1}{2}\sin z_1(2j-(p-r))[1+(-1)^{p-r}],
\end{aligned}$$

allows us to write

$$\sum_{k=1}^{n} \cos 2z_k(2j-(p-r)) = -\frac{1}{2}[1+(-1)^{p-r}].$$

Therefore,

$$\sum_{k=1}^{n} \cos 4jz_k \cos 2(p-r)z_k = -\frac{1}{2}[1+(-1)^{p-r}],$$

and we obtain the following equation for the second sum (13):

$$\sum_{k=1}^{n} \sin^{2p} z_k \cos 4\pi j z_k = \frac{1}{2^{2p}}\left\{-\sum_{r=0}^{p-1}(1+(-1)^{p-r})\binom{2p}{r} - \binom{2p}{p}\right\}.$$

Combination this equation with (13) and (15) results in

$$\left.\frac{\partial^{2p}\bar{T}_j}{\partial t^{2p}}\right|_{t=0} = -2^{2p-3}\binom{2p}{p}\beta.$$

Thus, in the initial state, the derivative of energy $\bar{T}_j(t)$ with respect to time does not depend on the number of the particle j. This result is explained by the fact that the function $\bar{T}_j(t)$ is not continuous at the point $(0, j)$ due to (11). Therefore, we will use the exact representation (12) for the energy $\bar{T}_j(t)$ to calculate its derivative with respect to j at $t = +0$:

$$\left.\frac{\partial\bar{T}_j}{\partial j}\right|_{t=0} = \frac{\beta}{2(n+1)}\sum_{k=1}^{n} z_k \sin 4jz_k = \frac{\pi\beta}{4(n+1)^2}\sum_{k=1}^{n} k \sin\frac{2\pi jk}{(n+1)}.$$

The summation of this series is presented in reference books:

$$\sum_{k=1}^{n} k\sin x = \frac{\sin(n+1)x}{4\sin^2\frac{x}{2}} - (n+1)\frac{\cos\frac{2n+1}{2}x}{2\sin\frac{x}{2}}.$$

Inserting $x = 2\pi j/(n+1)$ gives

$$\sum_{k=1}^{n} k\sin\frac{2\pi jk}{(n+1)} = -(n+1)\frac{\cos\frac{(2n+1)\pi j}{(n+1)}}{2\sin\frac{\pi j}{(n+1)}} = -\frac{n+1}{2}ctg\frac{\pi j}{(n+1)}.$$

Hence, the derivative of energy $\bar{T}_j(t)$ with respect to particle number is

$$\left.\frac{\partial \bar{T}_j}{\partial j}\right|_{t=0} = -\frac{\pi\beta}{8(n+1)}ctg\frac{\pi j}{(n+1)}. \tag{16}$$

Since average kinetic energy is a measure of temperature, it is sufficient to set microscopic inhomogeneous initial conditions for the occurrence of spatial non-uniformity of temperature distribution.

4 General Presentation for Average Kinetic Energy

To analyze the behavior of the average kinetic energy over a finite time interval, we should to calculate the sums in (12). We take into account the summation formula for trigonometric functions and the representation obtained by Jacobi [6]:

$$\cos(2\omega_k t) = \cos(4t\sin z_k) = J_0(4t) + 2\sum_{p=1}^{\infty} J_{2p}(4t)\cos(2pz_k).$$

Then

$$\sum_{k=1}^{n}\left(-\cos\frac{2\pi jk}{n+1} + \cos 2\omega_k t\right) = 1 + nJ_0(4t) - 2\sum_{p=1}^{\infty} J_{4p}(4t).$$

The sum of the Bessel functions is determined by the ratio:

$$\sum_{p=1}^{\infty} J_{4p}(4t) = \frac{1}{2}[\cos^2 2t - J_0(4t)]. \tag{17}$$

After calculations, we get the following result for the first three sums in the right part of (12):

$$\sum_{k=1}^{n}\left(1 - \cos\frac{2\pi jk}{n+1} + \cos 2\omega_k t\right) = \left[n + \sin^2 2t + (n+1)J_0(4t)\right]. \tag{18}$$

The sum generated by the last contribution to (12) is equal to:

$$\sum_{k=1}^{n} \cos \frac{2\pi jk}{n+1} \cos 2\omega_k t = -J_0(4t) + 2\sum_{p=1}^{\infty} CC \cdot J_{2p}(4t),$$
$$CC = \sum_{k=1}^{n} \cos \frac{2\pi jk}{n+1} \cos \frac{\pi pk}{n+1} \tag{19}$$
$$= \frac{1}{2}\sum_{k=1}^{n} \cos \frac{\pi k}{n+1}(p-2j) + \frac{1}{2}\sum_{k=1}^{n} \cos \frac{\pi k}{n+1}(p+2j).$$

If we apply the formula for the sum of the trigonometric series in (19), then

$$CC = \frac{1}{2}\frac{\cos \frac{\pi(p-2j)}{2} \sin \frac{\pi n(p-2j)}{2(n+1)}}{\sin \frac{\pi(p-2j)}{2(n+1)}} + \frac{1}{2}\frac{\cos \frac{\pi(p+2j)}{2} \sin \frac{\pi n(p+2j)}{2(n+1)}}{\sin \frac{\pi(p+2j)}{2(n+1)}}. \tag{20}$$

It can be seen that, at a certain ratio between p and j, the value of the denominator may be small ($\sin \frac{\pi(p\pm 2j)}{2(n+1)} \to 0$), however, it is compensated by the smallness of the numerator $\sin \frac{\pi n(p\pm 2j)}{2(n+1)}$, what leads to the need to calculate expressions (20) for such values of the parameters p and j. Therefore, we distribute the numbers in (19) into four disjoint sets: (A), (B), (C), (D) - and write

$$\sum_{p=1}^{\infty} CC \cdot J_{2p}(4t) = \sum_{p\in(A)}^{\infty} CC \cdot J_{2p}(4t) + \sum_{p\in(B)}^{\infty} CC \cdot J_{2p}(4t) \tag{21}$$
$$+ \sum_{p\in(C)}^{\infty} CC \cdot J_{2p}(4t) + \sum_{p\in(D)}^{\infty} CC \cdot J_{2p}(4t).$$

We define the first set (A) from the condition that $p-2j$, $p+2j$ are not divided by $n+1$, then for CC use (19), (17), and obtain:

$$\sum_{p\in(A)}^{\infty} CC \cdot J_{2p}(4t) = -\sum_{N=1}^{\infty} J_{4N}(4t) = -\frac{1}{2}[\cos^2 2t - J_0(4t)]. \tag{22}$$

The second set (B) includes those p for which $p-2j$ is divided by $n+1$ and $p+2j$ is not divided by $n+1$. In this case, p are parametrized by $p = 2j+m(n+1)$, where m is a integer number, and $p+2j = 4j+m(n+1)$ is not divided by $n+1$, if $j \neq (n+1)/4$. Then

$$CC = \sum_{k=1}^{n} \cos \frac{2\pi jk}{n+1} \cos \frac{\pi k}{n+1}[2j+m(n+1)] = \sum_{k=1}^{n} (-1)^{km}\left(\cos \frac{2\pi jk}{n+1}\right)^2. \tag{23}$$

Let m be even, then

$$CC = \sum_{k=1}^{n} \left(\cos \frac{2\pi jk}{n+1} \right)^2 = \frac{n}{2} + \frac{1}{2} \sum_{k=1}^{n} \cos \frac{4\pi jk}{n+1} = \begin{cases} \frac{n-1}{2}, \; j \neq \frac{n+1}{2}; \\ n, \; j = \frac{n+1}{2}. \end{cases}$$

For odd m, we get

$$CC = \sum_{k=1}^{n} (-1)^k \left(\cos \frac{2\pi jk}{n+1} \right)^2 = \frac{1}{2} \sum_{k=1}^{n} (-1)^k \left[1 + \cos \frac{4\pi jk}{n+1} \right]$$
$$= \begin{cases} \frac{(-1)^n - 1}{2}, \; j \neq \frac{n+1}{4}; \\ \frac{n}{2} + \frac{(-1)^n - 1}{2}, \; j = \frac{n+1}{4}. \end{cases}$$

After calculations, we can write for the set (B):

$$\sum_{p \in (B)}^{\infty} CC \cdot J_{2p}(4t) = \left(1 - \delta_{j, \frac{n+1}{4}}\right) \left(\sum_{\substack{p=2j+m(n+1), \\ even\ m}}^{\infty} CC \cdot J_{2p}(4t) + \sum_{\substack{p=2j+m(n+1), \\ odd\ m}}^{\infty} CC \cdot J_{2p}(4t) \right)$$
$$= \left(1 - \delta_{j, \frac{n+1}{4}}\right) n \delta_{j, [\frac{n+1}{2}]} \delta_{\frac{n+1}{2}, [\frac{n+1}{2}]} \sum_{even\ m}^{\infty} J_{2(1+m)(n+1)}(4t)$$
$$+ \frac{n-1}{2} \left(1 - \delta_{j, \frac{n+1}{4}}\right) \left(1 - \delta_{j, \frac{n+1}{2}}\right) \sum_{even\ m}^{\infty} J_{2(2j+m(n+1))}(4t)$$
$$+ \frac{(-1)^n - 1}{2} \left(1 - \delta_{j, \frac{n+1}{4}}\right) \sum_{odd\ m}^{\infty} J_{2(2j+m(n+1))}(4t). \quad (24)$$

The third set (C) is given by the condition: $p+2j$ is divisible by $n+1$ whereas $p-2j$ is not divisible by $n+1$. We parametrize p by $p = -2j + q(n+1)$ with the natural q, then the corresponding sum in (21) is calculated in a manner similar to the set (B) and is equal to

$$\sum_{p \in (C)}^{\infty} CC \cdot J_{2p}(4t) = \left(1 - \delta_{j, \frac{n+1}{4}}\right) n \delta_{j, [\frac{n+1}{2}]} \delta_{\frac{n+1}{2}, [\frac{n+1}{2}]} \sum_{even\ m}^{\infty} J_{2(1+m)(n+1)}(4t)$$
$$+ \frac{n-1}{2} \left(1 - \delta_{j, \frac{n+1}{4}}\right) \left(1 - \delta_{j, \frac{n+1}{2}}\right) \sum_{even\ m}^{\infty} J_{2(-2j+m(n+1))}(4t)$$
$$+ \frac{(-1)^n - 1}{2} \left(1 - \delta_{j, \frac{n+1}{4}}\right) \sum_{odd\ m}^{\infty} J_{2(-2j+m(n+1))}(4t). \quad (25)$$

The fourth set (D) is determined from the requirement that the numbers $p - 2j$ and $p + 2j$ are divisible by $n + 1$. Then values p can be parametrized by $p = 2j + r(n + 1)$, $p = -2j + s(n + 1)$ with natural r and s. As a consequence, we have

$$p = \frac{r+s}{2}(n+1), \; j = \frac{s-r}{4}(n+1). \quad (26)$$

Since $1 \le j \le n$, then from the last relation (26) there follow the restrictions on r, s:

$$\alpha)\ s - r = 1, \quad \beta)\ s - r = 2, \quad \gamma)\ s - r = 3. \tag{27}$$

If the condition $\alpha)$ is fulfilled, we obtain from (26) the values $p = r(n+1) + (n+1)/2$ and $j = (n+1)/4$, i.e. solution for j exists only for $n+1$, which is divisible by 4. Value CC (19) is

$$CC = \sum_{k=1}^{n} \cos\frac{\pi k}{2} \cos\frac{\pi k}{n+1}[r(n+1) + (n+1)/2] = \sum_{k=1}^{n} (-1)^{kr}[\cos(\pi k/2)]^2. \tag{28}$$

For even r the sum is calculated on the base of equation:

$$\sum_{k=1}^{n}\left[\cos\frac{\pi k}{2}\right]^2 = \frac{1}{2}\sum_{k=1}^{n}(1+\cos\pi k) = \frac{n}{2} + \frac{1}{2}\sum_{k=1}^{n}(-1)^k = \frac{n}{2} + \frac{(-1)^n - 1}{4}.$$

For odd r we get

$$\sum_{k=1}^{n}(-1)^k\left[\cos\frac{\pi k}{2}\right]^2 = \frac{1}{2}\sum_{k=1}^{n}(-1)^k + \frac{1}{2}\sum_{k=1}^{n}\cos k\pi(-1)^k = \frac{(-1)^n - 1}{4} + \frac{n}{2}.$$

Therefore,

$$CC = \frac{n}{2} + \frac{(-1)^n - 1}{4}. \tag{29}$$

For the condition $\beta)$ in (27) we find $p = (r+1)(n+1)$, $j = (n+1)/2$ and

$$\cos\frac{\pi pk}{n+1} = \cos\pi k(r+1) = (-1)^{k(r+1)}, \ \cos\frac{2\pi jk}{n+1} = \cos\pi k = (-1)^k.$$

Then and

$$CC = \sum_{k=1}^{n}(-1)^{kr} = \begin{cases} n \text{ for even } r, \\ \dfrac{(-1)^n - 1}{2} \text{ for odd } r. \end{cases} \tag{30}$$

For the latter condition $\gamma)$ in (27) we get $p = (r+3/2)(n+1)$ and $j = 3(n+1)/4$. Since

$$\cos\frac{\pi pk}{n+1} = \cos[\pi k(r+1) + \pi k/2] = (-1)^{k(r+1)} \cos\frac{\pi k}{2}, \ \cos\frac{2\pi jk}{n+1} = \cos\frac{\pi k}{2},$$

then from here and from (190, it follows that $CC = \sum_{k=1}^{n}(-1)^{k(r+1)}\cos^2\frac{\pi k}{2}$ and coincides with (28), whose result is given by Eq. (29).

Basing on (28) - (300, we can consequently formulate the result for the set (D):

$$\sum_{p\in(D)}^{\infty} CC\cdot J_{2p}(4t) = \left(\frac{n}{2}+\frac{(-1)^n-1}{4}\right)\delta_{j,\left[\frac{n+1}{4}\right]}\delta_{\frac{n+1}{4},\left[\frac{n+1}{4}\right]}\sum_{r=1}^{\infty} J_{(1+2r)(n+1)}(4t)$$
$$+\ \delta_{j,\left[\frac{n+1}{2}\right]}\delta_{\frac{n+1}{2},\left[\frac{n+1}{2}\right]}\left[n\sum_{\text{even } r}^{\infty} J_{2(1+r)(n+1)}(4t)+\frac{(-1)^n-1}{2}\sum_{\text{odd } r}^{\infty} J_{2(1+r)(n+1)}(4t)\right]$$
$$+\ \delta_{j,\left[\frac{3(n+1)}{4}\right]}\delta_{\frac{n+1}{4},\left[\frac{n+1}{4}\right]}\left(\frac{n}{2}+\frac{(-1)^n-1}{4}\right)\sum_{r=1}^{\infty} J_{2(r+3(n+1)/2)}(4t). \quad (31)$$

Summarizing the results of calculations given in this section, we write the Eq. (12) for the temperature distribution in a crystal with an arbitrary number of particles n:

$$\bar{T}_j = \frac{\beta}{8(n+1)}\left(n+1+nJ_0(4t)\right)+\frac{\beta}{8}F_j,\ F_j=\frac{1}{n+1}\Phi_j. \quad (32)$$

The function Φ_j is defined by the equation

$$\Phi_j = -2\sum_{p\in(B)}^{\infty} CC\cdot J_{2p}(4t) - 2\sum_{p\in(C)}^{\infty} CC\cdot J_{2p}(4t) - 2\sum_{p\in(D)}^{\infty} CC\cdot J_{2p}(4t), \quad (33)$$

in which the individual terms are given by Eqs. (24), (25), (31).

5 Reduction of the Obtained Equations

In the general case, the structure (33) is cumbersome; therefore, to analyze it, we exclude from consideration points j that can take values $(n+1)/4$, $(n+1)/2$, $3(n+1)/4$. Then F_j (32) is

$$F_j = -\frac{n-1}{n+1}\sum_{evenm}^{\infty}\left[J_{2(2j+m(n+1))}(4t)+J_{2(-2j+m(n+1))}(4t)\right]. \quad (34)$$

From the viewpoint of physics, it is of interest to consider the case of $t \gg 1$, which corresponds to times exceeding the time of particle oscillation in the cell, while not taking into account the influence of the crystal boundaries. In dimensionless variables, the time of the perturbation propagation through the crystal $\tau \sim n$, therefore the effects of reflection from the boundaries do not affect the temperature distribution, provided that $t \ll \tau$. This means that $n \gg 1$ and inequalities $t \ll j+n+1$, $t \ll -j+2(n+1)$ are valid for the functions in (34) for all values j. Hence, the Bessel functions (34) can be replaced by their asymptotic representations:

$$J_\nu\left(\frac{\nu}{ch\alpha}\right) \sim \frac{\exp\left(\nu th\alpha-\nu\alpha\right)}{\sqrt{2\pi\nu th\alpha}} \sim \frac{\exp[-\nu(\alpha-1)]}{\sqrt{2\pi\nu}},$$

since $\nu/x = ch\alpha \gg 1$. It is easy to show that the contribution of these terms in (34) does not exceed $\frac{1}{\sqrt{n}} \ll \frac{1}{\sqrt{t}}$. Then (32) can be written in the form

$$\bar{T}_j = \frac{\beta}{8}\left(1 + J_0(4t) + ...\right). \tag{35}$$

A numerical analysis of the Eq. (12) was carried out. Plots of the time dependent behavior of the kinetic energy normalized by $\beta/2$ are presented in Fig. 1. The calculations were carried out with $n = 500$ and different values of j. The graphs with $j = 1$ and $j = 2$ are indicated by a stroke and a short stroke consequently. The continuous line corresponds to the case of $j = 3$. The absolute value $\left|\frac{\partial \bar{T}_j}{\partial j}\bigg|_{t=0}\right|$ is seen to decrease (Fig. 1), which corresponds to the behavior defined by formula (16). The amplitude of the presented functions decreases with time, while their behavior becomes slightly different from each other for different j. This means that the influence of particle number on temperature becomes insignificant with time.

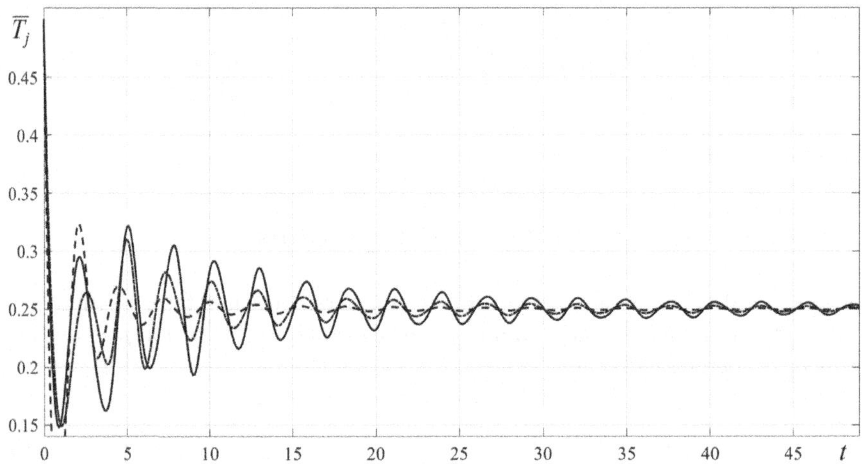

Fig. 1. Kinetic energy with time for $j = 1$ (stroke), $j = 2$ (short stroke), $j = 3$ (continuous line).

The following graphs illustrate the accuracy of the obtained formulae. Figure 2 the continuous line is drawn by means of the formula (35), and the stroke line is drawn on the base of the formula (12) for $j = 2$. Comparison of graphs shows that the exact solution decreases with respect to time faster than the approximate one, and the energy output to a stationary value occurs faster as well.

In Fig. 3 we presented empty squares which correspond to the values obtained on the formula (12) for $j = 100$. From here we can see that they lie on the continuous line drawn by means of the formula (35).

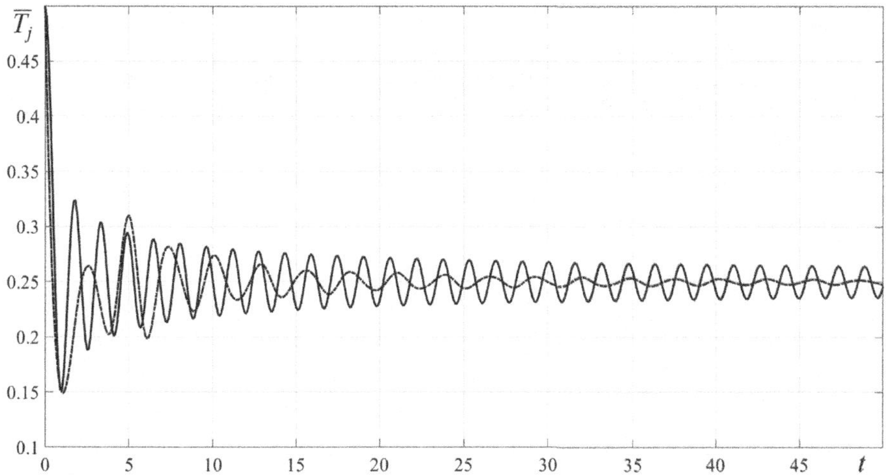

Fig. 2. Kinetic energy with time for $j = 2$.

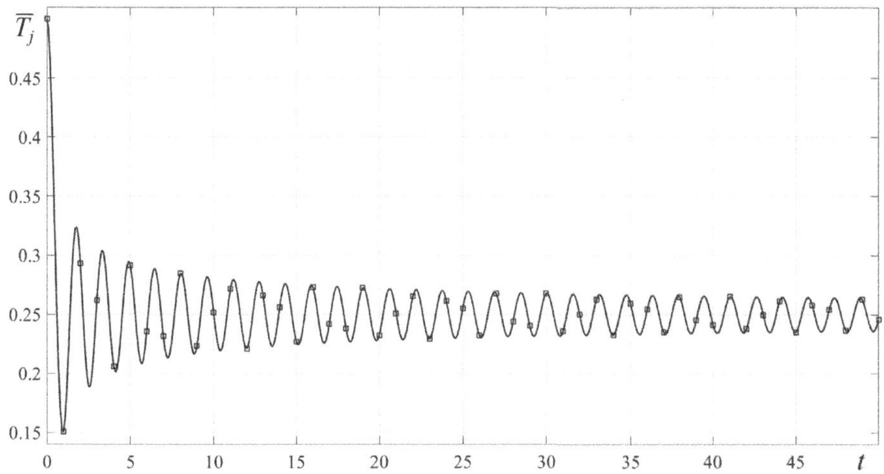

Fig. 3. Kinetic energy with time for $j = 100$.

Acknowledgements. The author Guzev M.A. grateful to Beijing High-caliber Talent from Overseas (BHT0 201612129-WD) for financial support.

References

1. Allen, M.P., Tildesley, A.K.: Computer Simulation of Liquids. Clarendon Press, Oxford (1987)
2. Krivtsov, A.M.: Heat transfer in infinite harmonic one dimensional crystals. Dokl. Phys. **60**(9), 407–411 (2015)

3. Hemmer, P.C.: Dynamic and stochastic types of motion in the linear chain. Thesis, Trondheim, Norway (1959)
4. Lepri, S., Livi, R., Politi, A.: Thermal conduction in classical low-dimensional lattices. Phys. Rep. **377**, 1–80 (2003)
5. Kubo, R., Toda, M., Hashitsume, N.: Statistical Physics II. Springer Series in Solid State Sciences, vol. 31, Springer, Heidelberg (1991)
6. Watson, G.N.: A Treatise on the Theory of Bessel Functions. University Press, Cambridge (1944)

Dynamics of the Microresonator in the Regime of Supercritical Compression

Vasilisa Igumnova[1(✉)], Lev Shtukin[1,2(✉)], Alexey Lukin[1(✉)], and Ivan Popov[1(✉)]

[1] Peter the Great St.Petersburg Polytechnic University (SPbPU), St.Petersburg 195251, Russia
igumnovavs@mail.ru, lvtvsh4749@gmail.com, {lukin_av,popov_ia}@spbstu.ru
[2] Institute for Problems in Mechanical Engineering of the Russian Academy of Sciences, St.Petersburg 199178, Russia

Abstract. In this work we investigate nonlinear dynamics of an electrostatically actuated microbeam resonator, located between two stationary electrodes, in the regime of supercritical compression. Longitudinal movement of the elastic fastening creates a longitudinal force in the elastic element of the microresonator. The equations of motion of the resonator are supplemented by equations of electrical circuits containing sources of electromotive force and capacitors of variable capacitance formed by fixed electrodes and the elastic element of the resonator. Equilibrium positions depending on the longitudinal displacement of the elastic fastening mechanism are obtained for various configurations of the electric field. With different switched on sources of constant electromotive force either two or three critical values of the force are possible, which differ from the Euler force. A numerical experiment demonstrating the possibility of the occurrence of a self-oscillatory regime was performed.

Keywords: Microresonator · Electric field · Bifurcations of equilibrium forms · Critical force · Self-oscillations

1 Introduction

At the moment, the development of MEMS is of great interest and has a wide range of applications. The use of micromechanical systems in modern technology is associated with high sensitivity to changes in system parameters: mass, pressure, acceleration, temperature.

The problem of axial compression of a rod begins with the works of L. Euler [1], who solved two static limit problems: the critical load and possible forms of instability were found in the linear approximation, and with a nonlinear approach all possible equilibrium forms of the rod loaded at the ends were found (Euler elastics). In the work of M.A. Lavrenty'va and A.Yu. Ishlinsky [2], taking into account the inertial forces of transverse motion, a compressive load, significantly exceeding the Eulerian load, was considered, and it was established that the

D. A. Indeitsev and A. M. Krivtsov (Eds.): APM 2019, LNME, pp. 139–151, 2020.
https://doi.org/10.1007/978-3-030-49882-5_14

greatest growth rate of the transverse deflection amplitude has a form with a large number of waves in the longitudinal direction.

In [5,6], the finite velocity of propagation of longitudinal waves in the rod is taken into account. A short-term longitudinal impact is considered under the assumption that the impact time is shorter than the travel time of the longitudinal wave along the doubled rod length and a prolonged impact [3].

In the article [7] the authors consider MEMS accelerometer based on double-ended-tuning fork resonators. The dynamic performance was demonstrated by single-frequency and hybrid-frequency vibration tests, and the results showed that device is suitable for detecting low-frequency vibration (0.5–5 Hz). Device showed its superiority in mixed acceleration measurement, which makes it a potentially attractive option for geophone or seismometer applications.

In [8] proposes a resonant accelerometer with sensitivity enhancement and adjustment mechanisms based on microelectromechanical systems (MEMS). Accelerometer utilizes a fishbone-shaped resonator as sensing element to enhance and adjust sensitivity. The preliminary tilt experiment verifies that this prototype has potential usage in tilt easurement.

In this work we research the dynamics of a microresonator consisting of an elastic element in the form of a beam located between stationary electrodes under the action of a compressive longitudinal force. When turning on the electric field, the loss of stability occurs if a compressive longitudinal force greater than Euler force was established. The free oscillations of the resonator under supercritical compressive loads are investigated, and a positive feedback scheme for the excitation of self-oscillating regimes is proposed. When conducting numerical experiments, the effect of variations of different parameters—the gain, the time constant, and the limit level—on the frequency stability of the oscillator was investigated.

2 The Scheme and Principle of Operation of the Microresonator

The Fig. 1. presents the microresonator circuit.

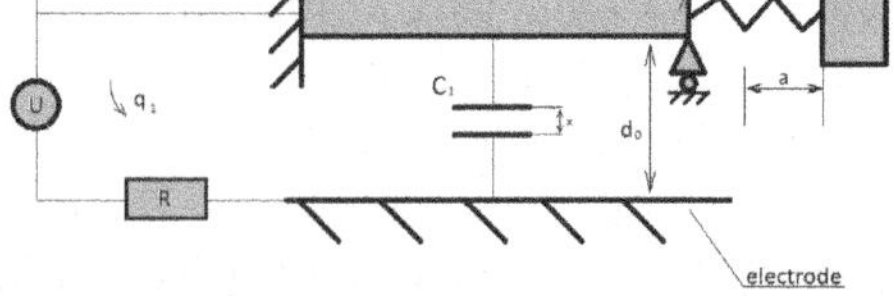

Fig. 1. Microresonator

The resonator consists of an elastic element in the form of a beam located between the stationary electrodes. One end of the beam is rigidly clamped, and the other one is elastically fixed in the longitudinal direction. The elastic element and the electrodes form two capacitors with variable capacitance. In the space between the elastic element and the fixed electrodes, an electric field is created using an electromotive force (EMF) source.

3 Equations of Motion

Consider the oscillations of an elastic element located between two fixed electrodes. In general case, these equations can be written as:

$$L(w) + R\ddot{w} = F_e(w) \tag{1}$$

where w – transverse deflection, $L(w)$ – is the differential operator of elasticity, R – is the inertia operator, $F_e(w)$ – is the transverse force that is created by the electric field.

The equations of motion of an elastic element are derived using the Galerkin method. The solution is represented as a series expansion with respect to certain coordinate functions W_k with time dependent coefficients $x_k(t)$:

$$w(x,t) = x_1(t)W_1 + x_2(t)W_2 + ... \tag{2}$$

The eigenforms of the elastic element are taken as the coordinate functions, without regard to its nonlinear elastic properties and in the absence of the electric field, which satisfy the equation:

$$L_0 W_k - \lambda^2 R W_k = 0, k = 1, 2, ... \tag{3}$$

where $L_0(W)$ – is the linearized operator $L(W)$.

Leaving in the decomposition only one addend corresponding to the first form, we get:

$$w(x,t) = x_1(t)W_1 \tag{4}$$

The application of the Galerkin method to Eq. (1) with regard to Eq. (4) leads to the equation of the elastic element oscillations:

$$m\ddot{x} + b\dot{x} + P(x) - \frac{1}{2}\frac{q_1^2}{C_1^2}\frac{dC_1}{dx} + \frac{1}{2}\frac{q_2^2}{C_2^2}\frac{dC_2}{dx} = 0 \tag{5}$$

where $C_1 = C_0\frac{d_0}{d_0-x}$, $C_2 = C_0\frac{d_0}{d_0+x}$ – capacitance on the respective capacitors, $q_1 = C_1v_1$, $q_2 = C_2v_2$ – charges in respective contours, v_1, v_2 – voltage on the respective capacitors, C_0 – undeformed capacitor capacitance, P – longitudinal compressive force, depending on the offset of the end of the spring and on the deflection of the beam.

Equation (5) should be supplemented with the equations of the second Kirchhoff law for two electrical circuits:

$$\begin{cases} U + u_{out} - R\dot{q}_1 - v_1 = 0 \\ U - R\dot{q}_2 - v_2 = 0 \end{cases} \tag{6}$$

The output signal of the amplifier depends on the input signal as follows:

$$u_{out} = F(u_{in}) = F(R\dot{q}_2) = F(U - v_2) \tag{7}$$

The characteristic of the amplifier in general has the form:

$$F(u) = V_0 \frac{2}{\pi} arctg(\frac{\pi}{2} K \frac{u}{V_0}) \tag{8}$$

where K – gain at low input signals, V_0 – voltage limiting signal at the output of the amplifier. This characteristic makes it possible to take into account an almost linear segment with small input signals, smooth nonlinearity with large input signals and the limited output signal by some level depending on the amplifier supply voltage.

Introducing dimensionless parameters: $\xi = \frac{x}{d_0}$ – dimensionless deflection, $\tau = \lambda t$ – dimensionless time, $\eta_1 = \frac{v_1}{U}$ and $\eta_2 = \frac{v_2}{U}$ – dimensionless voltage, respectively, on the first and second capacitors.

Then Eqs. (5) and (6) can be rewritten in dimensionless form:

$$\begin{cases} \xi'' + 2n\xi' + (1-\alpha)\xi + \gamma\xi^3 - \beta^2\eta_1^2\frac{1}{(1-\xi)^2} + \beta^2\eta_2^2\frac{1}{(1+\xi)^2} = 0 \\ V_0\frac{2}{\pi}arctg(\frac{\pi}{2}K(1-\eta_2)) + \delta(\frac{1}{1-\xi}\dot{\eta}_1 - \frac{1}{(1-\xi)^2}\eta_1\dot{\xi}) + \eta_1 = 1 \\ \delta(\frac{1}{1+\xi}\dot{\eta}_2 - \frac{1}{(1+\xi)^2}\eta_2\dot{\xi}) + \eta_2 = 1, \end{cases} \tag{9}$$

where the physical meaning of each of the introduced dimensionless factors is the following:

1) $\beta^2 = \frac{1}{2}\frac{C_0U^2}{md_0^2\lambda^2}$ - the ratio of the energy of the electric field to the maximum energy of oscillations with an amplitude equal to the total gap.
2) $\delta = RC_0\lambda = 2\pi\frac{RC_0}{T}$ - the ratio of the capacitor charge time constant to the period natural frequency. With a small value of this factor, we can not see the signal at the output, and with a large value, the capacitor will not have time to recharge in one oscillation period.
3) $v_0 = \frac{V_0}{U}$ - the ratio of the limiting level of the output signal of the amplifier to the voltage value of the sources of constant emf.
4) K - gain. By choosing a certain gain value, it is possible to obtain a buildup of oscillations at small amplitudes.

4 Research of Equilibrium Positions

The motion of the resonator in a dimensionless form under the action of an electric field only in one of the gaps:

$$\xi'' + 2n\xi' + (1-\alpha)\xi + \gamma\xi^3 - \beta^2\eta^2\frac{1}{(1-\xi)^2} = 0 \tag{10}$$

In Fig. 2 shows the bifurcation of the equilibrium positions described by the Eq. (10)

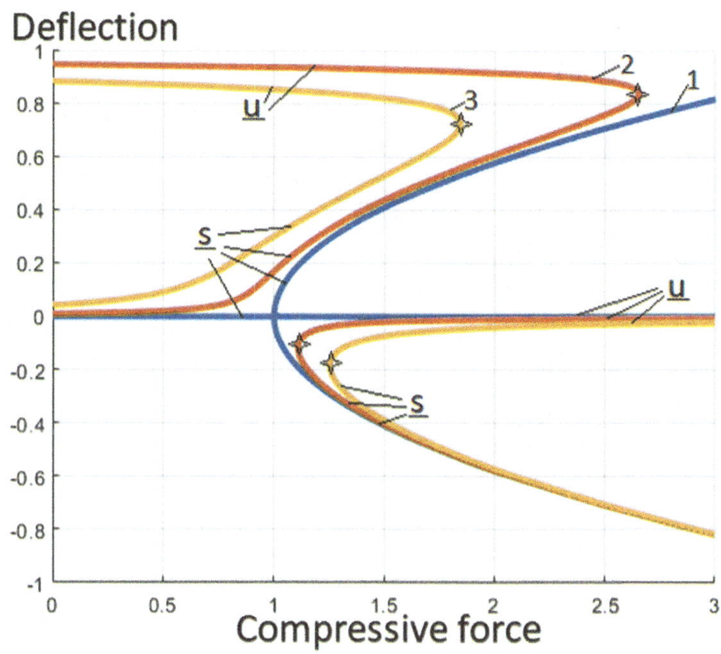

Fig. 2. Bifurcation of equilibrium positions for different values of the electric field in one gap $\beta = 0$ (1), $\beta = 0.05$ (2), $\beta = 0.1$ (3)

In the presence of an electric field, two or four equilibrium positions with alternation of stable and unstable positions are possible. The equilibrium position is stable with a small compressive load, remains stable even when the critical value of the force is equal to unity is exceeded (Euler force in the absence of an electric field). Two new critical force values appear on the diagram, both of which are greater than Euler's strength. With a longitudinal displacement less than the first critical there are two equilibrium positions, a position with a smaller deflection is stable, with a large - is unstable. With a longitudinal displacement greater than the first critical and less than the second critical equilibrium position - four, two stable and two unstable. With a longitudinal displacement greater than the second critical equilibrium position two, one with a smaller deflection is unstable, the other with a large deflection is stable.

Now consider the situation when a symmetric electric field acts. The equation of motion of the beam in this case is described by the equation:

$$\xi'' + 2n\xi' + (1-\alpha)\xi + \gamma\xi^3 - \beta^2\eta^2\frac{4\xi}{(1-\xi^2)^2} = 0 \tag{11}$$

In Fig. 3 shows the bifurcation of the equilibrium positions described by the Eq. (11)

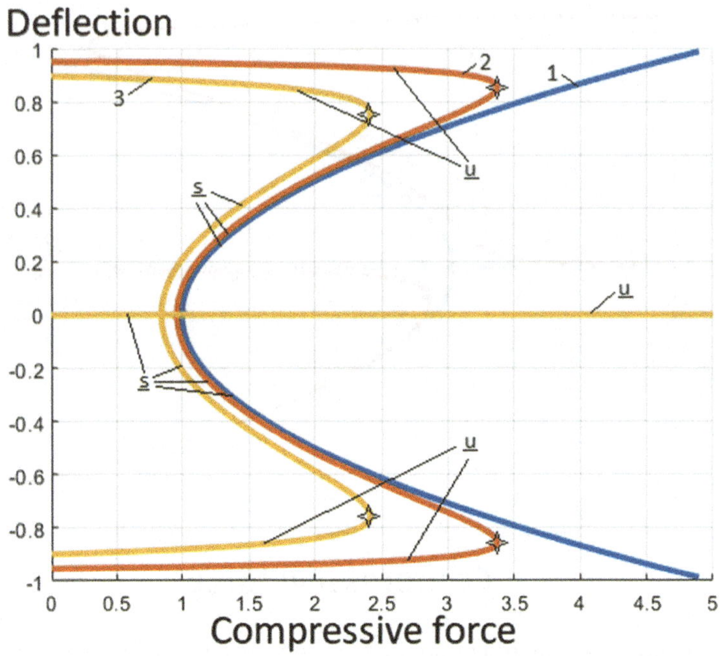

Fig. 3. Bifurcation of equilibrium positions at various values of a symmetric electric field $\beta = 0$ (1), $\beta = 0.05$ (2), $\beta = 0.1$ (3)

The bifurcation diagrams have a symmetrical appearance. Depending on the compressive force, we have one, three or five equilibrium positions with alternation of stable and unstable positions. In the presence symmetric electric field in each of the gaps critical values is two. The first value is close to the Euler force, the second is of greater importance. The equilibrium position corresponding to zero deflection exists at any value of the longitudinal displacement, it is stable at the longitudinal displacement less than the first critical and unstable at a larger longitudinal displacement. With a longitudinal displacement less than the first critical, two equilibrium positions are added, both unstable. With a longitudinal displacement greater than the first critical and less than the second critical there are five equilibrium positions, two of them are stable, three are unstable. With a longitudinal displacement greater than the second critical value, the equilibrium position is only one, it corresponds to zero deflection and it is unstable.

Finally, we consider the motion of a resonator, when the electric field acts on both sides and with different strengths. The differential equation has the form:

$$\xi'' + 2n\xi' + (1-\alpha)\xi + \gamma\xi^3 - \beta^2\eta_1^2\frac{1}{(1-\xi)^2} + \beta^2\eta_2^2\frac{1}{(1+\xi)^2} = 0 \quad (12)$$

In Fig. 4. shows the bifurcation of the equilibrium positions described by the Eq. (12)

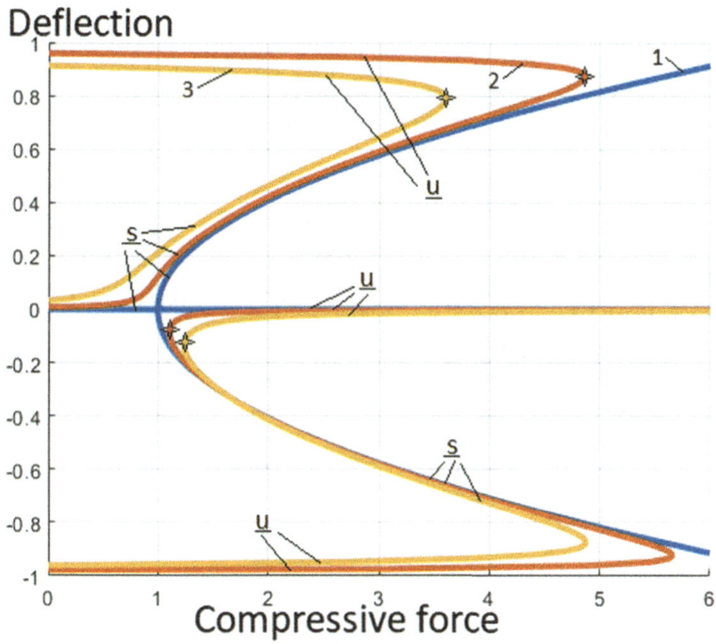

Fig. 4. Bifurcation of equilibrium positions for different values of the electric field in two gaps $\beta = 0$ (1), $\beta = 0.05$ (2), $\beta = 0.1$ (3)

Under the action of different electric field in each of the gaps, there are three critical values. They are all greater than the value corresponding to Euler force. With a longitudinal displacement less than the first critical there are three equilibrium positions, one with a small deflection - stable, two with a large deflection - unstable. With a longitudinal displacement greater than the first critical and less than the second critical equilibrium position five, two stable and three unstable. With a longitudinal displacement greater than the second critical and less than the third critical equilibrium position three, one stable and two unstable. With a longitudinal displacement greater than the third critical equilibrium position is one and it is unstable.

5 Analysis of Free Oscillations

The motion of the system under the action of a field in one gap is described by the equation:

$$\xi'' + (1-\alpha)\xi + \gamma\xi^3 - \beta^2\eta^2\frac{1}{(1-\xi)^2} = 0 \tag{13}$$

Writing the energy integral of the Eq. (13), we can get the phase portrait of the system, which is shown in Fig. 5.

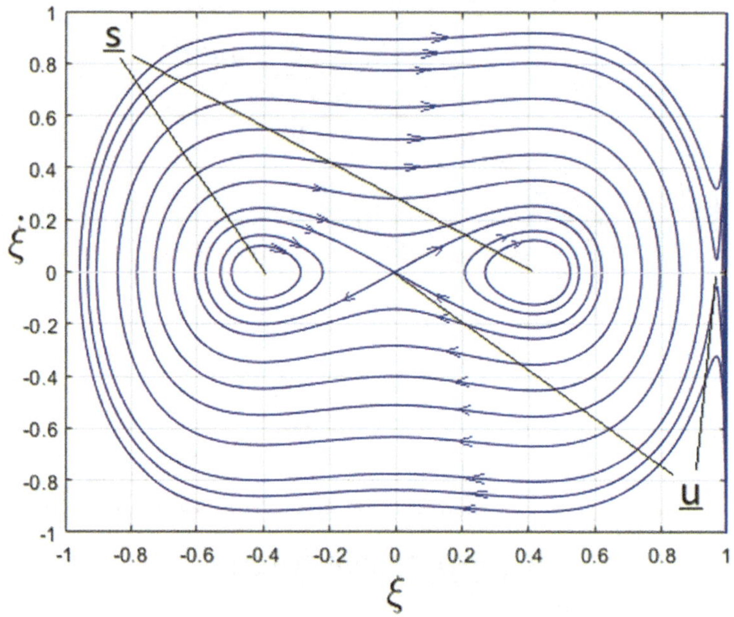

Fig. 5. Phase portrait with an included electric field in one of the gaps ($\alpha = 1.5$)

In Fig. 5 it is clear that with a force exceeding the critical value (Euler's force), the system has four equilibrium positions, two of which are stable and two unstable. With large initial conditions, it is possible to jump from the oscillatory regime around three positions to the so-called "sticking" regime, that is, when the beam touches the stationary electrode and remains there.

The movement of the system under the action of a symmetric field is described by the equation:

$$\xi'' + (1-\alpha)\xi + \gamma\xi^3 - \beta^2\eta^2\frac{4\xi}{(1-\xi^2)^2} = 0 \tag{14}$$

Similarly, to obtain the phase portrait of the system (14), the energy integral was written. Figure 6 shows the phase portrait.

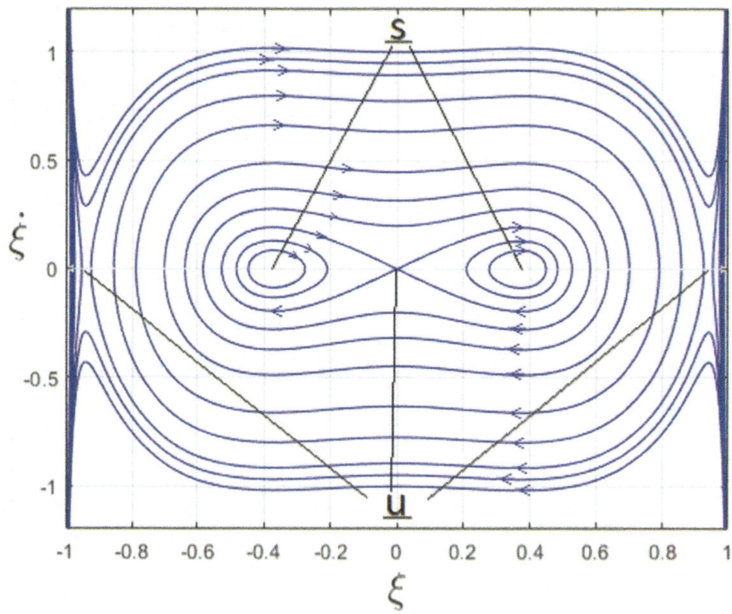

Fig. 6. Phase portrait with an included symmetric electric field ($\alpha = 1.5$)

From Fig. 6 it can be seen that in the supercritical compression mode with the symmetric electric field switched on there are five equilibrium positions, two of which are stable and three are unstable. Equilibrium positions, both stable and unstable, are symmetric about the zero position. Sticking mode is also possible.

Finally, the equation of motion under the action of a field in two gaps takes the form:

$$\xi'' + (1 - \alpha)\xi + \gamma\xi^3 - \beta^2\eta_1^2\frac{1}{(1-\xi)^2} + \beta^2\eta_2^2\frac{1}{(1+\xi)^2} = 0 \qquad (15)$$

Figure 7 shows the phase portrait of the system (15).

According to Fig. 7, it can be seen that with a compressive force exceeding the critical value, taking into account the effect of different electric fields in both gaps of the resonator, there are five equilibrium positions, two of which are stable and three unstable. It is possible to switch to the "sticking" regime in both gaps.

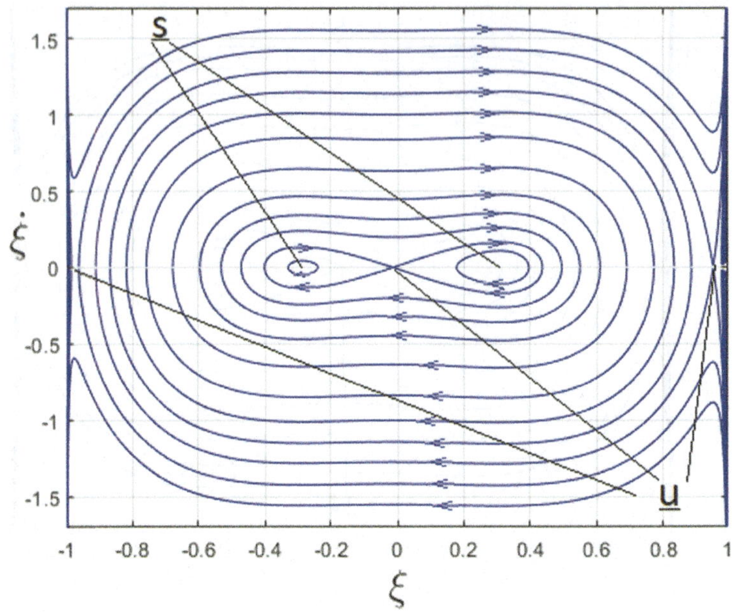

Fig. 7. Phase portrait with an included electric field in two gaps ($\alpha = 1.5$)

6 Self-oscillating Regime

In Fig. 8 shows the electromechanical system for the excitation of self-oscillations.

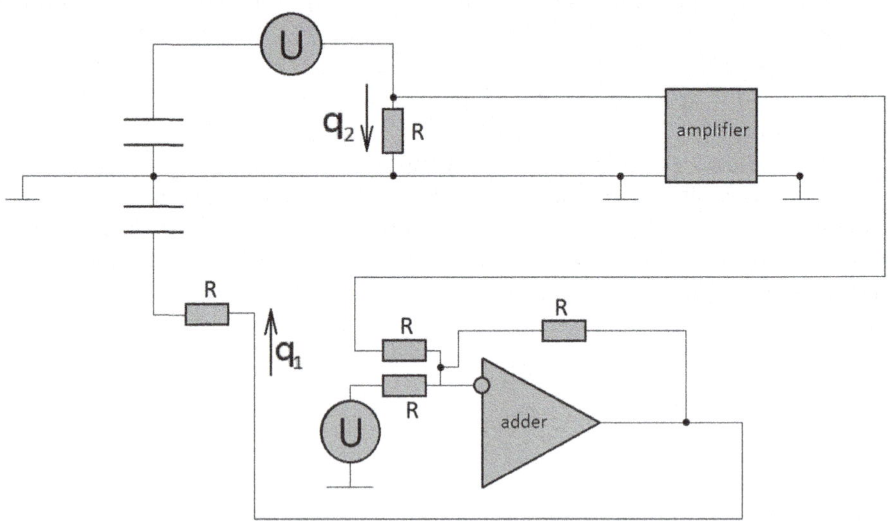

Fig. 8. Electromechanical system

The electromechanical model consists of an elastic element in the form of a beam located between two stationary electrodes. The elastic element and stationary electrodes form two parallel-plate capacitors. A voltage from a constant electromotive source is applied to the capacitor plates. When the beam is bent, the capacitance of the capacitor changes and, accordingly, a charge exchange current arises, depending on the rate of change of the deflection. The upper electrode is supplied through a resistor R with a voltage from a constant emf source U. Through the resistor R, the sum voltage is fed to the lower electrode from a source of constant emf U and the output signal of the amplifier u_{out}.

The system of equations for the excitation of self-oscillations:

$$\begin{cases} \xi'' + 2n\xi' + (1-\alpha)\xi + \gamma\xi^3 - \beta^2\eta_1^2\frac{1}{(1-\xi)^2} + \beta^2\eta_2^2\frac{1}{(1+\xi)^2} = 0 \\ V_0\frac{2}{\pi}arctg(\frac{\pi}{2}K(1-\eta_2)) + \delta(\frac{1}{1-\xi}\dot{\eta_1} - \frac{1}{(1-\xi)^2}\eta_1\dot{\xi}) + \eta_1 = 1 \\ \delta(\frac{1}{1+\xi}\dot{\eta_2} - \frac{1}{(1+\xi)^2}\eta_2\dot{\xi}) + \eta_2 = 1, \end{cases} \quad (16)$$

The numerical experiment carried out in solving the system of Eqs. (16) showed that a self-oscillatory regime is possible, corresponding to oscillations around three equilibrium positions and around one equilibrium position (Fig. 9).

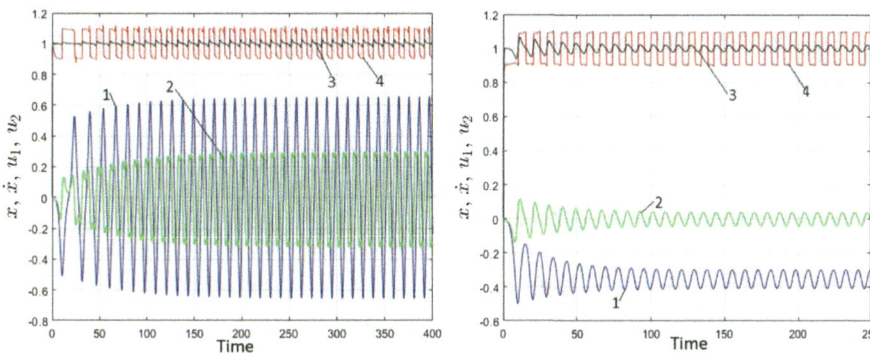

Fig. 9. Oscillogram of the self-oscillation process. 1 – oscillation amplitude, 2 – oscillation speed, 3 – voltage drop across the second resistor (input signal), 4 – voltage drop across the first resistor (output signal)

Figure 9 shows the process of buildup of self-oscillations and the subsequent exit to a limited level of the amplitude of oscillations.

7 The Influence of System Parameters on the Stability of Self-oscillation Frequency

When conducting numerical experiments, the effect of different parameters variations the gain, the time constant, and the limit level—on the frequency stability of the oscillator was investigated. Figure 10 shows these dependencies.

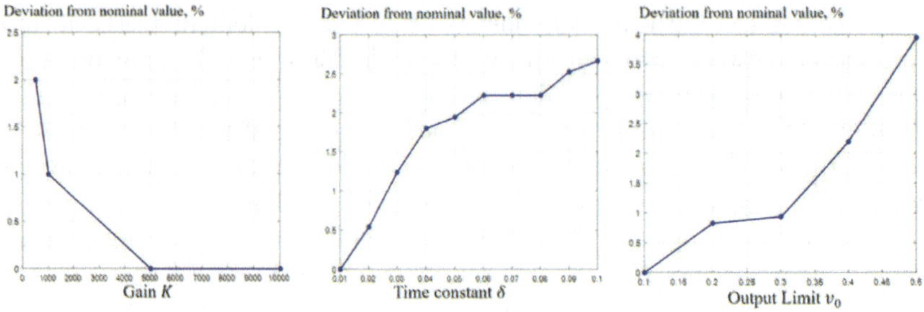

Fig. 10. The dependence of the system parameters on the stability of self-oscillations

The gain must be taken large (5000) and its further increase will not lead to a change in frequency. The time constant δ is better to take a small one. With a small value of this multiplier, we can not see the signal at the output, but we obtain the frequency stability of the oscillator, and with a large value the capacitor will not have time to recharge over one oscillation period, therefore, at large values of the time constant, the frequency stability of the oscillator decreases. To ensure frequency stability, it is necessary to take small values of v_0.

Consider the effect of compressive force on the frequency and amplitude of self-oscillations, which is shown in Fig. 11.

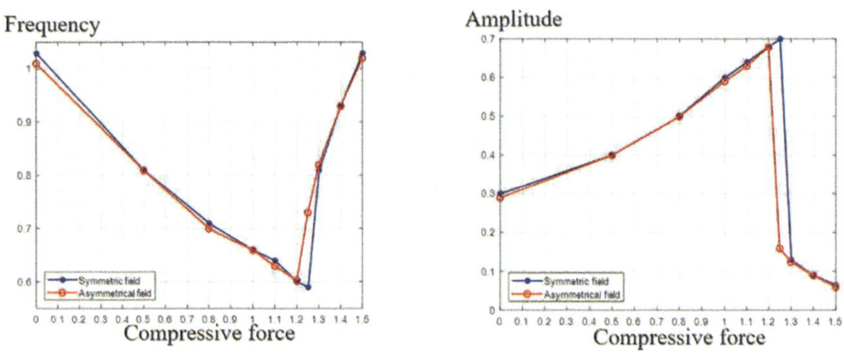

Fig. 11. Influence of compressive force on the frequency and amplitude of self-oscillations

As can be seen from Fig. 11, with an increase in the compressive force, the oscillation frequency decreases, but after the critical value of the force is exceeded, the frequency begins to increase. With an increase in the compressive force, the amplitude of the oscillations increases, but after exceeding the critical value of the force, the amplitude begins to fall. For an asymmetric field, the value of the critical force is slightly less than for a symmetric field. In the supercritical zone, the slope of the graph is steeper than in the subcritical zone.

8 Conclusions

A model of a microresonator consisting of an elastic element in the form of a beam located between stationary electrodes under the action of a compressive longitudinal force is proposed.

Equilibrium positions are obtained in the presence of a longitudinal compressive force, switched on by one or two sources of constant emf depending on the longitudinal displacement of the mechanism of elastic fastening. With different switched on sources of constant electromotive force either two or three critical values of the force are possible, which differ from the Euler force.

Phase portraits of systems with different field switching are built, using which the number of equilibrium positions, as well as their stability, are analysed.

A positive feedback scheme for the excitation of self-oscillatory regime was proposed. The possibility of the oscillation regime both around three equilibrium positions and around one was shown. When conducting numerical experiments, the effect of various parameters—gain, time constant, and limiting level—on the frequency stability of the oscillator was considered. The influence of the compressive force on the frequency and amplitude of self-oscillations was investigated.

The work was carried out at the support of RFBR Grant 17-01-00414.

References

1. Eulero, L.: Methodus inveniendi lineas curvas maximi minimive proprietate gaudentes, sive solution problematis isoperimetrici latissimo sensu accepti. Lausannae, Genevae. MDCCXLIV (1744)
2. Lavrent'ev, M.A., Ishlinsky, A.J.: Dynamical modes of stability loss of elastic systems. DAN SSSR **64**, 779–782 (1949)
3. Morozov, N.F., Tovstik, P.E., Tovstik, T.P.: Rod stability with long-term axial compression. Prob. Strength Plast. **77**, 40–48 (2015)
4. Morozov, N.F., Tovstik, P.E.: On the rod dynamic stability loss under the load less that the Eulerian one. DAN **453**, 282–285 (2013)
5. Morozov, N.F., Tovstik, P.E.: Rod dynamics in longitudinal impact. Bull. St. Petersburg Univ. **1**, 105–111 (2009)
6. Morozov, N.F., Tovstik, P.E.: Rod dynamics in a short-term longitudinal impact. Bull. St. Petersburg Univ. **3**, 126–130 (2013)
7. Wang, S., Wei, X., Zhao, Y., Jiang, Z., Shen, Y.: A MEMS resonant accelerometer for low-frequency vibration detection. Sens. Actuators A **283**, 151–158 (2018)
8. Ding, H., Wang, W., Ju, B., Xie, J.: A MEMS resonant accelerometer with sensitivity enhancement and adjustment mechanisms. J. Micromech. Microeng. **27**, 115010 (2017)

Design and Simulation of an Acoustic Metamaterial Plate Incorporating Tunable Shape Memory Cantilever Absorbers

Hua-Liang Hu, Yi-Ting Wu, Ji-Wei Peng, and Chun-Ying Lee(✉)

Graduate Institute of Manufacturing Technology,
National Taipei University of Technology, Taipei 10608, Taiwan
leech@ntut.edu.tw

Abstract. Metamaterials are materials having artificially tailored internal structure and unusual physical and mechanical properties. Due to their unique characteristics, metamaterials possess great potential in engineering applications. This study proposes a tunable metamaterial for the applications in vibration or acoustic isolation. For the state-of-the-art structural configurations in metamaterial, the geometry and mass distribution of the crafted internal structure is employed to induce the local resonance inside the material. Therefore, a stopband in the dispersion curve can be created because of the energy gap. For the conventional metamaterial, the stopband is fixed and unable to be adjusted in real-time once the design is completed. Although the metamaterial with distributed resonance characteristics has been proposed in the literature to extend its working stopband, the efficacy is usually compromised. In this study, the incorporation of tunable shape memory materials (SMM) via phase transformation into the metamaterial plate is proposed. Its theoretical finite element formulation for determining the dynamic characteristics is established. The effect of the configuration of the SMM cantilever absorbers on the metamaterial plate for the desired stopband in wave propagation is simulated by using finite element model. The result demonstrates the tunable capability on the stopband of the metamaterial plate under different activation controls of the SMM absorbers. The result of this study should be beneficial to precision machinery and defense industries which have desperate need in vibration and noise isolation.

1 Introduction

Metamaterials are artificially designed and fabricated material structures which demonstrate special and/or peculiar properties in contrast to conventional engineering materials. Due to their special properties, such as negative reflectivity, negative mass and negative Poisson's ratio, they have potential applications in electromagnetic wave cloaking, acoustic and vibration inhibition of structure [1–4].

D. A. Indeitsev and A. M. Krivtsov (Eds.): APM 2019, LNME, pp. 152–162, 2020.
https://doi.org/10.1007/978-3-030-49882-5_15

The bandgap of the metamaterial in acoustic transmission presents a useful tool for isolating the targeted acoustic wave. Several tunable mechanisms for the bandgap of acoustic metamaterial have been proposed, such as the use of piezoelectric actuator [5], the volume control of Helmholtz resonator [6], the change in the size parameters of a kagome-sphere lattice [7], the use of distributed vibration absorbers [8,9], and the incorporation of chiral elastic lattice [10], etc. They all demonstrate the wide bandgap characteristics of their proposed designs. Although by changing the frequency dispersion function of the structure or the parameters of the electric circuits, the changeable bandgap can provide more flexibility of the metamaterial to adapt to different design circumstances, the time-varying excitation source presents another challenge to real-time adaptability of the metamaterial. For example, the speed of the machine in operation can change which induces the variation in associated vibration and noise excitation.

Smart materials, such as piezoelectric ceramics, magnetorheological fluid, electrorheological fluid, shape memory materials, are a class of materials which properties can be tuned according to the demand [11]. If smart material is incorporated into the metamaterial, the bandgap property can then be tuned by controlling the property of its constituent smart material. In our previous study, absorbers consisted of the shape memory material (shape memory alloy and shape memory polymer) as the structural member were employed in the design of an acoustic metamaterial beam [12–16]. The activation on the phase change of the shape memory material by controlling the heating current tuned the frequency of the bandgap and obtained the required isolation of the designated excitation. As a continuation of the previous work, the SMM absorbers were used in a plate structure. The design and dynamic characteristics of this acoustic metamaterial plate were simulated through finite element modeling.

2 Design of the SMM Absorber and Metamaterial Plate

Figure 1 presents the proposed tunable absorber made of SMM spring. The spring can be in helical configuration or cantilevered beam configuration. The absorber with helical spring is able to provide more structural flexibility and more vibration modes at low frequency range [14]. Therefore, it can present the acoustic metamaterial plate with wide bandwidth of the stopband. Nevertheless, the fabrication of the helical spring is more complicated comparing with its cantilever counterpart. The SMM cantilever is close to a single degree of freedom system because its second natural frequency is nearly four times higher than its fundamental frequency [17]. For acoustic frequency, this SMM cantilever is easier to design in the required frequency span. However, due to its lower mass, the effect of vibration isolation can be limited [18]. As required by the design of metamaterial, these absorbers should be arranged and attached to base plate with lattice configuration.

The design of an exemplary SMM cantilever absorber is shown in Fig. 2. An insulated fixture pad was employed to mount the cantilevers and to secure on the

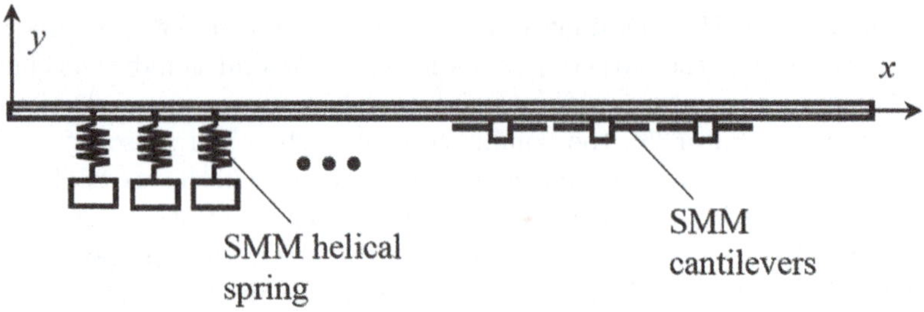

Fig. 1. The schematic diagram of the proposed design of SMM absorbers

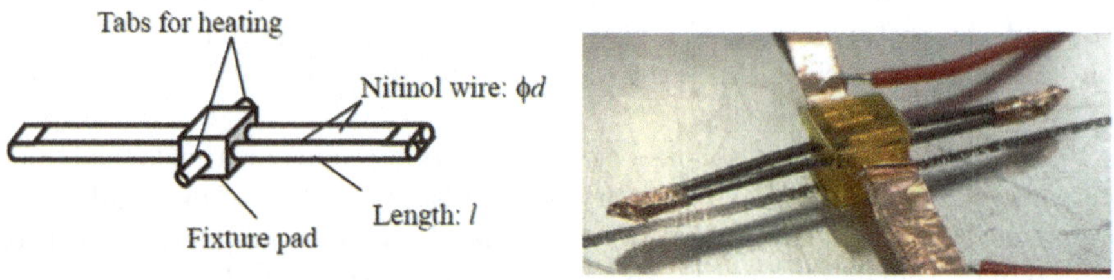

Fig. 2. The design of an SMM cantilever absorber

base plate. The cantilevers were arranged in parallel and symmetric configuration. The parallel cantilevers were to complete an electrical circuit for the heating current and the symmetric arrangement was to minimize the rotation loading at the fixture pad. On the sides of the fixture pad there were two conducting end tabs for introducing the electric current and forming the control circuit for the SMM absorbers. If a NiTinol cantilever with diameter d and length l was used, the following material properties can be applied: mass density $\rho = 6450\,\mathrm{kg/m^3}$, Young's modulus of martensite phase $E_M = 24\,\mathrm{GPa}$, and Young's modulus of austenite phase $E_A = 76\,\mathrm{GPa}$ [11]. The fundamental frequency of this cantilever absorber can be written as:

$$\omega_a = (1.875)^2\sqrt{\frac{EI}{\rho A l^4}}$$

Accordingly, the fundamental frequencies at low temperature (martensite) and elevated temperature (austenite) are 1499 Hz and 2668 Hz, respectively. For the following numerical analyses, the fundamental frequencies for the SMM absorber in the martensitic and austenitic phases were taken as 1500 Hz and 2500 Hz for simplicity, respectively.

3 Theoretical Formulation

No matter it is helical SMM absorber or cantilever absorber, its simplification into a single degree-of-freedom system can facilitate the theoretical formulation

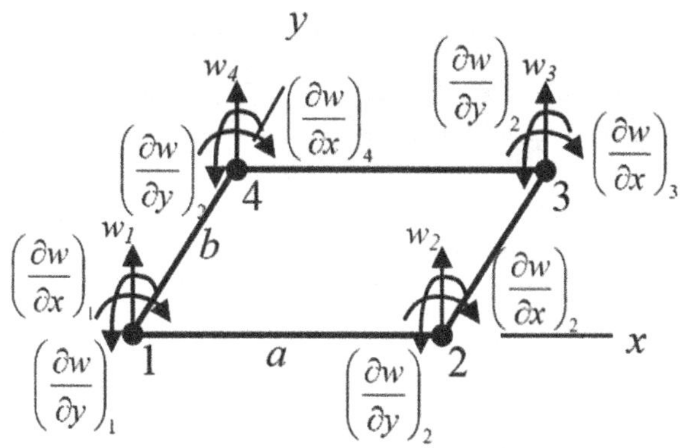

Fig. 3. Schematic diagram of a rectangular element used in this study

without losing the important physical nature. Starting from the base plate, a rectangular plate element as shown in Fig. 3 with three degrees-of-freedom at each corner node was used. An interpolation function $w(x, y)$ for the transverse deflection of the plate can be written as:

$$\begin{aligned} w(x,y) = {} & \alpha_1 + \alpha_2 x + \alpha_3 y + \alpha_4 x^2 + \alpha_5 y^2 + \alpha_6 xy + \alpha_7 x^3 \\ & + \alpha_8 y^3 + \alpha 9 x^2 y + \alpha_{10} xy^2 + \alpha_{11} x^3 y + \alpha_{12} xy^3 \end{aligned} \tag{1}$$

The plate element used has 3 degrees-of-freedom at each node and totally has 12 degrees-of-freedom. The derivation of the elemental stiffness matrix and consistent mass matrix follows the finite element method and the results can be found in related textbook [17]. Therefore, the detail was not presented herein. If the damping of the base plate is to be considered, the structural damping matrix can be obtained by assuming proportional damping for simplicity. With the derived element matrices, the system matrices for the base plate can be established by considering the connectivity of the elements. Using the Principle of Virtual Work, the unconstrained system equations can be obtained as:

$$M_s \ddot{X}_s + C_s \dot{X}_s + K_s X_s = F_s \tag{2}$$

If the SMM absorber with single degree-of-freedom was attached to the ith node, as schematically shown in Fig. 4, an additional degree-of-freedom of the absorber's displacement was added accordingly. Depending on the number of attached absorbers, the system matrices augmented with the corresponding dimensions. The following modifications were performed to consider the effect of attached absorbers:

$$M = M_s + \Delta M_s, C = C_s + \Delta C_s, K = K_s + \Delta K_s \tag{3}$$

$$\Delta M_s = \begin{bmatrix} 0 & 0 \\ 0 & m_a \end{bmatrix}, \Delta C_s = \begin{bmatrix} c_a & -c_a \\ -c_a & c_a \end{bmatrix}, \Delta K_s = \begin{bmatrix} k_a & -k_a \\ -k_a & k_a \end{bmatrix} \tag{4}$$

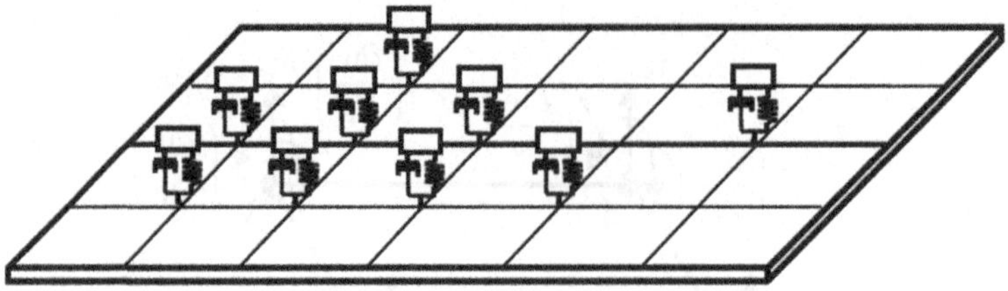

Fig. 4. The schematic diagram of the metamaterial plate installed with absorbers at the nodes

In the above equations, the first degree-of-freedom corresponded to that of w_i at i^{th} node of the base plate while the second degree-of-freedom related to added degree-of-freedom of w_{ai} from the absorber. Figure 4 shows an example of the base plate having absorbers attached to the nodes. After finishing the amendment of the system matrices with all the attached absorber, the system equations need to consider the boundary constraints on the base plate. By imposing the boundary conditions on the corresponding nodes, the constrained system governing equation can be written as:

$$M\ddot{X} + C\dot{X} + KX = F \tag{5}$$

The frequency response of the system equation, Eq. (5), can be obtained by performing sweep sine analysis. A sinusoidal input force function was applied at certain input node and the corresponding response at designated output node can be calculated. The frequency response of the system over a desired frequency span can be calculated by increasing the sinusoidal frequency step-by-step. The system characteristics can also be examined through the time response. In this study, the following Newmark's method [17] was adopted:

$$\dot{X}^{t+\Delta t} = \dot{X}^t + \left[(1-\delta)\ddot{X}^t + \delta\ddot{X}^{t+\Delta t}\right]\Delta t \tag{6}$$

$$X^{t+\Delta t} = X^t + \dot{X}^t\Delta t + \left[(\frac{1}{2}-\alpha)\ddot{X}^t + \alpha\ddot{X}^{t+\Delta t}\right]\Delta t^2 \tag{7}$$

$$M\ddot{X}^{t+\Delta t} + C\dot{X}^{t+\Delta t} + KX^{t+\Delta t} = F^{t+\Delta t} \tag{8}$$

In the above equations, α and δ are the parameters for controlling the integration accuracy and stability. In this study, a trapezoidal rule with $\alpha = \frac{1}{4}$ and $\delta = \frac{1}{2}$ was employed as in usual practice. It should be mentioned that the purpose of this study is mainly to simulate steady state system response as the absorber is switched between two controllable phases- martensite and austenite. The transient response and the physical model from the hysteresis of the phase transformation was not modeled and considered herein.

4 Results and Discussion

4.1 A Rectangular Plate in Simply-Supported Boundary Condition

A rectangular plate made of aluminium (Young's modulus 72.4 GPa, Poisson's ratio 0.33, mass density 2780 kg/m^3) with 187.5 mm × 150 mm and thickness of 3 mm was discretized into 4 × 5 elements. Firstly, consider the plate was in simply-supported edges. The frequency response function of this simply-supported plate was shown in Fig. 5. The first three natural frequencies were extracted from the peaks of the frequency response function and listed in Table 1. Along with these finite element results, the analytical ones calculated from the formula of vibration textbook [18] were also included. The percentage errors of these first three frequencies from the finite element simulation were all smaller than 4%. Therefore, the accuracy of the finite element program was validated. Also presented in Fig. 5 are the frequency response spectra of the simply-supported plate with absorbers attached to each node inside the discretized domain. There were 12 internal nodes inside the simply-supported boundary.

Table 1. The natural frequencies of a simply-supported rectangular plate

Mode	1st	2nd	3rd
Analytical results (Hz)	536.0	1163.6	1516.6
FEM results (Hz)	523.0	1117.0	1475.0
Error (%)	−2.43	−4.00	−2.74

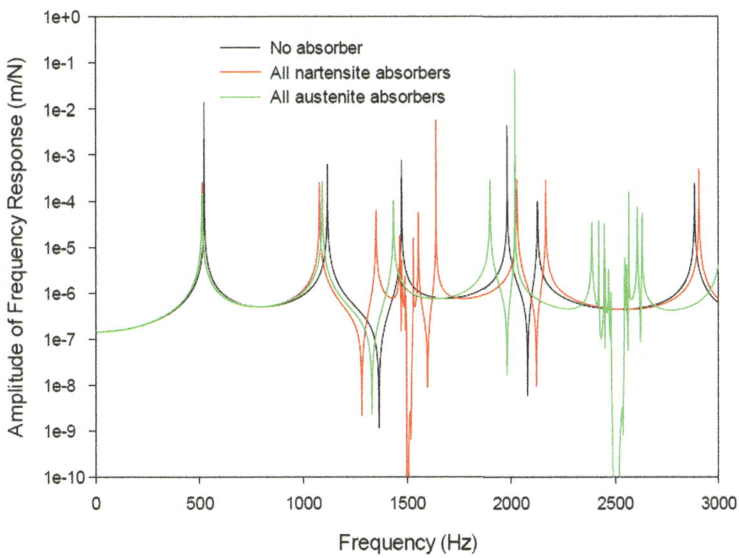

Fig. 5. The frequency response spectra of a rectangular plate in simply-supported boundary condition with/without absorbers

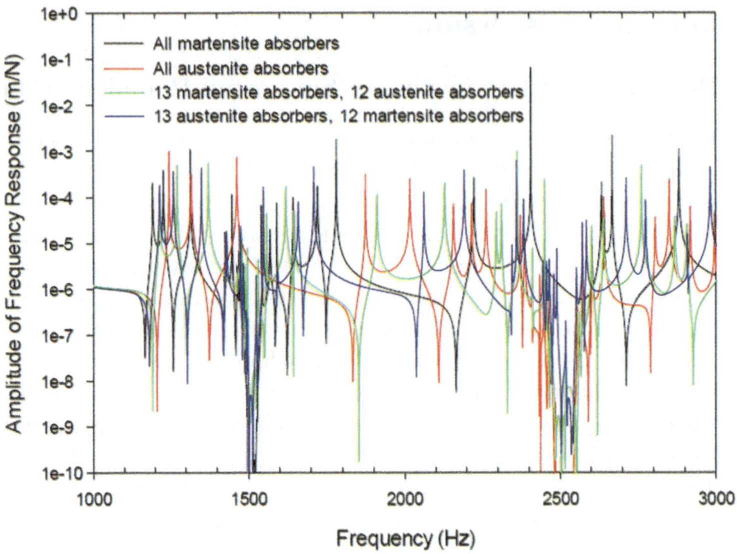

Fig. 6. The frequency response spectra of a cantilevered plate with 25 absorbers activated in different phase states

Therefore, 12 single degree-of-freedom absorbers were used in this metamaterial plate. As mentioned previously, each absorber can be tuned in martensite state ($\omega_a = 1500$ Hz) or austenite state ($\omega_a = 2500$ Hz). It can be clearly seen that for each tuned state, the plate showed a stopband corresponding to the natural frequency of the absorber. In other words, if the absorbers were excited from martensite to austenite phase, the stopband frequency can be shifted from 1500 to 2500 Hz.

4.2 A Rectangular Plate in Cantilevered Boundary Condition

The same plate with the same finite element discretization but with cantilevered boundary condition along the left edge was considered in the following. There were 25 nodes for the installation of one absorber at each node. For the absorbers activated in martensite or austenite state, the frequency response spectra, as seen in Fig. 5, showed a bandgap at 1500 Hz and 2500 Hz, respectively. Similar stopband performance was observed for this plate in cantilevered configuration comparing with those in simply-supported configuration. If the absorbers were activated differently in two grouped regions, their frequency response spectra were also presented in Fig. 5. In these cases, both stopbands at 1500 Hz and 2500 Hz occurred but with less effectiveness. More absorbers at right frequency installed could demonstrate more damping on the targeted signal. Figure 7 presents the frequency response spectra of the cantilevered plate with the martensite and austenite absorbers arranged in different configurations. The alternate configuration represented the 25 absorbers were activated in spatially alternating

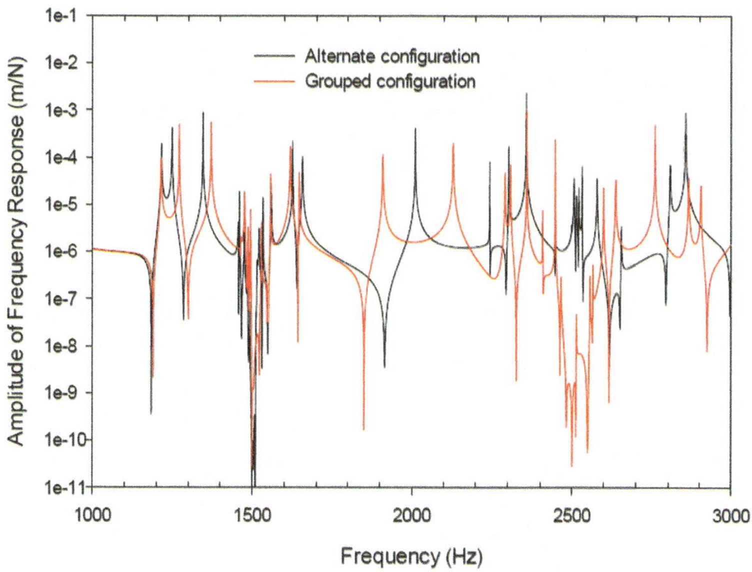

Fig. 7. The frequency response spectra of a cantilevered plate with 25 absorbers activated in different configuration states

nodes, i.e. one martensite absorber was surrounded by neighboring austenite absorbers. On the other hand, the grouped configuration denoted the 13 martensite absorbers and 12 austenite absorbers were arranged in proximity neighborhood. The same activated absorbers were placed in closer spacing for the grouped configuration. It is clearly seen that both configurations still revealed the bandgap at 1500 Hz. However, at higher frequency, the grouped configuration showed better bandgap formation than its alternate counterpart. Therefore, for inducing the bandgap at higher frequency, closer spacing of the installed absorbers is required. The distribution effect of the absorbers on the stopband characteristics were further explored in this cantilevered plate. With only 12 austenite absorbers installed at the discretized nodes, Fig. 8(a) presents the distribution of 6 different configurations. Their corresponding frequency response spectra near the stopband were also shown in Fig. 8(b). Because only 12 out of 25 available nodes were installed with absorbers, the stopband at designated 2500 Hz was not effective as that presented in Fig. 6 for all nodes installed. Nevertheless, among the 6 different distribution configurations, the more grouped arrangement demonstrated better stopband characteristics. For the alternate arrangement of the absorbers, nearly no stopband was observed in the frequency response spectrum, which was also mentioned in Fig. 7 previously. Figure 9 presents the time responses of the cantilevered plate with/without absorbers installed and subjected to sinusoidal excitation at 2500 Hz. Without absorber, the base plate vibrated in the same frequency as the excitation and showed no decay in vibration amplitude with time because no damping for the base plate

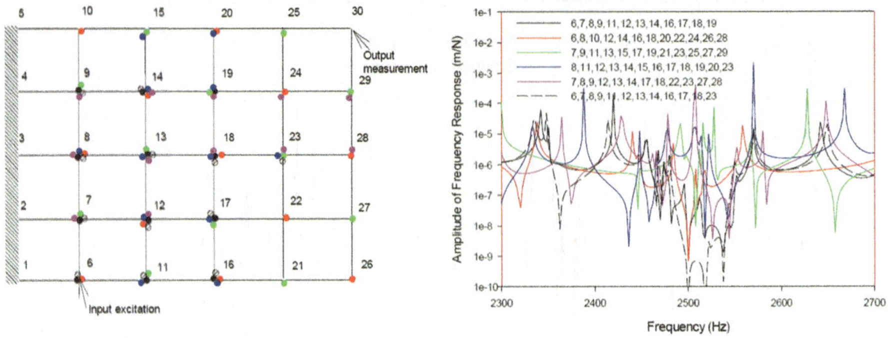

(a) (b)

Fig. 8. A cantilevered plate with 12 absorbers installed in different distributions: (a) the configurations of absorber distribution; (b) the corresponding frequency response spectra

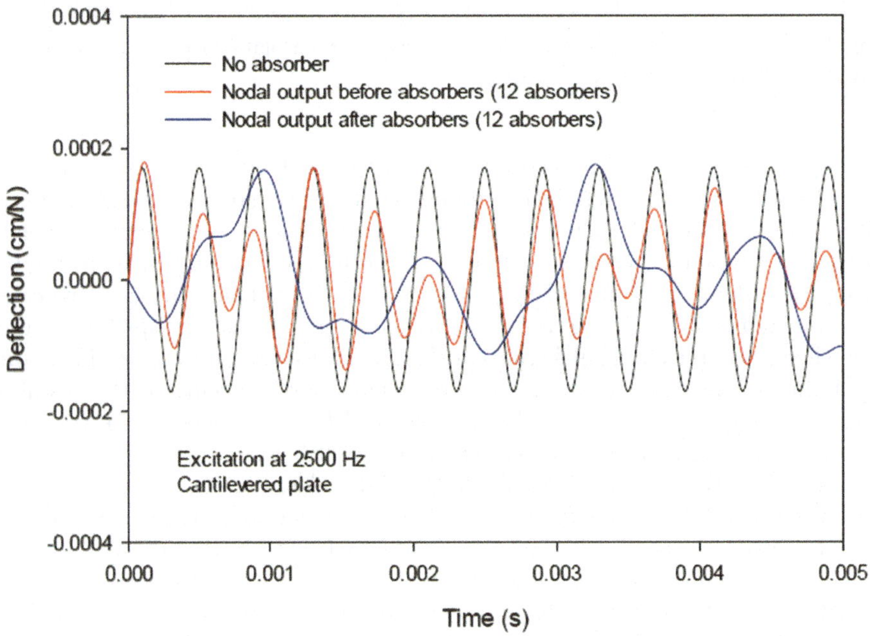

Fig. 9. The time responses of the cantilevered plate with and without absorbers installed and subjected to sinusoidal excitation at 2500 Hz

was assumed. As 12 absorbers were installed in grouped configuration, the simulated time responses at the measured locations before and after the grouped absorber domain in the propagation path looked different in nature. Before the absorbers in the propagation path, although a disturbance from the installed absorbers altered the time response, a significant constituency of the input excitation at 2500 Hz was clearly seen. Since a small damping was assumed in the absorbers, slight reduction in the time response with time was observed. The time response at node after the absorbers showed totally different result. The constituent from the 2500-Hz excitation in the time response was barely seen. The result demonstrates the tuned absorbers are able to trap the vibration at the designed frequency and prevent it from propagating downstream.

5 Conclusion

A simulation on the design and dynamic characteristics of the proposed acoustic metamaterial plate with installation of SMM absorbers was performed in this study. The cantilevered beam configuration of the SMM absorber was easy to implement on an existing structure. The simulation using finite element formulation demonstrated the tunable stopband of this metamaterial plate in different boundary conditions. In other words, by controlling the cantilever absorber in martensite or austenite phase the working stopband frequency can be tuned in lower or higher frequency domain. The arrangement of the SMM absorbers on the base plate also plays an important role in demonstrating the function of this metamaterial plate. A more grouped arrangement of the absorbers was better in enforcing the stopband especially at higher design frequency. The experimental measurement on this tunable metamaterial plate can be subject for further study.

Acknowledgements. The partial financial support from Ministry of Science and Technology, Taiwan under Grant No. MOST 106-2221-E-027-054- is gratefully acknowledged.

References

1. Srivastava, A.: Elastic metamaterials and dynamic homogenization: a review. Int. J. Smart Nano Mater. **6**(1), 41–60 (2015)
2. Zhu, R., Liu, X.N., Hu, G.K., Yuan, F.G., Huang, G.L.: Microstructural designs of plate-type elastic metamaterial and their potential applications: a review. Int. J. Smart Nano Mater. **6**(1), 14–40 (2015)
3. Liu, Z., Zhang, X., Mao, Y., Zhu, Y., Yang, Z., Chan, C.: Locally resonant sonic materials. Science **289**, 1734–1736 (2000)
4. Pendry, J.B.: Negative refraction makes a perfect lens. Phys. Rev. Lett. **85**, 3966–3969 (2000)
5. Airoldi, L., Ruzzene, M.: Design of tunable acoustic metamaterials through periodic arrays of resonant shunted piezos. New J. Phys. **13**, 113010 (2011)

6. Li, J.-B., Wang, Y.-S., Zhang, C.: Tuning of acoustic bandgaps in phononic crystals with Helmholtz resonators. J. Vib. Acoust. **135**, 031015-1 (2013). 9 pages
7. Liu, Y., Sun, X.-Z., Jiang, W.-Z., Gu, Y.: Tuning of bandgap structures in three-dimensional Kagome-sphere lattice. J. Vib. Acoust. **136**, 021016-1 (2014). 8 pages
8. Pai, P.F.: Metamaterial-based broadband elastic wave absorber. J. Intell. Mater. Syst. Struct. **21**, 517–528 (2010)
9. Pai, P.F., Peng, H., Jiang, S.: Acoustic metamaterial beams based on multi-frequency vibration absorbers. Int. J. Mech. Sci. **79**, 195–205 (2014)
10. Zhu, R., Liu, X.N., Hu, G.K., Sun, C.T., Huang, G.L.: A chiral elastic metamaterial beam for broadband vibration suppression. J. Sound Vib. **333**, 2759–2773 (2014)
11. Gandhi, M.V., Thompson, B.S., Thompson, B.D.: Smart Materials and Structures. Springer, Dordrecht (1992)
12. Lee, C.Y., Zhuo, H.C., Hsu, C.W.: Lateral vibration of a composite stepped beam consisted of SMA helical spring based on equivalent Euler-Bernoulli beam theory. J. Sound Vib. **324**(1–2), 179–193 (2009)
13. Lee, C.Y., Chen, C.C., Yang, T.H., Lin, C.J.: Structural vibration control using a tunable hybrid shape memory material vibration absorber. J. Intell. Mater. Syst. Struct. **23**(15), 1725–1734 (2012)
14. Lee, C.-Y., Pai, C.-A.: Design and implementation of tunable multi-DOF vibration absorber made of hybrid shape memory helical springs. J. Intell. Mater. Syst. Struct. **27**(8), 1047–1060 (2016)
15. Lee, C.-Y., Chen, C.-Y.: Experimental application of a vibration absorber in structural vibration reduction using tunable fluid mass driven by micropump. J. Sound Vib. **348**(1), 31–40 (2015)
16. Hu, H.-L., Peng, J.-W., Lee, C.-Y.: Dynamic simulation of a metamaterial beam consisting of tunable shape memory material absorbers. Vibration **1**, 81–92 (2018)
17. Bathe, K.-J.: Finite Element Procedures in Engineering Analysis. Prentice-Hall, Inc., Englewood Cliffs (1982)
18. DenHartog, J.P.: Mechanical Vibrations, 4th edn. Dover Publication, New York (1985)

A New Class of Optimization Problems Related to Structural Control by Contact Interaction

István Páczelt[1(✉)] and Zenon Mróz[2]

[1] Institute of Applied Mechanics, University of Miskolc, Miskolc-Egyetemváros 3515, Hungary
paczelt@freemail.hu

[2] Institute of Fundamental Technological Research, A. Pawińskiego 5B, 02-106 Warsaw, Poland
zmroz@ippt.pan.pl

Abstract. For some structures under service loads there is a need of precise control of local boundary displacement and/or its tangential gradient by an additional loading of one or two punches. Such problems exist in design of robot grippers or mechanical tools used in element assembling or in other mechanical processes. The punch interaction is assumed to be executed by a discrete set of pins or by a continuously distributed contact pressure. The optimal contact force and pressure distribution are defined in terms of assumed control function, for which contact shape is specified for both discrete and continuous punch action. For beam or plate structures three classes of control are considered. First, requiring by punch action the fixed load F_Q and displacement u_Q^* at a specified position, second, requiring the load-displacement $F_Q = F_Q(u_Q^*)$ evolution by the varying punch load and third, provide deflection and slope control at a specified position by a coordinated action of two punches. The reciprocal motion of a transverse pin attached to the beam is induced by varying punch forces. The punch position is specified by satisfying constraints on maximum punch pressure and equivalent Mises stress on the contact interface. Several illustrative examples are presented to illustrate punch control for different boundary supports and three control classes.

Keywords: Contact problem · Displacement and slope control · Optimal pressure distribution · Optimal contact shape

1 Introduction

The design parameters in structural optimization are usually defined as material moduli, structure size, shape and topology characteristic parameters, supports, loads, inner links, reinforcement, cf. Banichuk [1]. The mathematical programming technique has been used by many authors for shape optimization of structures. Referring to contact problems, usually the peak contact pressure and the

D. A. Indeitsev and A. M. Krivtsov (Eds.): APM 2019, LNME, pp. 163–178, 2020.
https://doi.org/10.1007/978-3-030-49882-5_16

interface stress concentration have been minimized by using special mathematical programming techniques and contact shape sensitivity analysis [2]. In [3], and [4,5] several classes of contact optimization problems have been considered with account for wear process.

A class of contact optimization problems for kinematical constraints has been treated in the paper by Páczelt [6]. Several classes of optimization problems have also been considered in the paper [5]: for axisymmetric punch shapes of arbitrary meridian profile the contact shape optimization problems were treated for specified punch displacement, prescribed punch load and for steady wear state conditions. In some examples the effective Mises stress was required to be below a prescribed ultimate stress. The numerical solutions have been obtained by applying a special iteration process using also the concept of partially controlled contact pressure [7]. The monographs of Goryacheva [3] and Wriggers [8] are essential sources for analytical and numerical methods of solution of contact problems, including wear analysis. The finite element analysis is useful for solving contact problems, cf. Szabó and Babuska [9].

In work [10] a new class of optimization problems was formulated. It is required that at the same boundary point (or points) of a structure, the displacement and force are prescribed. To achieve this condition, the punch contact pressure action is applied at some location on the structure boundary. Then, the punch force and its location should be specified, combined with specified contact pressure distribution and required contact shape, satisfying the constraint set on the maximal contact pressure and the stress level at contacting material interfaces. In this paper the problem is extended. Namely, the beam is investigated in a way, that the normal displacement, slope and normal force are prescribed at the same section of the beam. In this case two punches execute action on the beam. For specified geometry and material parameters an interval can be set, in which the position of the punch/punches - at adequate loads - can ensure the desired displacement and maximal/minimal slope value. The illustrative cases can be seen in Example 2.2 and in Figs. 7, 9 in Sects. 3, 4 concerning the above mentioned problems. The prescribed displacement and slope at the cross section Q can be chosen at any time and they can be reached by adequate positioning and punch action control. This fact ensures the existence and uniqueness of our solution. This problem can exist in the design of robot elements, such as clippers and gardening or plantation tools used for mechanical processing. In other words, the problem is reduced to a local displacement and its slope control in a structure subjected to service loads, such as an assembling robot gripper [11].

It is assumed that materials of the contacting bodies are linearly elastic, displacements and strains are small. The supporting constraints set on a structural element are most important in specifying a proper controlling punch action.

In mechanical engineering practice fairly abundant operation called "pick and place" is related to lifting the cylinder and placing in a new position, Fig. 1a. In this case, the cylinder is compressed along its diameter by two plates inducing normal contact displacements u_Q^* nonlinearly related to contact forces F_Q. In the assembling process of mechanical elements a typical operation is to place

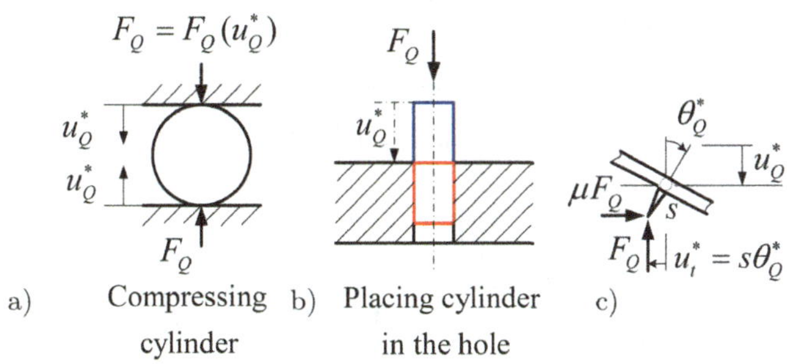

Fig. 1. Typical robot gripper operations: a) compressing cylinder by required normal forces b) placing cylinder into a structural element hole. In both cases the load-displacement relation $F_Q = F_Q(u_Q^*)$ results from contact interaction. c) tool displacement control.

one body (cylinder) into a hole of another body, Fig. 1b. In this case the cylinder must execute translation u_Q^* under increasing axial force F_Q induced by friction between cylinder and hole. In both operations the prescribed force and displacement values are required at the same point. In Fig. 1c the control of normal and tangential tool displacement is presented.

In the paper three classes of displacement control problems will be discussed. First, it is assumed that punch should apply the fixed loading assuring the required values of F_Q and u_Q^* at a specified position Q, where the interaction of the gripper with an object occurs. The optimal punch action inducing such control is discussed in Sect. 2. The second class of control requires the varying punch load in order to follow the relation $F_Q = F_Q(u_Q^*)$ resulting from gripper interaction with a structural element. Such varying load control is considered in Sect. 3 by assuming the action of two punches. In Sect. 4, the most advanced control by two punches applying fixed or varying loads is considered by requiring the control of both: normal displacement and its slope. Then a working tool attached to the beam can execute not only normal displacement, but also tangential displacement required by an executed technological process.

2 Control of Beam Deflection at Loading Point Q

Consider a beam shown in Fig. 2 with two support conditions at the left end A and with free right end B. In the first case the cantilever beam is built-in at its end A (it is referred to as *Beam I*). In the second case, the beam is allowed to slide vertically at its support A with constrained rotation (it is referred to as *Beam II*). In both cases the beam is loaded at point Q by the force F_Q, inducing the deflection $u_n^{(2)}$ in the $-z$ direction. To keep the deflection of Q at the required value u_Q^*, the discrete or continuous punch action is applied within the specified contact zone region Ω defined as an interval $L_1 \leq x \leq L_4$ of the length $l_c = L_4 - L_1$. The punch is allowed to translate in the normal direction $\mathbf{n}$ to the beam and exert contact pressure.

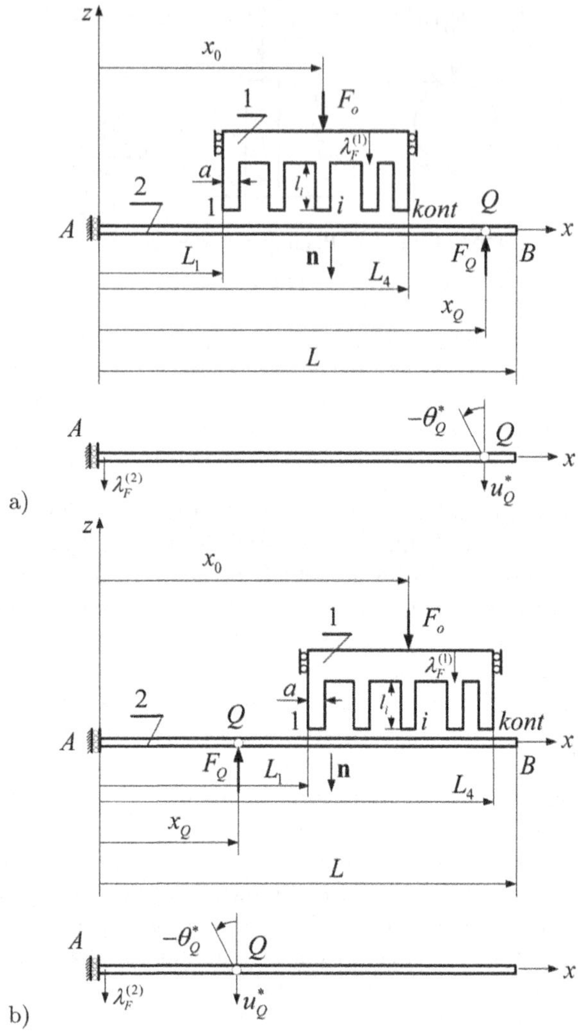

Fig. 2. Beam and punch system for transmission of load F_0 interacting with the force F_Q to induce vertical displacement u_Q^* and slope θ_Q^*. Beam 2 is allowed for rigid body displacement $\lambda_F^{(2)}$ and elastic deformation, a) $x_0 \leq x_Q$, b) $x_0 \geq x_Q$.

The beam cross section area is $A_b = a_b h_b$, inertia moment equals $I = a_b h_b^3/12$, and Young modulus is denoted by E. Two schemes correspond to a) $x_0 \leq x_Q$ and b) $x_Q \leq x_0$. In our analysis the strength condition will be applied in specifying beam heights. An alternative design of varying heights is also considered. The contact pressure distribution is assumed in the form

$$p_n = c(x) p_{\max}, \quad c(x) \leq 1, \quad \text{for} \quad x \in \Omega \tag{1}$$

where $c(x)$ is the control function and $p_{\max}$ is the maximal pressure. Usually the control function is assumed and the constraint is set on the maximal pressure which depends on punch position and the contact zone length. Determination of the initial gap between punch and beam corresponding to pressure distribution (1) constitutes the objective of numerical analysis. For the first class of control problem two designs are considered. First, the punch center location is assumed as fixed and second, specifying punch position with account for the beam stress constraint $\sigma_{\max} \leq \sigma_u$, where σ_u is the ultimate stress value. The problem is solved in two main steps. First, using the stamp equilibrium condition at given control function $c(x)$ and constraint of normal displacement at the point Q. In this way $p_{\max}$ is specified, see Eq. (3). Second, from contact condition between the stamp and beam the initial gap between the contacting bodies is determined.

Consider first the case of punch action executed by a set of punch pins, as shown in Fig. 2, with thickness a, width b and cross section area $A = ab$. The forces between punch and beam are specified by Eq. (1), thus

$$P_j = Ac(x_j)p_{\max}, \quad j = 1, 2, \ldots, \text{kont} \tag{2}$$

For *Beam II* under applied load $F_0 = F_Q$ on the punch, the maximal pressure results from the equilibrium condition

$$p_{\max} = \frac{F_0}{\sum\limits_{j=1}^{kont} c(x_j)A} \tag{3}$$

but for *Beam I*, using equation $u_Q^* = \sum\limits_{j=1}^{\text{kont}} H^{(2)}(x_Q, x_j)P_J + u_{n,load}^{(2)}$ we find

$$p_{\max} = \frac{u_Q^* - u_{n,load}^{(2)}}{\sum\limits_{j=1}^{\text{kont}} H^{(2)}(x_Q, x_j)c(x_j)A} \tag{4}$$

where $H^{(2)}(x, s)$ is the influence Green function (see Appendix A), $u_{n,load}^{(2)} < 0$ is the displacement at point Q induced by the load F_Q. In the equation for u_Q^* the first term provides the deflection due to load F_0, and the second term provides the deflection due to load F_Q. The *Beam II* displacement in the normal direction $\mathbf{n}$ along the axis $-z$ equals

$$u_Q^* = \lambda_F^{(2)} + \sum_{j=1}^{kont} H^{(2)}(x_Q, x_j)P_J + u_{n,load}^{(2)} \tag{5}$$

because the beam end A executes the sliding displacement $\lambda_F^{(2)}$. Using the influence function for calculation of $u_{n,load}^{(2)}$, we can write

$$u_{n,load}^{(2)} = -H^{(2)}(x_Q, x_Q)F_Q = -H_{Q,Q}^{(2)}F_Q \tag{6}$$

and the support displacement can be determined from (5)

$$\lambda_F^{(2)} = u_Q^* - \left(\sum_{j=1}^{kont} H^{(2)}(x_Q, x_j) \frac{c(x_j)}{\sum\limits_{j=k}^{kont} c(x_k)} - H_{Q,Q}^{(2)} \right) F_Q \tag{7}$$

The contact condition (gap after deformation) between the punch and beam is

$$d_i = u_{in}^{(2)} - u_{in}^{(1)} + g_i^{(0)} = 0, \qquad i = 1, \ldots, kont \tag{8}$$

Using the influence Green functions, Eq. (8) can be expressed as follows

$$\lambda_F^{(2)} + \left(\sum_{j=1}^{kont} H(x_i, x_j) - H_{i,Q}^{(2)} \right) F_Q - \lambda_F^{(1)} + g_i^{(0)} = 0 \tag{9}$$

where

$$H(x_i, x_j) = \left(H^{(1)}(x_i, x_j) + H^{(2)}(x_i, x_j) \right) \frac{c(x_j)}{\sum\limits_{k=1}^{kont} c(x_k)} \tag{10}$$

Discretizing (8)–(10) we can write

$$\mathbf{d} = \left({}^{iter}\mathbf{He} - \mathbf{h}_Q^{(2)} \right) F_Q + \mathbf{e}\lambda_F^{(2)} - \mathbf{e}^{(iter)}\lambda_F^{(1)} + {}^{(iter)}\mathbf{g}^{(0)} = \mathbf{0} \tag{11}$$

where $\mathbf{h}_Q^{(2),T} = \left[H_{1,Q}^{(2)}, H_{2,Q}^{(2)}, \ldots, H_{i,Q}^{(2)}, \ldots H_{kont,Q}^{(2)} \right]$, $\mathbf{e}^T = [1, \ldots, 1, \ldots, 1]$. Because F_Q and $\lambda_F^{(2)}$ are known (see (7)), then ${}^{iter}\mathbf{u}$ can be calculated, namely

$$^{iter}\mathbf{u} = \left({}^{iter}\mathbf{He} - \mathbf{h}_Q^{(2)} \right) F_Q + \mathbf{e}\lambda_F^{(2)} \tag{12}$$

and from the following equation

$$^{iter}\mathbf{u} - \mathbf{e}\,{}^{(iter)}\lambda_F^{(1)} + {}^{(iter)}\mathbf{g}^{(0)} = \mathbf{0} \tag{13}$$

one can easily find ${}^{(iter)}\lambda_F^{(1)}$. When specifying ${}^{(iter)}\mathbf{g}^{(0)}$, suppose ${}^{(iter)}g_1^{(0)} = 0$, then ${}^{(iter)}\lambda_F^{(1)} = {}^{(iter)}u_1$ and from (13) we determine the initial gap ${}^{(iter)}\mathbf{g}^{(0)}$.

2.1 Example: Beam *I* for the Discrete Punch Action

Geometric parameters: $a_b = 20\,\text{mm}$, $h_b = 70\,\text{mm}$, $L = 950\,\text{mm}$, $x_Q = 850\,\text{mm}$, $L_1 = 220\,\text{mm}$, $L_4 = 280\,\text{mm}$. Punch pins cross section $A = ab = 5{\cdot}20 = 100\,\text{mm}^2$, $l_i = 50\,\text{mm}$, $i = 1, \ldots, kont$, $kont = 5$. Material parameters: Young modulus $E = 2 \cdot 10^5\,\text{MPa}$, $\sigma_u = 150\,\text{MPa}$. The punch centre position $x_0 = 250\,\text{mm}$ corresponds to the position parameter $\xi = x_0/x_Q = 0.556$. The specified vertical displacement at the point Q is $u_Q^* = 2\,\text{mm}$ and the required force values are

$F_Q = 4\,\mathrm{kN}$, $F_Q = 5\,\mathrm{kN}$, $F_Q = 6\,\mathrm{kN}$. The influence functions for punch are calculated as $H_{i,j}^{(1)} = l_i/AE$ and for the beam according to Appendix A.

In the starting position of the punch, at $x_0 = 250\,\mathrm{mm}$, $l_c = L_4 - L_1 = 60\,\mathrm{mm}$ the maximum stress is $\sigma_{\max} = 536\,\mathrm{MPa}$. However, for the optimal solution the punch should be moved in the right direction. The optimal deflection form, initial gap form and $\sigma_{\max}$ are presented in Fig. 3 for three values of force F_Q. A special "second type iteration" method [5] was used for solution of these optimization problems. At the optimal solution the values of load F_0 and pressure $p_{\max}$, position length L_1, rigid body displacement $\lambda_F^{(1)}$ are collected in Table 1. The load factor efficiency now is $f_0 = F_0/F_Q \approx 2$–2.5.

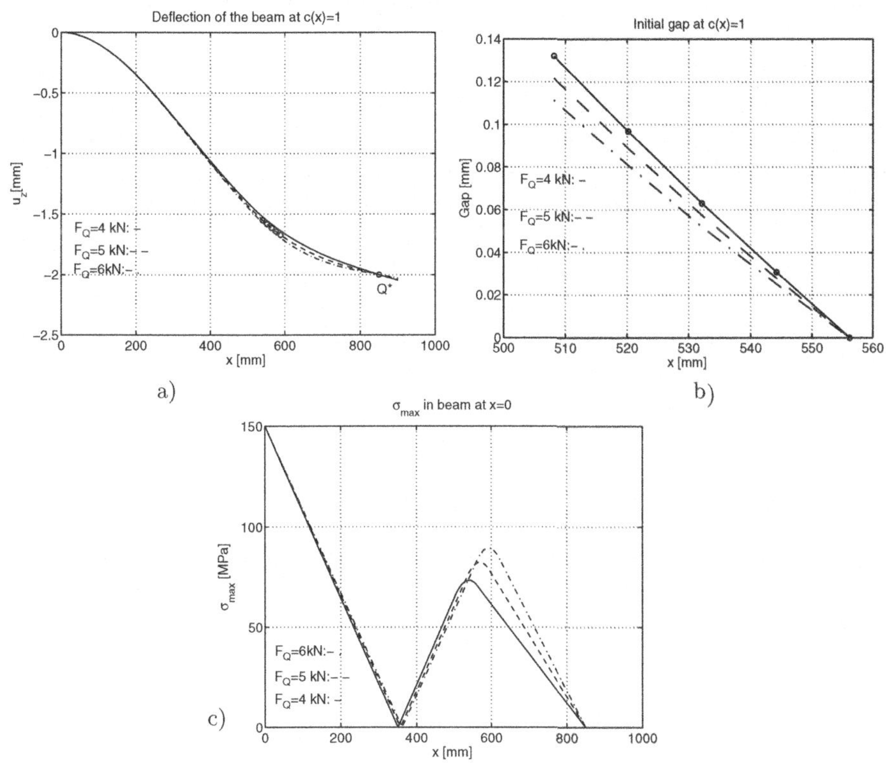

Fig. 3. Beam I deflection form for $c(x) = 1$, a) for prescribed forces $F_Q = 4\,\mathrm{kN}$, $F_Q = 5\,\mathrm{kN}$, $F_Q = 6\,\mathrm{kN}$, b) Initial contact gap, c) $\sigma_{\max}$ distribution along beam length. /Position of punch is denoted by o at $F_Q = 5\,\mathrm{kN}$, see Fig. 3a/

2.2 Example: Beam *II* for the Continuous Punch Action

Geometric data and material parameters are the same before. It is also supposed that $c(x) = 1$, that is contact pressure is constant, $p_n(x) = F_0/(b(L_4 - L_1)) =$

Table 1. Punch loads, maximal contact pressures, punch positions, and rigid body displacements for optimal design solutions.

F_Q [kN]	F_0 [kN]	$p_{\max}$ [MPa]	L_1 [mm] $x_0 = L_1 + l_c/2$	$\lambda_F^{(1)}$ [mm]
4	10.99	21.98	502.12	1.57
5	11.87	23.74	534.59	1.68
6	12.78	25.56	560.76	1.76

4.1666 MPa. The force at point Q is equal to $F_Q = 5$ kN, the required vertical displacement is $u_n^* = 2$ mm. The punch centre position is assumed at $x_0 = 250$ mm (for $L_1 = 220$ mm, $L_4 = 280$ mm). The plane stress state is assumed in the elastic punch. The normal displacement $u_n^{(1)} = (x, p_n)$ was calculated by the finite element method, using p-version technique, Szabó and Babuska [9]. Solving the contact problem, it is found $\lambda_F^{(1)} = 9.26$ mm, $\lambda_F^2 = 9.9$ mm and the beam maximal bending stress is $\sigma_{\max} = 183.67$ MPa. Assuming the admissible stress value $\sigma_u = 150$ MPa, the punch action must be moved in the right direction. Using the iteration method of Ref. [5] for solution of this optimization problem, we obtain the punch centre position at $x_0 = 360$ mm, the contact zone length $l_c = 60$ mm and the position factor $\xi_c = x_0/x_Q = 0.4235$. The rigid body displacements are $\lambda_F^{(1)} = 7.71$ mm, $\lambda_F^{(2)} = 8.88$ mm. The deflection curves and the maximal normal stresses of initial and optimized designs are shown in Fig. 4.

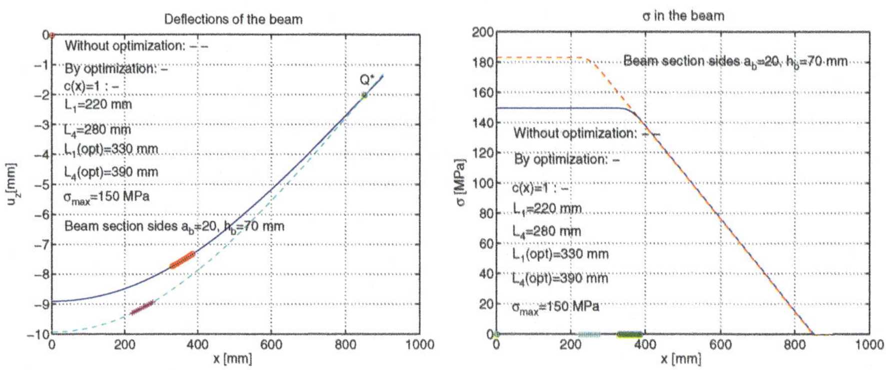

a) b)

Fig. 4. a) Beam II deflection forms of the initial and optimized designs, b) maximal normal stresses along the beam axis x for two designs. The punch position is marked by o or $+$.

3 Control of Beam Deflection and Slope by the Action of Two Concentrated Loads

Consider now the beam of Fig. 5 with the sliding support at left end, loaded by two concentrated loads $F_0^- = F_Q^-$ and $F_0^+ = F_Q^+$ at the distances x_0^- and x_0^+. The specified contact force at Q located at the distance x_Q now is $F_Q = F_Q^- + F_Q^+$. At the support the transverse force equals $F_b = 0$. Denote the load position factors by $\xi_0^- = x_0^-/x_Q$ and $\xi_0^+ = x_0^+/x_Q$. The case of concentrated loads can be treated analytically and its solution provides an input to the analysis of distributed two punches action.

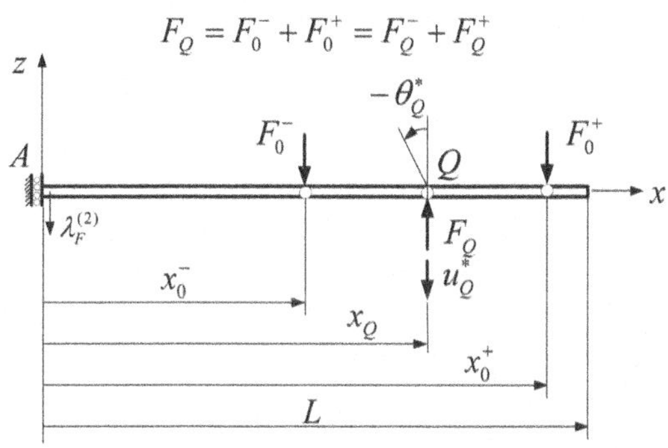

Fig. 5. Beam II is loaded by two forces F_0^-, F_0^+ inducing at the point Q the required force $F_Q = F_Q^- + F_Q^+$. The exact normal displacement value (deflection) at Q is required, $w = u_Q^*$, also the slope at this point is required, $\theta = \mathrm{d}w/\mathrm{d}x = \theta_Q^*$.

The normal deflection at the point Q (along $-z$-axis) due to force F_Q (can be expressed from the formulae of Appendix A) equals $w_Q = -\frac{F_0}{3EI}x_Q^3 = -|w_Q|$. Deflection in the direction $-z$ will be denoted by w. Consider first the case of single force action at $x_0^- \le x_Q$, or $\xi_0^- \le 1$. The deflection form now is expressed as follows

$$w^- = -\frac{F_Q^-}{2EI}(x_Q - x_0^-)x^2 + w_b^-, \qquad 0 \le x \le x_0^-,$$

$$w^- = -\frac{F_Q^-}{2EI}\left[\left[x_Q x - (x_0^-)^2\right]x + \frac{1}{3}\left[(x_0^-)^3 - x^3\right]\right] + w_b^-, \qquad x_0^- \le x \le x_Q$$

$$w^- = -\frac{F_Q^-}{2EI}\left[\left[x_Q^2 - (x_0^-)^2\right]x + \frac{1}{3}\left[(x_0^-)^3 - x_Q^3\right]\right] + w_b^-, \qquad x_Q \le x \tag{14}$$

For the single force action at $x_0^+ \ge x_Q$, or $\xi_0^+ \ge 1$, we have

$$w^+ = \frac{F_Q^+}{2EI}(x_0^+ - x_Q)x^2 + w_b^+, \qquad 0 \le x \le x_Q,$$

$$w^+ = \frac{F_Q^+}{2EI}\left[(x_0^+ x - x_Q^2)x - \frac{1}{3}(x^3 - x_Q^3)\right] + w_b^+, \qquad x_Q \le x \le x_0^+$$

$$w^+ = \frac{F_Q^+}{2EI}\left[\left[(x_0^+)^2 - x_Q^2\right]x - \frac{1}{3}\left[(x_0^+)^3 - x_Q^3\right]\right] + w_b^+, \qquad x_0^+ \le x \tag{15}$$

When two forces load the beam, then

$$w = w^- + w^+, \quad F_Q = F_Q^- + F_Q^+, \quad w_b = w_b^- + w_b^+ \tag{16}$$

where w_b denotes the beam translation at the support. Supposing the contact interaction load F_Q to be attained by the combined two forces action, we can write (Fig. 6)

$$F_Q = F_Q^- + F_Q^+ = k^-(u_Q^*)^m + k^+(u_Q^*)^m. \tag{17}$$

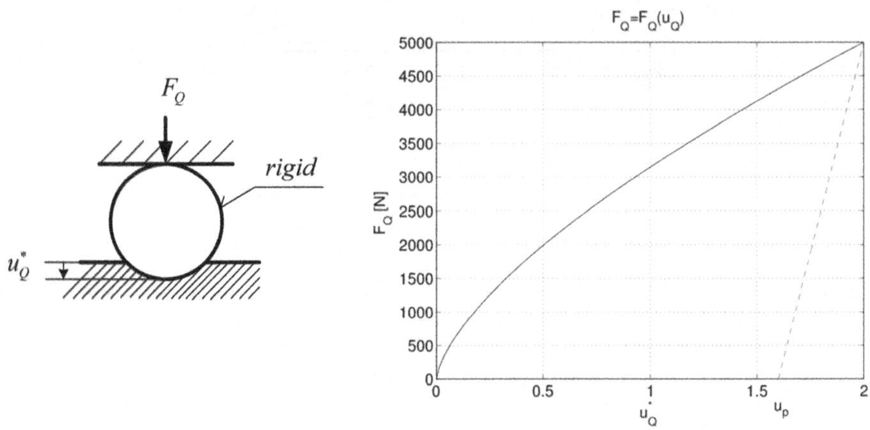

Fig. 6. Contact of a rigid sphere with an elastic-plastic body: non-linear indentation function $F_Q = F_Q(u_Q^*) = k(u_Q^*)^m$.

Requiring $w = u_Q^*$ at Q, the beam translation w_b is then determined, thus

$$w_b = u_Q^* + \frac{k^-(u_Q^*)^m}{3EI}x_Q^3\left[1 - \frac{1}{2}(3(\xi_0^-)^2 - (\xi_0^-)^3)\right] - \frac{1}{2}\frac{k^+(u_Q^*)^m}{EI}x_Q^3(\xi_0^+ - 1) \tag{18}$$

If the separate load action is assumed, the beam deflections at the points x_0^-, x_0^+ are

$$w_0^- = u_Q^* + \frac{k^-(u_Q^*)^m}{3EI}x_Q^3\left[1 - 3(\xi_0^-)^2 + 2(\xi_0^-)^3\right], \quad \xi_0^- \le 1 \tag{19}$$

and

$$w_0^+ = u_Q^* + \frac{k^+(u_Q^*)^m}{3EI}x_Q^3\left[(\xi_0^+)^3 - 3\xi_0^+ + 2\right], \quad \xi_0^+ \ge 1. \tag{20}$$

At force $F_Q^- = k^-(u_Q^*)^m$ and $F_Q^+ = k^+(u_Q^*)^m$ deflection at the point Q is

$$\left|w_Q^{\mp}\right| = \frac{k^{\mp}(u_Q^*)^m}{3EI} x_Q^3 \tag{21}$$

Using (14) and (15) we can derive the formulae for displacements at the points x_0^-, x_0^+ due to two forces

$$\begin{aligned} w_0^- = w_0^-(u_Q^*) = &\frac{k^-(u_Q^*)^m}{3EI} x_Q^3 \frac{3}{2}\left[(\xi_0^-)^3 - (\xi_0^-)^2\right] \\ &+ \frac{k^+(u_Q^*)^m}{3EI} x_Q^3 \frac{3}{2} (\xi_0^-)^2 (\xi_0^+ - 1) + w_b \end{aligned} \tag{22}$$

$$\begin{aligned} w_0^+ = w_0^+(u_Q^*) = &\frac{k^-(u_Q^*)^m}{3EI} x_Q^3 \frac{1}{2}\left[3\xi_0^+(\xi_0^-)^2 - 3\xi_0^+ - (\xi_0^-)^3 + 1\right] \\ &+ \frac{k^+(u_Q^*)^m}{3EI} x_Q^3 \frac{1}{2} (1 + 2(\xi_0^+)^3 - 3\xi_0^+) + w_b \end{aligned} \tag{23}$$

If the $k^{\mp} = k^{\mp}(\tau)$ are changing in time, then the forces $F_Q(\tau) = F_Q^-(\tau) + F_Q^+(\tau) = F_0$ and the deflection also varies in time, thus

$$w(\tau) = w^-(\tau) + w^+(\tau) \tag{24}$$

At the point Q the beam deflection is required to preserve the value $u_Q^*(\tau)$. This value is reached through the varying displacements $w_0^-(\tau)$, $w_0^+(\tau)$. Alternatively, the force-displacement relation $F_Q = F_Q(u_Q^*)$ can be attained by one varying force action.

Consider now the advanced beam control by requiring the specified deflection and slope value to be preserved at the interaction point Q. Such control can be attained by the action of two punches at x_0^+ and x_0^-. Considering the concentrated load action, from Eqs. (14) and (15) we obtain the expression of slope value

$$w'_Q = \frac{\mathrm{d}w}{\mathrm{d}x} = -\frac{F_Q^- x_Q^2}{2EI}\left[1 - (\xi_0^-)^2\right] + \frac{F_Q^+ x_Q^2}{2EI} 2(\xi_0^+ - 1) = \theta_Q^* \tag{25}$$

Satisfying the condition $F_Q = F_0^- + F_0^+ = F_Q^- + F_Q^+$, from (25) the values of two loads per unit beam width are obtained, thus

$$F_0^- = \frac{2F_Q x_Q^2(\xi_0^+ - 1) - 2EI\theta_Q^*}{x_Q^2\left[2\xi_0^+ - (\xi_0^-)^2 - 1\right]}, \quad F_0^+ = \frac{F_Q x_Q^2(1 - (\xi_0^-)^2) + 2EI\theta_Q^*}{x_Q^2\left[2\xi_0^+ - (\xi_0^-)^2 - 1\right]}. \tag{26}$$

Denoting the load fractions by $f_0^- = F_Q^-/F_Q$ and $f_0^+ = F_0^+/F_Q$, the diagram of evolution of the slope angle θ_Q^* on load fractions and their positions is presented in Fig. 7. The relations (26) can now be written as follows

$$f_0^- = \frac{2(\xi_0^+ - 1) - 2\beta\theta_Q^*}{2\xi_0^+ - (\xi_0^-)^2 - 1}, \quad f_0^+ = \frac{\left[1 - (\xi_0^-)^2\right] + 2\beta\theta_Q^*}{2\xi_0^+ - (\xi_0^-)^2 - 1}, \quad \beta = \frac{EI}{F_Q x_Q^2}. \tag{27}$$

If we take account the influence on pin (see Fig. 8) for deflection and slope, $(EI)_{mod} = EI - sF_Q x_Q$ will be taken instead of EI.

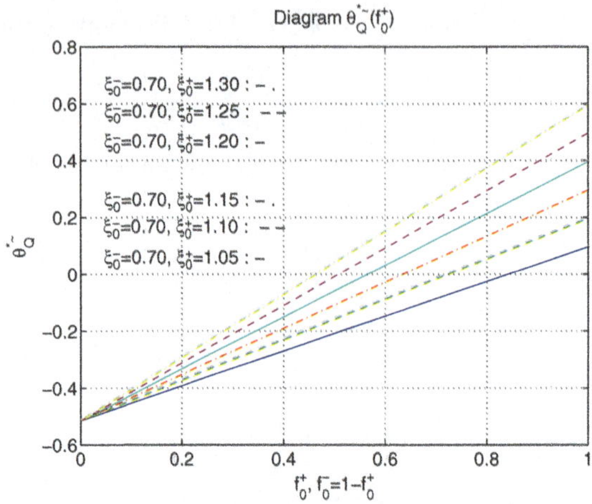

a)

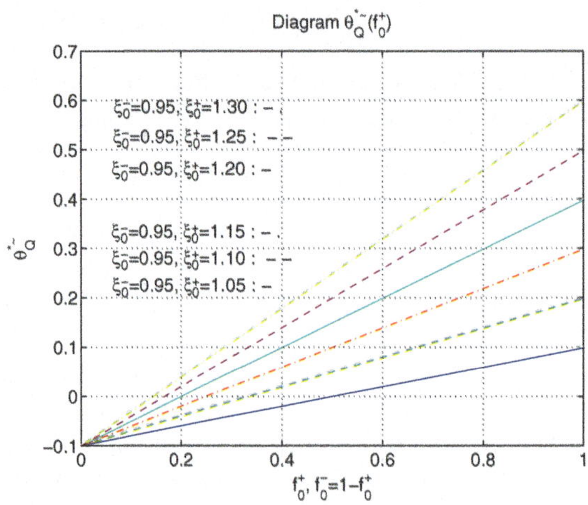

b)

Fig. 7. Dependence of the beam slope $\theta_Q^{*\sim} = 2\beta\theta_Q^*$ on two load fractions f_0^-, f_0^+ their positions ξ_0^-, ξ_0^+ and beam stiffness–54load ratio $\beta = EI/F_Q x_Q^2$, a) $\xi_0^- = 0.70$, b) $\xi_0^- = 0.95$.

4 Control of Displacement and Slope in a Beam with a Transverse Tool at Point Q

Consider now the beam with transversely attached rigid pin of length s, Fig. 8. At the end of the pin only vertical force is supposed. The frictional case will be investigated in a separate paper. Now the control of beam is aimed to induce a required trajectory in the plane $x-z$. Assume that first the normal deflection u_Q^* is attained at vanishing slope $\theta_Q^* = 0$ and next the end of pin executes reciprocal translation of amplitude $2\tilde{a}$ along the x-axis at fixed value of deflection. The derived formulae (26) or (27) allow us to execute such pin motion by varying two loads. First, setting $\theta_Q^* = 0$ in (26) the values of F_0^- and F_0^+ are obtained

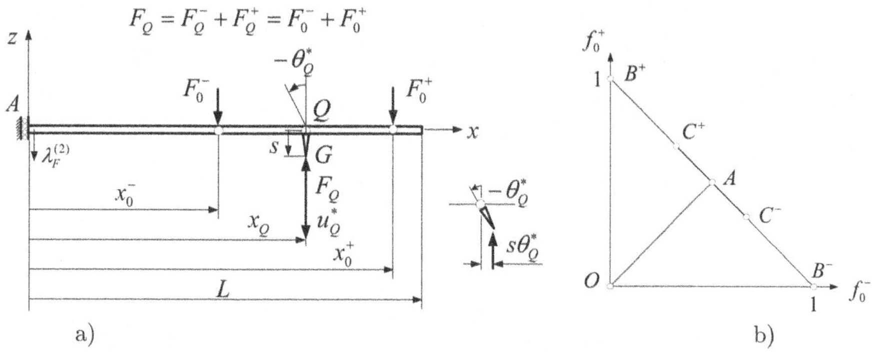

Fig. 8. *Beam II* with a rigid transverse tool at Q loaded by two forces F_0^-, F_0^+ such that $F_0^- + F_0^+ = F_Q$. The control of displacement u_Q^* and slope θ_Q^* at Q provides the control of tool end displacement.

and the value of F_Q assures the transverse displacement value at Q. The ratio of forces equals

$$\frac{f_0^-}{f_0^+} = \frac{F_0^-}{F_0^+} = \frac{2(\xi_0^+ - 1)}{1 - (\xi_0^-)^2}, \qquad f_0^- + f_0^+ = 1. \tag{28}$$

In the load diagram, Fig. 8b, the initial force state is represented by the point A. Next, the loads vary following the loading path $f_0^- + f_0^+ = 1$ in the force plane. For the assumed symmetric sliding of the pin end relative to the initial position, the force reversal states are represented by the points C^+ and C^- in Fig. 8b. Note that $AC^-/AC^+ = AB^-/AB^+$. Then the maximal and minimal values of the slope are reached at B^- and B^+ for the single load actions. From (27) it follows that

$$(\theta_Q^*)^- = -\frac{f_0^-\left[2\xi_0^+ - (\xi_0^-)^2 - 1\right] - 2(\xi_0^+ - 1)}{2\beta},$$

$$(\theta_Q^*)^+ = \frac{f_0^+\left[2\xi_0^+ - (\xi_0^-)^2 - 1\right] - (1 - (\xi_0^-)^2)}{2\beta} \tag{29}$$

and the total slope equals $\theta_Q^* = (\theta_Q^*)^- + (\theta_Q^*)^+$. The extreme slope values are obtained for $f_0^- = 1$, $f_0^+ = 0$ and $f_0^+ = 1$, $f_0^- = 0$, thus

$$(\theta_Q^*)_{\min} = -\frac{1 - (\xi_0^-)^2}{2\beta}, \quad (\theta_Q^*)_{\max} = \frac{\xi_0^+ - 1}{\beta} \tag{30}$$

The related tool end displacements are $\Delta x^- = -(\theta_Q^*)_{\min} s$, $\Delta x^+ = -(\theta_Q^*)_{\max} s$ and the maximal translation amplitude is $(\Delta x)_{\tilde{a}} = \left[2\xi_0^+ - 1 - (\xi_0^-)^2\right] s/\beta$.

When the specified tool sliding amplitude is less than the maximal value, the loading program is executed along the portion of the loading path shown in Fig. 8b. Non-linear effects related to large beam deflections, large slope variation at Q and the effect of contact friction are not considered here. The slope dependence on the parameter β is shown in Fig. 9 and the deflection curve is presented in Fig. 10.

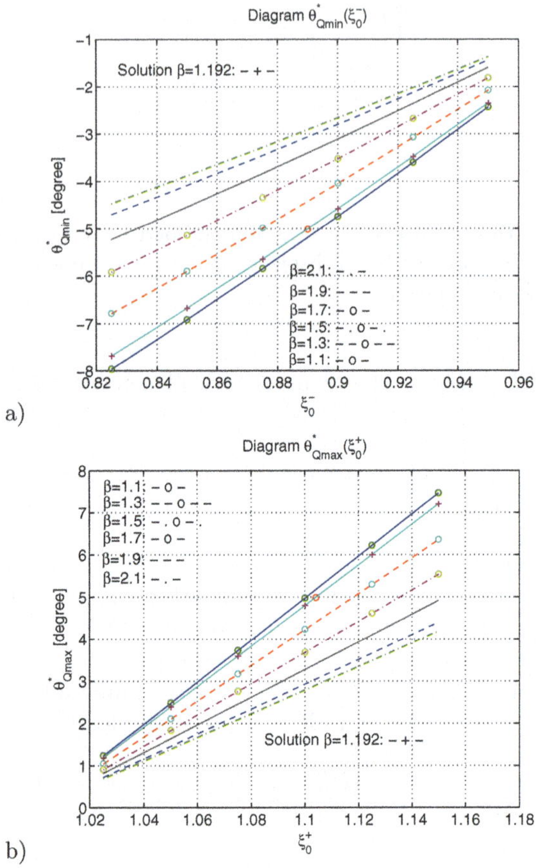

Fig. 9. Minimum (a) and maximum (b) of the slope for varying parameter β.

The formulae presented are valid also for the case $\xi_c^- = 0$, when the load F_0^- is applied at the sliding support. A similar design is used in the atomic force microscope, where a cantilever beam with a sliding support at the left end and a sharp transverse pin at the right end are used. The load is then applied at the left support and its translation is controlled in order to generate proper contact force of pin. The present design is more complex requiring the pulsating load application of two punches in order to generate the reciprocal sliding motion of pin. Such design and punch action can be used in developing a wear testing apparatus at both micro and macro-scales.

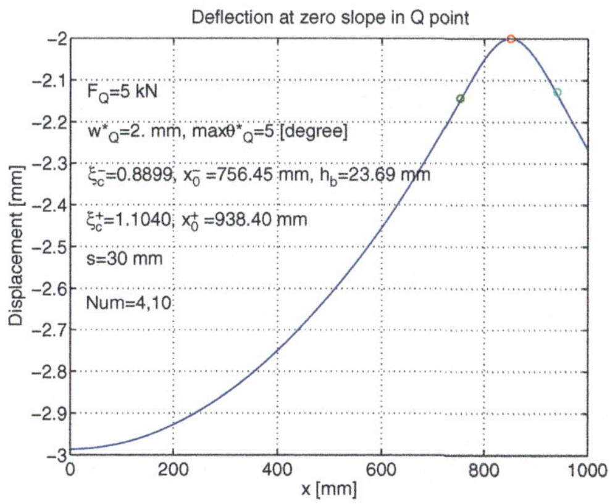

Fig. 10. Deflection of the beam II for loads $F_0^- = F_0^+$, $f_0^- = f_0^+ = 0.5$, $\beta = 1.2271$, $\beta_{mod} = 1.1918$.

5 Conclusions

The optimal design of punch action in order to control normal displacement at a loaded boundary point in a structural element has been discussed in the paper and illustrated by the specific examples of beam deflection control. A new class of optimization problems has been analysed, when at one boundary point both displacement and force are specified. It has been shown that support constraint can affect essentially the punch load and the beam deflected form. The problem formulation was also extended to control both the deflection and its slope at a loaded boundary by the action of two punches. For varying punch loads the control was aimed to induce reciprocal tool sliding on the contact surface.

Acknowledgements. The present research was partially supported by the Hungarian Academy of Sciences, by the grant National Research, Development and Innovation Office – NKFIH: K115701.

Appendix A

Using Betti theorem, the influence (Green) function for cantilever beam (Fig. 11) has the form

$$H^{(2)}(x_i, x_j) = H^{(2)}_{i,j} = \frac{1}{6EI}\left[(3x_j x_i^2 - x_i^3) + \langle x_i - x_j \rangle^3\right]$$

where I is the inertia moment of cross section, E is the Young modulus,

$$\langle x_i - x_j \rangle^3 = \Big\{(x_i - x_j)^3, \;\; if\; x_i > x_j; \quad and \quad 0 \;\; if\; x_i \leq x_j\Big\}.$$

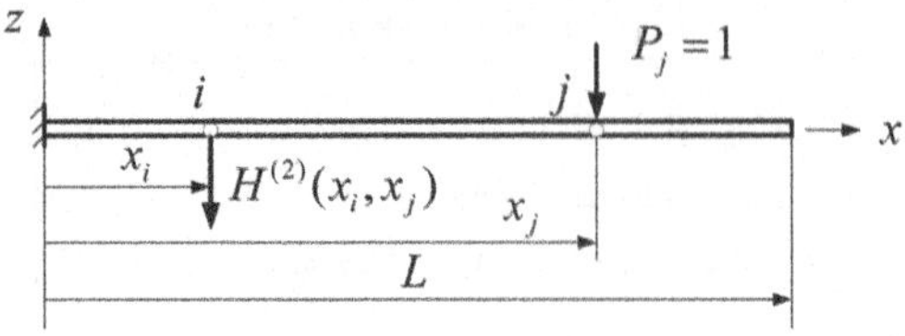

Fig. 11. Load for calculation of the influence function of the cantilever beam.

References

1. Banichuk, N.V.: Introduction to Optimization of Structures. Springer, London (2011)
2. Mróz, Z., Bojczuk, D.: Shape and topology sensitivity analysis and its application to structural design. Arch. Appl. Mech. **82**(10–11), 1541–1555 (2012)
3. Goryacheva, I.G.: Contact Mechanics in Tribology. Kluwer Academic Publishers, Dordrecht (1998)
4. Páczelt, I., Mróz, Z.: Solution of wear problems for monotonic and periodic sliding with p-version of the finite element method. Comput. Methods Appl. Mech. Eng. **249–252**, 75–103 (2012)
5. Páczelt, I., Baksa, A., Mróz, Z.: Contact optimization problems for stationary and sliding conditions. In: Neittaanmaki, P., et al. (eds.) Mathematical Modeling and Optimization of Complex Structures. Computational Methods in Applied Sciences, vol. 40, pp. 281–312. Springer, Cham (2016). https://doi.org/10.1007/978-3-319-23564-6_16
6. Páczelt, I.: Some optimization problems of contact bodies within the linear theory of elasticity. In: Nemat-Nasser, S. (ed.) Variational Methods in the Mechanics of Solids, pp. 349–356. Pergamon Press, Oxford (1980)
7. Páczelt, I.: Iterative methods for solution of contact optimization problems. Arch. Mech. **52**(4–5), 685–711 (2000)
8. Wriggers, P.: Computational Contact Mechanics. Wiley, New York (2002)
9. Szabó, B., Babuška, I.: Finite Element Analysis. Wiley-Intersience, New York (1991)
10. Páczelt, I., Mróz, Z.: Optimized punch contact action related to control of local structure displacement. Struct. Multidisc. Optim. https://doi.org/10.1007/s00158-019-02300-z
11. Monkman, G., Hesse, S., Steinmann, R., Schunk, H.: Robot Grippers. Wiley-VCH, Berlin (2006)

On Crack Propagation in a Two-Component Thermally Reinforced Pipe

Alexander Maksimov[1] and Yulia Pronina[2(✉)]

[1] Kerch State Maritime Technological University, Ordzhonikidze str. 82, Kerch 298309, Russia
[2] Saint Petersburg State University, Universitetskaya emb. 7/9, St. Petersburg 199034, Russia
y.pronina@spbu.ru

Abstract. One of the effective approaches to the retardation of cracks in pipes is based on thermal reinforcement of rolled sheet. For the manufacture of pipes it was proposed to use thermally reinforced steel sheet with periodic strengthened strips. These barriers in the crack path reduce the rate of its propagation and change its trajectory. The direction of crack propagation may be controlled by the configuration of these barriers. This paper presents experimental and theoretical results concerning the direction of crack propagation near the boundary of a strengthened part of a steel sheet depending on the angles of inclination of this boundary with respect to the initial crack trajectory. The competition between deflection and penetration at the interface is investigated for different ratios of the strength limits of the strengthened material and the matrix. The problem is studied using known expressions for the stress components in the vicinity of crack tips with non-singular terms associated with load biaxiality, and the generalised maximum tensile stress criterion.

Keywords: Thermally reinforced sheet · Pipeline · Direction of crack propagation

1 Introduction

The problems of safe and reliable operation of trunk oil and gas pipelines are of great importance from different points of view. One of the most expensive types of accidents on trunk pipelines is extended destruction when the length of a longitudinal crack in a pipe can reach hundreds of meters [1]. The losses induced by cracks propagation in pipelines may be minimized by creating conditions for cracks arrest and controlling their trajectories. Many scientists have made significant contributions to the problems of preventing cracks propagation in pipelines. Among the fundamental ways of preventing cracks propagation is increasing the fracture toughness of pipe materials by doping. However, with an increase in the number of alloying elements, the weldability and affordability of the steel suffer [10]. Design solutions that can contribute to these problems include, e.g., the use of linings, stiffeners and separately mounted couplings. Another possible way to prevent extended destruction was developed by the authors of [4] (see also [1]). They suggested the formation of annular ribs (corrugations) on tube

D. A. Indeitsev and A. M. Krivtsov (Eds.): APM 2019, LNME, pp. 179–184, 2020.
https://doi.org/10.1007/978-3-030-49882-5_17

blanks by cyclic heat changes and shortening the pipe. Heat treatment can also significantly improve the mechanical properties of steel. Based on these features, A. Maksimov developed an approach to the retardation of cracks in pipes using thermal reinforcement of rolled sheet [8]. For the manufacture of pipes it was proposed to use thermally reinforced steel sheet with periodic strengthened strips. These barriers in crack path reduce the rate of its propagation and change its trajectory. The direction of crack propagation may be controlled by the configuration of such barriers. This paper presents experimental and theoretical results concerning the prediction of the direction of crack propagation near the boundary of a strengthened part of a steel sheet depending on the angles of inclination of this boundary with respect to the initial crack trajectory. The competition between deflection and penetration at the interface is investigated. Crack behavior near the interface joining two materials with different elastic properties was investigated earlier by a number of authors using the integral equation methods [5]. Here, due to the specific of the considered thermally reinforced material, the problem is studied using known closed-form expressions for the stress-components in the vicinity of crack-tips with non-singular terms [3] and the generalised maximum tensile stress criterion. The results can be adapted to other criteria [11].

2 Experimental Results

To study the ability to control the direction of crack propagation by thermal reinforcement of rolled sheets, a series of tensile tests of St3sp steel plane samples with initial straight cuts were carried out [9, 10]. The samples were first heated to the austenitization temperature, after that one part of each sample was rapidly cooled by immersion in 7% aqueous solution of salt. The angle α between the immersion line and the initial cut was 0, 20, 30, 45, 60 and 90°. As a result, two regions separated by the immersion line were obtained, each with a homogeneous microstructure. These regions differed in hardness about 1.7 times. The elastic moduli of the matrix and the thermally strengthened material remained the same.

Tensile forces were applied perpendicularly and symmetrically to the initial cut. Forces increased gradually until the sample ruptured. Tensile tests revealed that when the initial cut was perpendicular to the phase boundary, or at $\alpha = 60°$ the crack passed to the strengthened area. At a smaller angle α of inclination, the crack did not pass through the interface but traveled along it [9].

3 Calculation Model

3.1 Fracture Criterion

Consider the problem under study within the framework of linear fracture mechanics, assuming that the crack approaches the interface quasistatically. Taking into account dynamic effects may alter the results [2].

To determine the direction of crack propagation we will use the maximum tensile stress criterion in the following generalized form. *The crack extends (originates) perpendicular to the direction upon which the ratio of tensile stress to the strength limit attains maximum value.*

3.2 Stress Field in an Intact Pipe

Consider a linearly elastic long thin-walled pipe made of thermally reinforced rolled sheet with periodic strengthened strips. The middle radius and thickness of the tube are denoted by R and h, correspondingly. The angle of inclination of the strengthened strips to the generatrix of the pipe is denoted by α. The elastic moduli of the matrix material and the strengthened strips are the same, while the strength limits of the matrix, σ_M^*, and the strengthened material, σ_A^*, differ by k times: $k = \sigma_A^*/\sigma_M^*$ (determined by their hardness ratio). Therefore, up to the onset of destruction, conventional formulae of linear elasticity for homogeneous materials can be applied.

The tube is subjected to internal pressure p. The action of the ends of the pipe is not taken into account. Consider a cylindrical coordinate system (ρ, θ, z) with the z-axis coincide with the axis of the tube. Since the elastic moduli of both strengthened and virgin materials are the same, for a thin-walled pipe under internal pressure stress components can be approximately defined by the well-known formulae

$$\sigma_{\theta\theta} = pR/h, \sigma_{zz} = \widetilde{\nu}pR/h = \widetilde{\nu}\sigma_{\theta\theta}, \sigma_{\rho\rho} = o(\sigma_{\theta\theta}, \sigma_{zz}), \tag{1}$$

ignoring the stress variation across the wall thickness (which can be essential under mechanochemical corrosion conditions [12]). Here, $\widetilde{\nu} = 1/2$ for the tube with plugs, $\widetilde{\nu} = \nu$ (Poison's ratio) in the case of plane strain, and $\widetilde{\nu} = 0$ if there is no longitudinal stresses.

The fact that the hoop stress $\sigma_{\theta\theta}$ is at least twice the longitudinal stress σ_{zz} explains that the crack originates from a random microdefect in non-strengthened material along the tube generatrix, i.e., perpendicular to the direction of the maximum ratio of tensile stress to the corresponding strength limit.

3.3 Stress Field Near the Crack Tips

As a first approximation, it can be assumed that the stress state in the immediate vicinity of the crack tips is plane and the remote loads parallel and perpendicular to the crack edges are equal to the longitudinal σ_{zz} and hoop $\sigma_{\theta\theta}$ stress-components in the pipe (1), correspondingly. For definiteness, consider the right end of the crack. Since the elastic constants of the matrix and strengthened parts of the material are the same, we can apply formulae for homogeneous material [3]. Passing in Eqs. (30) from [3] to the polar coordinates and using (1), we obtain the expression for the circumferential stress around the crack tip

$$\sigma_{\varphi\varphi}(r, \varphi) = \frac{K_I}{\sqrt{2\pi r}}\cos^3\frac{\varphi}{2} - \sigma_{yy}^{\infty}(1 - \widetilde{\nu})\sin^2\varphi, \tag{2}$$

where (r, φ) are local polar coordinates with origin at the current crack-tip, $\varphi \in (-\pi, \pi)$; σ_{yy}^{∞} is the remote stress perpendicular to the crack, while $\sigma_{xx}^{\infty} = \tilde{\nu}\sigma_{yy}^{\infty}$ is the remote stress parallel to the crack; K_I is the crack-tip stress intensity factor.

It can be verified that for $\tilde{\nu} \leq 1/2$, the circumferential stress $\sigma_{\varphi\varphi}$ attains maximum value at $\varphi = 0$. This corresponds to the observed straight-line crack growth in a homogeneous material or at some distance from the interface. Therefore, the crack extends straight along its original direction until it impinges the interface at an angle α.

Based on the foregoing, as a computational model for determining the direction of propagation of the crack near the boundary of the strengthened region, consider a two-component infinite plane with a rectilinear crack of length $2l$ located along the x-axis and impinging the interface at an angle α. The plane is subjected to the biaxial tensile loads σ_{yy}^{∞} and $\sigma_{xx}^{\infty} = \tilde{\nu}\sigma_{yy}^{\infty}$, where $\tilde{\nu} \leq 1/2$.

4 Deflection Versus Penetration

4.1 Deflection/ Penetration Criterion

In accordance with the accepted fracture criterion, it can be shown that, depending on the angle α and the ratio $k = \sigma_A^*/\sigma_M^*$, there are two possible initial ways of the crack extension near the interface. If at given α and k,

$$\sigma_{\varphi\varphi}(r, 0)/\sigma_A^* > \sigma_{\varphi\varphi}(r, \alpha)/\sigma_M^*$$

then the crack will pass through the interface, otherwise it will deflect along the interface. Using (2), this criterion may be rewritten in the following form. Crack growth direction is determined by comparing the values of the expressions

$$k \quad \text{and} \quad \frac{1 + \sin^2\alpha\,(1 - \tilde{\nu})\,\sigma_{yy}^{\infty}/\sigma_M^*}{\cos^3(\alpha/2)} = K. \tag{3}$$

If $k < K$ then the crack will penetrate the strengthened material, otherwise it will deflect along the interface.

These results are applicable to a single crack. With regard to effects of multiple cracks interaction, the reader is referred to works of the authors of [6, 7] and others.

4.2 Calculation Results

Figure 1 shows the dependence of K defined by (3) on $|\alpha|$ (in degrees) for $\sigma_{yy}^{\infty}/\sigma_M^* = 0$ (solid line), 0.5 (dashed lines), and 1 (dotted lines), at $\tilde{\nu} = 0;\ 0.25;\ 0.5$ (lines A, B, C, correspondingly). Dash-dotted horizontal line 1 corresponds to $k = 1.7$ (observed in the experimental samples) and line 2 to $k = 2$. Let $|\alpha^*|$ denote the abscissa of the point of intersection of the curve $K(|\alpha|)$ with the horizontal line $k = \sigma_A^*/\sigma_M^*$ corresponding to the given initial data. If $|\alpha| < |\alpha^*|$ then the crack will deflect along the boundary of the strengthened material, otherwise it will penetrate the interface.

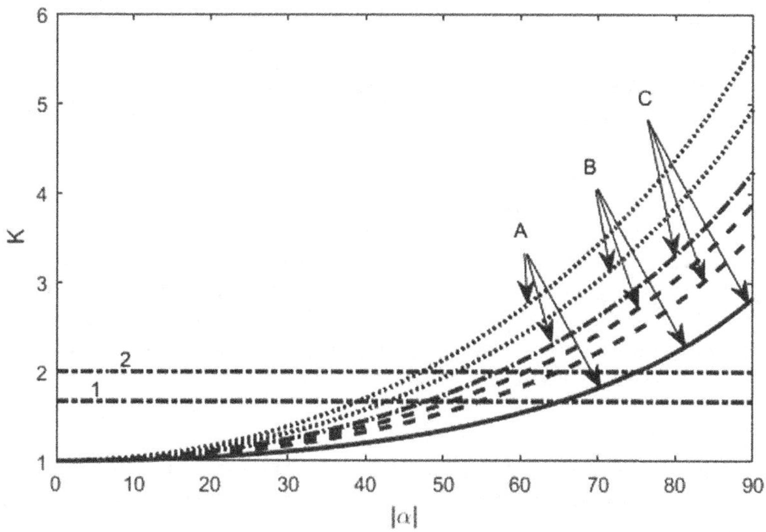

Fig. 1. Dependence of K on $|\alpha|$ (in degrees) for $\sigma_{yy}^{\infty}/\sigma_{M}^{*} = 0$ (solid line), 0.5 (dashed lines), and 1 (dotted lines), at $\widetilde{\nu} = 0$ (lines A), $\widetilde{\nu} = 0.25$ (lines B), and $\widetilde{\nu} = 0.5$ (lines C). Horizontal line 1 corresponds to $k = 1.7$, line 2 to $k = 2$.

To minimise the length of the destroyed part of a pipe, it is most reasonable to set the angle $|\alpha|$ as large as possible, at which the crack will deflect along the interface. This angle is determined by the value of $|\alpha^{*}|$ with taking into account a certain safety factor. As one can see, the greater the ratio $k = \sigma_{A}^{*}/\sigma_{M}^{*}$ and/or the coefficient $\widetilde{\nu}$, the greater the angle $|\alpha^{*}|$. Curves A, corresponding to uniaxial tension, give the smallest (i.e. guaranteed) value of $|\alpha^{*}|$ among all possible for various $\widetilde{\nu}$, other parameters being equal. Dotted lines give the smallest (guaranteed) value of $|\alpha^{*}|$ among all possible for fixed k and $\widetilde{\nu}$. The solid line can also be obtained by the formulae for a crack in an infinite plane without non-singular terms entered in [3]. This curve yields the over-estimated value of the optimal angle $|\alpha^{*}|$ for any initial data.

The calculation results for $k = 1.7$ are consistent with the results of experiments: at $|\alpha|$ equal to 90° and 60° the crack penetrates the strengthened material, while at $|\alpha| \leq 45°$ the crack deflects along the interface.

5 Conclusion

Thus, it has been shown that thermal reinforcement of rolled sheet can help to control the main crack trajectory, causing it to deviate from the straight path. Ignoring the non-singular terms in expressions for stress-components in the vicinity of the crack-tips can lead to errors in the control of the crack trajectory.

References

1. Albaghdadi, B.M.H., Cherniavsky, A.O.: Comparison of fracture criteria for cracked steel pipelines by finite element method. Bull. South Ural State Univ. Ser. Mech. Eng. Ind. **17**(1), 13–20 (2017)
2. Bragov, A.M., Konstantinov, A.Y., Lomunov, A.K., Petrov, Y.V., Basalin, A.V.: Dynamic deformation and fracture toughness of pipe steel. Procedia Struct. Integrity **13**, 1811–1816 (2018). https://doi.org/10.1016/j.prostr.2018.12.355
3. Eftis, J., Liebowitz, H.: On the modified Westergaard equations for certain plane crack problems. Int. J. Fract. Mech. **8**(4), 383–392 (1972). https://doi.org/10.1007/BF00191100
4. Gokhfel'd, D.A., Cherniavsky, O.F., Il'in, A.V.: Sposob poluchenija gofrov na trubnyh zagotovkah (The method of obtaining corrugations on tube blanks). USSR Inventor's Certificate N 1506719 (1989). (in Russian)
5. He, M.-Y., Hutchinson, J.W.: Crack deflection at an interface between dissimilar elastic materials. Int. J. Solids Struct. **25**(9), 1053–1067 (1989)
6. Kachanov, M., Mishakin, V.V.: On crack density, crack porosity, and the possibility to interrelate them. Int. J. Eng. Sci. **142**, 185–189 (2019). https://doi.org/10.1016/j.ijengsci.2019.06.010
7. Lapin, R.L., Kuzkin, V.A., Kachanov, M.: On the anisotropy of cracked solids. Int. J. Eng. Sci. **124**, 16–23 (2018). https://doi.org/10.1016/j.ijengsci.2017.11.023
8. Maksimov, A.B.: Sposob uprochnenija listovogo prokata (Method of reinforcing of steel rolled products). Patent of Ukraine № 63929.2011 (2011). (in Rassian)
9. Maksimov, A.B.: The destruction of thermally reinforced steel. Izvestiya Vysshyh Uchebnyh Zavedenii. Series: Chernaya Metallurgiya **5**, 12–14 (2014). (in Russian)
10. Maksimov, A., Erokhina, I.: Thermal reinforcement of rolled sheet. Steel Translation **47**(8), 571–575 (2017). https://doi.org/10.3103/S0967091217080101
11. Nosikov, A.I., Semenov, A.S., Melnikov, B.E., Rayimberdiyev, T.P.: Prediction of short fatigue crack propagation on the base of non-local fracture criterion. Mater. Phys. Mech. **31** (1–2), 44–47 (2017)
12. Pronina, Y., Sedova, O., Grekov, M., Sergeeva, T.: On corrosion of a thin-walled spherical vessel under pressure. Int. J. Eng. Sci. **130**, 115–128 (2018). https://doi.org/10.1016/j.ijengsci.2018.05.004

Rock Fracture During Oil Well Perforation

Evgenii Riabokon[1(✉)], Mikhail Turbakov[1], Evgenii Kozhevnikov[1], Vladimir Poplygin[1], and Marian Wiercigroch[1,2]

[1] Department of Oil and Gas Technologies, Perm National Research Polytechnic University, Perm, Russian Federation
riabokon.evgenii@gmail.com
[2] Centre for Applied Dynamics Research, University of Aberdeen, Aberdeen, UK

Abstract. Rock stress fields in the near wellbore of an oil well during perforation are modelled and studied numerically. The main parameters that influence the stress field of rocks during perforation are identified. The dynamic stress field formed in the near wellbore zone causes degradation of reservoir properties and annular zones of rock fracture formed around a production wellbore. The model of porous and permeable reservoir was built considering mechanical parameters such as Poisson's ratio, Young's modulus, coupling along the crack system, uniaxial compression strength and compressibility of cracks. The dynamics of rock fracture and stress state change were simulated using Finite Element Method. It is shown in the simulation that the type of perforation has a strong influence on the mechanism of rock fracture and distance of fracturing zone from the well. It is investigated in experiments that pressure drawdown between reservoir and bottomhole zones affects the formation stress state and as a result distribution of permeability and cracks opening.

Keywords: Rock · Oil well · Reservoir · Jet · Perforation · Modelling · Fracture

1 Introduction

The stress state of reservoir rocks caused by the rock weight and the pressure of saturating fluids is heterogeneous in the natural conditions. The stress in rock formations is generated mainly by the geostatic pressure, which is balanced by rock matrices and pore pressures. Well drilling changes this stress field while rock formations are replaced with fluid when a pressure field is created in a well. The change in pressure in the pores of the rock surrounding the well can be explained by a difference between reservoir and bottomhole pressures and softening of the rocks when wetted with drilling fluids, which leads to redistribution of fluid pressure in pores of a reservoir and to a change in the stress state of the entire matrix. Rock fracture in a near wellbore zone causes softening of reservoirs and a change in reservoir properties. At the same time, annular zones of rock fracture around the wellbore can appear. To assess resistance of rocks to fracture, a relatively simple analytical method can be used, which is based on

D. A. Indeitsev and A. M. Krivtsov (Eds.): APM 2019, LNME, pp. 185–192, 2020.
https://doi.org/10.1007/978-3-030-49882-5_18

elastic models of rock media and various criteria for fracture of rocks. Description of fracture of the porous media of such plastic rocks as clay, shale, weak unconsolidated sandstone by an elastic model is not enough due to a significant plastic fracturing during well drilling [1–4]. Therefore, when assessing the stability of the borehole walls and slot channels in clays and weak layered rocks, an elastoplastic transversely isotropic medium with a Coulomb-Mohr fracture criterion and considered shifts in stratification layers are the most suitable. Modelling of a reservoir rock fracture can show the dynamics in a stress state using various perforation techniques. During the perforation of a naturally fractured formation rock fracture zones occur along fracture systems. The nature of the fracture zones is associated with the redistribution of stresses due to the formation of perforation slots, which is accompanied by a creation of large tangential stresses and the removal of normal stresses from the slot surface. High pressure drawdown (PD) created after the perforation jobs should act for a short time, which is necessary to create fractured zones, after which the well should operate at low PD. Otherwise, high pressure drop will lead to a significant decrease in the permeability of the reservoir due to the closure of cracks. Created fracture zones connecting the ends of a slot is an important geomechanical factor that contributes to an increase in reservoir productivity. In case there are no fracture zones created slots have a positive effect on the well productivity due to the creation of new channels in the bottomhole formation zone. The radius of the zone of compressed cracks and the radius of the zone of a sharp decrease in permeability are determined by the PD developed inside the formation and the compressibility characteristics of cracks.

2 Modelling of Changes in the Stress State of an Oil Field

The solution for viscoplastic displacement rates described by Eq. (1) is computed by the Finite Element Method (FEM),

$$\left\{\delta_T^{VP}\right\} = \begin{cases} 0\, at\, F_s \leq 0 \\ \frac{1}{\mu} F_s \left\{\frac{\partial Q_s}{\partial \sigma_T}\right\} at\, F_s > 0 \end{cases}, \quad \{\sigma_T\} = \begin{Bmatrix} \sigma_n \\ \tau_r \end{Bmatrix}, \quad \left\{\delta_T^{VP}\right\} = \begin{Bmatrix} \delta_n^{VP} \\ \delta_s^{VP} \end{Bmatrix} \tag{1}$$

where $\left\{\delta_T^{VP}\right\}$ is a displacement vector on the fracture surface, *VP* is viscoplasticity index, μ is viscosity of the material filling the interface, F_s is fracture threshold in the layer plane, Q_s is plastic potential corresponding to F_s in the associated law of plastic flow and $\{\sigma_T\}$ is stress vector on fracture surface [5–7].

Separately, the magnitude of the opening of the fracture zone is calculated depending on the initial stress state and the reservoir pressure determined by the PD. As a result of the solution, the stress and deformation tensor is obtained at an arbitrary point of the rock mass, taking into account its deformation due to an arbitrary number of cracks, as well as the fractured reservoir zone. This determines the permeability tensor of the fractured reservoir, which is the ultimate goal of the stress state calculations of the fractured reservoir. Note that the permeability in the direction of the crack plane can be determined by width of cracks, average distance between them, and, in the case of a vortex flow, the value of the relative roughness or the ideal crack width. The radii of the tubing, production casing are set in accordance with those used at the field. Rock fracture is specified as a system of horizontal cracks or two orthogonal fracture systems. The following Table 1 gives parameters input in the model.

Table 1. Input parameters of experiment.

Parameter	Symbol	Units	Value
Reservoir depth	H	m	2,080
Volume weight of rocks	γ	MN/m^3	0.022
Poisson's ratio	ν	–	0.25
Young's modulus	E	MPa	6,000
Coupling along a crack system	C	MPa	1.0
Uniaxial compression strength	$\sigma_{compress}$	MPa	29.6
Initial reservoir pressure	$P_{res.i}$	MPa	21.3
Azimuth strike of a crack system	α_τ	deg.	0
Crack angle of incidence	θ_τ	deg.	90
Angle of internal friction of a rock on a system of cracks	φ_τ	deg.	23
Residual angle of internal friction of a rock along a crack system	φ_w	deg.	25
Initial crack opening	b_0	μm	34
Compressibility of cracks	β_τ	MPa^{-1}	0.028
Angle of dilatancy	i_0	deg.	20
Length of a perforation slot	L_{slot}	m	0.5
Width of a perforation slot	W_{slot}	m	0.04

After the model (cluster of a reservoir) was built, at the stage of simulation different PDs were applied to the model for a certain period of time. During the experiment four options for calculating the stress state of an oil field were investigated. The main changing parameters in each simulation were PD at the bottom of the well, coupling of a media, Young's modulus of the array and type of perforation. In each scenario distant (reservoir) pressure in the right-hand upper corner of the reservoir cluster (one of the quarters with a production well in the center between them) (Fig. 1*a–d*) was greater than near (well bottomhole) pressure in the left-hand bottom corner. The dynamics of a rock fracture and reservoir stress state change was monitored. The downhole pressure of the well taking into account the PD on the formation and lateral rock pressure on the well circular contour were taken as boundary conditions. The stress states of the fractured fluid-conducting reservoir were computed using the ANSYS software package taking into account the distribution of reservoir pressure over the radius using a multiprocessor system for high-performance computing.

It was found that occurrence of fracture zones in rocks is likely to be as shown in Fig. 1*a*. During gun perforation, fracture zones are formed that are comparable in size to the well radius. With a PD of 3 MPa, small zones of fracture appear. The larger PDs, the larger the fracture zones as can been from Figs. 1*b*, *c* and *d*.

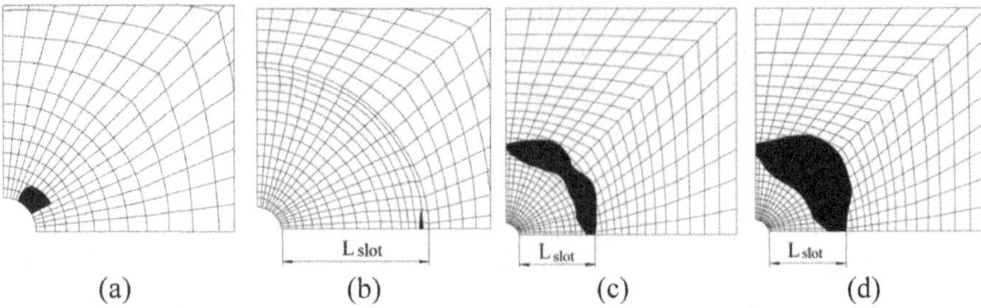

Fig. 1. Fracture zones around the production well; (a) gun perforation (PD of 5 MPa); (b, c, d) slot perforation (PD of 5, 7, 10 MPa); ■ – fracture zone; L_{slot} – slot length in the reservoir (m)

During the slot perforation small fracture zones occur at the ends of cracks with the PD of 5 MPa. With the PD of 6–7 MPa the fracture zones are connected (Fig. 1*b*, *c*, *d*). With further increase of the PD, the zones of fracture (in meters) become larger and reach maximum sizes at 10 MPa [8].

The nature of the fracture zones is associated with the redistribution of stresses due to the formation of perforation slots, which is accompanied by a creation of large tangential stresses and the removal of normal stresses from the slot surface. The formation of a fracture zone connecting the ends of the gap is an important geomechanical factor that contributes to an increase in reservoir productivity. In the absence of fractured zones, the created gaps have a positive effect on the well productivity due to the creation of new channels in the near wellbore zone. It is noted in [8] that the formation of annular zones of fractured rocks during slotted perforation is characteristic of a pore-type reservoir.

In the following experiment the procedure was the same. The relationship between distributions of permeability, crack opening and PD were investigated and corresponding functions were plotted.

The image for different values of PD is one of the outcomes of the calculation of the near wellbore zone. Crack permeability k_{perm} depends on fracture opening b_f, distance between cracks s_f, measured in m^2 and is defined as

$$k_{perm} = \frac{b_f^3}{12 s_f}. \tag{2}$$

A large PD decreases reservoir permeability due to closure of cracks as can be seen in Fig. 2. Consequently, a high PD should act for a short time, which is necessary to create the fractured zones, after which the well should work with a slight PD.

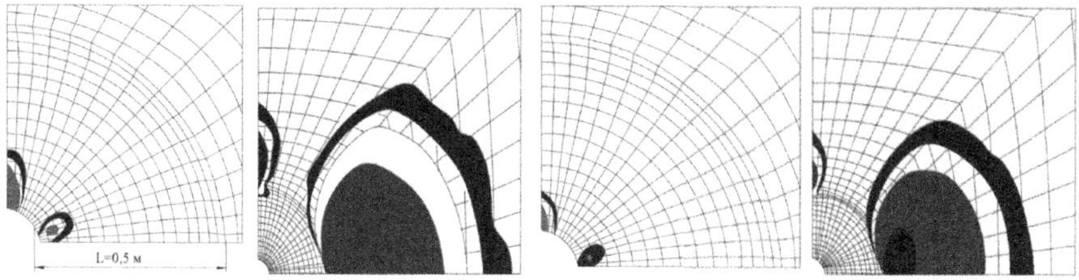

(a) (b) (c) (d)

Fig. 2. Distribution of permeability (a, b) and crack opening (c, d) in the zone of slot perforation with a PD of 3 MPa and 5 MPa; (a, b): ■ −0,02 μm^2; ▩ −0,06 μm^2; □−0,12 μm^2; (c, d): ■−2,9 μm; ▩−19,3 μm; □−35,7 μm

The formation of annular fracture zones depends on the PD and on the value of adhesion along the cracks grids. It means that a weak rock requires a small PD for annular fracture zones to be formed. Annular fracture zones can be formed using perforation and maintaining the necessary PD contributing to a steady increase in well production.

It is worth noting however that simple analytical methods based on elastic rock models and various criteria for rock fracture can be more difficult to assess the resistance of rocks to fracture for such rocks as plastic clay, shale, weak unconsolidated sandstone. For these rocks, the elastic model of the media is no longer accurate enough due to the development of significant plastic deformations during the drilling of wells. When assessing the stability of the borehole walls and slot channels in clays and weak layered rocks, an elastoplastic transversely isotropic medium with a Coulomb-Mohr fracture criterion and taking into account the possibility of shifts in stratification layers is the most suitable and can be implemented in the software package ISAMGEO. Plastic deformations can lead to overlap of the cross-section of perforation channels after perforation and reduce its effectiveness to 0.

Perforation has a strong influence on the distribution of crack opening and change in permeability. If the PD is large ($p_{res} << p_{lat}$), there are extensive zones of compression of cracks formed at the ends of the gap, and, accordingly, low permeability are formed. If the PD is small ($p_{res} > p_{lat}$) such zones do not arise, there are small zones of compression of cracks adjacent to the well. However, the transition from the unloaded media to the compressed one takes place very quickly and it is determined by the initial reservoir pressure and crack parameters. Crack compression zones and low permeability zones in the case of conventional perforations with $p_{res} > p_{lat}$ do not appear.

The radius of the zone of compressed cracks and the radius of the zone of a sharp decrease in permeability are determined by the PD created and the compressibility characteristics of the cracks. With the parameters adopted for calculations for a PD of 5 MPa, the area of compressed cracks is up to 1.8 m. With an increased PD, the cracks close at a distance of 30 m from the well seen in Fig. 3*a*.

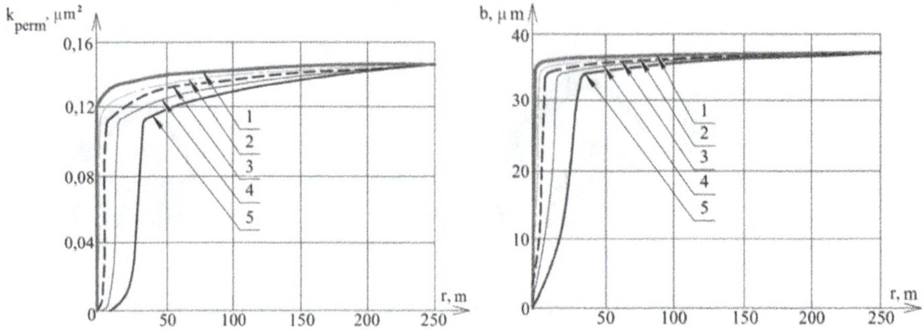

(a) (b)

Fig. 3. Change in permeability (a) and crack opening (b) with distance from the well; 1, 2, 3, 4, 5 – PD of 3 MPa, 5 MPa, 7 MPa, 9 MPa, 12 MPa

3 Jet Perforation

In a jet perforation a liquid that carries an abrasive material (e.g. sand) outflow from a nozzle which has a cone inner geometry and ensures pressure rise at the end of a nozzle that can be equal to 30 MPa (see Fig. 4). Thanks to the abrasive influence of the liquid-carrier the slots are cut in a casing, cement stone and the rock consequentially creating a channel with height of up to 0.25 m, depth up to 0.4 m and width 0.04 m.

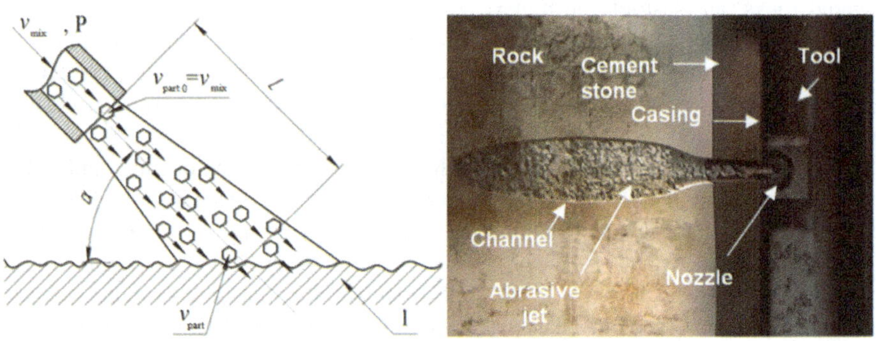

(a) (b)

Fig. 4. (a) Schematic of a jet perforation process, (b) cross-section of jet perforated well; *1* – impact surface; v_{part} – particle velocity; v_{mix} – mixture velocity; p – pressure; α – impact angle, l – jet length

As a result, the hydrocarbon flow area increases from 17.8 to 107.2% of the area of the open borehole, which increases the quality of the perforation of the reservoir (the skin effect is improved). The increase in the size of the formed slotted channels in height by the amount of stretching of the pipe string (with a simultaneous increase in their depth) is ensured by the natural elongation of the pipe under the influence of the abrasive fluid supplied through it exclusively by drawing the pipe without using any additional engine or ground mechanism, which reduces material and labor costs. After a pipe is divided by a slotted channel, its strength decreases, due to which the average pressure at the contact of the cement ring and column decreases greatly with increasing rigidity of the cement ring and rocks in the near wellbore zone [9–12].

To estimate the change in the elastic properties of the "environment-cement stone" system, Winkler's hypothesis is used, in which the strength factor is calculated as

$$K = \frac{E}{(1 - v)R}, \tag{3}$$

where E is the Young's modulus, v is the Poisson's ratio and is R is the radius. In the work it is assumed that the elastic modulus and the Poisson's ratio of cement stone and surrounding rocks are the same. If $E > 1.3 \cdot 10^5$ N/cm^2 (strong cement stone), the stress in the pipe after the perforations does not exceed 3% of the initial, the gap length of the channel does not affect the strength of the production tubing. In cases where the reservoir strength is substantially lower, it is recommended not to create long slotted channels. Longitudinal and transverse geometrical dimensions of the slotted channels will depend on each other. Relative geometric dimensions $\frac{l}{R}$ (relative length) of slot channel should not exceed 0.3 for thin-wall pipes and 0.55 for thick wall pipes.

4 Conclusions

In this work the rock stress fields in the near wellbore of an oil well during perforation have been modelled and computed numerically using the FEM. The following main conclusions can be drawn from study.

The change in the stress state of the rock during drilling is influenced by replacement of rock formations with a circulating mud, fluctuations in a circulating fluid temperature and osmosis. The largest changes in the stress state of the rock are observed at the boundary "well-reservoir". The change in the stress state of rocks in the near wellbore zone worsens reservoir properties of a formation and reduces an oil rate of a well.

Numerical modelling using FEM showed that in the process of slot perforation small rock fracture zones are formed and high permeability channels that create a system "well-reservoir", are formed. The nature of fracture zones associated with redistribution of stresses due to perforation slots formed. Perforation reduces normal and increases the tangential stresses at the surface of a crack. In such case, a high PD should act for a short time. The formation of a fracture zone connecting ends of a gap is an important geomechanical factor that contributes to an increase in reservoir productivity.

The elastic model of the rock formation is insufficient in the analytical assessment of the stability of plastic rocks (clay, shale, weak unconsolidated sandstone) due to development of large plastic deformations. Numerical modelling of the stress state of the plastic rock (clay) using an elastoplastic transversely isotropic medium with the

Coulomb-Mohr fracture criterion taking into account the possibility of shifts in the stratification layers showed an enlargement of the zone of plastic fracture of a rock in the near wellbore zone over the time and overlapping of slot perforation channels.

Acknowledgements. The investigation was carried out at the expense of the grant Agreement No. 19-19-00408 of the Russian Science Foundation (project No. 19-19-00408).

References

1. Zoback, M.D.: Reservoir Geomechanics. Cambridge University Press, Cambridge (2007)
2. Peirce, A.P., Bunger, A.P.: Interference fracturing: nonuniform distributions of perforation clusters that promote simultaneous growth of multiple hydraulic fractures. SPE J. **20**(5), 384–395 (2015)
3. Furui, K., Zhu, D., Hill, A.D., Davis, E.R., Buck, B.R.: Optimization of horizontal well-completion design with cased/perforated or slotted liner completions. SPE Prod. Oper. **22**(2), 248–253 (2007)
4. Guo, T., Qu, Z., Gong, D., Lei, X., Liu, M.: Numerical simulation of directional propagation of hydraulic fracture guided by vertical multi-radial boreholes. J. Nat. Gas Sci. Eng. **35**, 175–188 (2016)
5. Guo, T., Liu, B., Qu, Z., Gong, D., Xin, L.: Study on initiation mechanisms of hydraulic fracture guided by vertical multi-radial boreholes. Rock Mech. Rock Eng. **50**(7), 1767–1785 (2017)
6. Wang, Y., Ju, Y., Chen, J., Song, J.: Adaptive finite element-discrete element analysis for the multistage supercritical CO_2 fracturing and microseismic modelling of horizontal wells in tight reservoirs considering pre-existing fractures and thermal-hydro-mechanical coupling. J. Nat. Gas Sci. Eng. **61**, 251–269 (2019)
7. Kumar, D., Ghassemi, A.: A three-dimensional analysis of simultaneous and sequential fracturing of horizontal wells. J. Pet. Sci. Eng. **146**, 1006–1025 (2016)
8. Kashnikov, Y.A., Ashihmin, S.G., Kuchtinski, A.E., Vildeman, V.E., Tretyakov, M.P., Bazhukov, P.S.: Determination of fracture toughness of oil fields rocks. Oil Ind. **10**, 86–89 (2015)
9. Cai, C., Kang, Y., Wang, H.Y., Chen, H., Yuan, X., Cai, Y.: Mechanism of supercritical carbon dioxide (SC-CO_2) hydro-jet fracturing. J. CO2 Utilization **26**, 575–587 (2018)
10. Asadi, M., Shirazi, M.K., Ghalambor, A., Pratt, D.W.: Experimental and simulation analysis of jet-perforated rock damage. SPE Drilling Completion **16**(3), 176–181 (2001)
11. Halleck, P.M., Poyol, E., Santarelli, F.J.: Estimating perforation flow performance from variations in indentation hardness. SPE Drilling Completion **10**(4), 271–275 (1995)
12. Asadi, M., Ghalambor, A., Shirazi, M.K.: Assessment of jet perforated rock damage by finite element method. In: SPE - European Formation Damage Control Conference, Proceedings, pp. 25–32 (1997)

Cavity Flow of Nematic Liquid Crystals – A Parameter Study

Wilhelm Rickert(✉) and Wolfgang H. Müller

Institute of Mechanics, Chair of Continuum Mechanics and Constitutive Theory, Technische Universität Berlin, Einsteinufer 5, 10587 Berlin, Germany
rickert@tu-berlin.de, whmueller1000@gmail.com

Abstract. This paper presents a parameter study of the flow of nematic liquid crystals which possess both viscous and elastic properties. The well-known ERICKSEN–LESLIE theory is used. The underlying general equations are stated and subsequently simplified for non-isothermal and steady state conditions. The flow situation of a two-dimensional lid-driven cavity is analyzed. Hence, the equations are specialized for the case of two-dimensional flow. For numerical calculations, the complete boundary value problem is formulated and then expressed in dimensionless form. Several dimensionless parameters are identified and their impact on the solution is analyzed. Furthermore, the temperature rise due to viscous dissipation is studied, which is frequently ignored in the mechanics community. The finite element method is employed using the software package FEniCS. In particular, the numerical treatment of the constraints required in the theory, is analyzed. A convergence analysis is performed based on the constraints for rigid and incompressible nematic liquid crystals.

Keywords: Nematic liquid crystals · Microstructured material · Lid-driven cavity problem · Forced convection

1 Introduction

This short note presents a numerical analysis of the flow of fluids with internal rotational degrees of freedom, specifically nematic liquid crystals. These are commonly described in terms of the ERICKSEN–LESLIE theory on which a detailed overview is given in [16]. In this theory an additional degree of freedom of a continuum point is introduced by the vector $\boldsymbol{d}$, the so-called director, which captures the motion of the rigid rod-like molecules. Here, a two-dimensional cavity flow is analyzed, which is a classical benchmark problem. Usually in the literature two-dimensional problems of liquid crystal flows are analyzed, *e.g.*, [2,10,11], and in particular for cavity flow [1,14,18]. Most of these works focus on error estimates or on the evolution of defects rather than specific solutions. For inextensible nematics, which are customarily considered, the director field is of unit length, *i.e.*, $\|\boldsymbol{d}\| = 1$, which is called unit sphere constraint. For numerical calculations,

D. A. Indeitsev and A. M. Krivtsov (Eds.): APM 2019, LNME, pp. 193–205, 2020.
https://doi.org/10.1007/978-3-030-49882-5_19

this constraint is often enforced with a penalty parameter, see [1] and references therein. However, this approach relies heavily on the appropriate choice of the penalty factor. Therefore, we will make use of the LAGRANGE multiplier metod instead as it was introduced in the original works of ERICKSEN and LESLIE, which is often neglected, see [3,9]. Furthermore, non-isothermal conditions are considered. A general framework for the analysis of the temperature rise due to viscous dissipation is given in [5] or [8], but the numerical analysis of specific boundary value problems involving temperature are rarely found in literature, *e.g.*, [15].

2 Governing Equations

Mathematically speaking, our objective is to determine the fields of (a) linear velocity, $\boldsymbol{v}(\boldsymbol{x},t)$, (b) the director field, $\boldsymbol{d}(\boldsymbol{x},t)$, and (c) the temperature field, $T(\boldsymbol{x},t)$, in all points, $\boldsymbol{x}$, and at all times, t, within a region of space, $\mathcal{B}$, through which incompressible nematic crystal matter of constant mass density, ρ_0, and constant mass density of the director, ρ_0^{d}, (in units of $^{\mathrm{kg}}/_{\mathrm{m}}$) is flowing.

The determination of these fields relies on field equations for the primary fields. The field equations are based on balance laws and need to be complemented by suitable constitutive relations later. After introducing the substantial time derivative ($^{\mathrm{d}}/_{\mathrm{d}t}$), the balances of mass, linear momentum, director force and internal energy read as follows (see [9, pg. 268 and 276] or [12, pp. 354]):

$$
\begin{aligned}
0 = \nabla\cdot\boldsymbol{v}, \quad \rho_0\frac{\mathrm{d}\boldsymbol{v}}{\mathrm{d}t} = \nabla\cdot\boldsymbol{\sigma}, \quad \rho_0^{\mathrm{d}}\frac{\mathrm{d}^2\boldsymbol{d}}{\mathrm{d}t^2} = \nabla\cdot\boldsymbol{\pi}+\boldsymbol{g}, \\
\rho_0\frac{\mathrm{d}u}{\mathrm{d}t} = -\nabla\cdot\boldsymbol{q} + \boldsymbol{\sigma}\cdot\cdot(\nabla\otimes\boldsymbol{v}) + \boldsymbol{\pi}\cdot\cdot\left(\nabla\otimes\frac{\mathrm{d}\boldsymbol{d}}{\mathrm{d}t}\right) - \boldsymbol{g}\cdot\frac{\mathrm{d}\boldsymbol{d}}{\mathrm{d}t},
\end{aligned}
\tag{1}
$$

where body force, body couple, and volumetric heat supply have been neglected. The non-symmetric, second order CAUCHY stress tensor is denoted by $\boldsymbol{\sigma}$ and $\boldsymbol{\pi}$ is the so-called director stress tensor (in units of $\mathrm{kg/s^2}$, second order). The director density $\rho_0^{\mathrm{d}} = d_0^2\rho_0$ in units of kg/m includes an internal length parameter, d_0, which is characteristic to the length of the rod-shaped molecules. The symbol "$\cdot\cdot$" denotes a double contraction, $\boldsymbol{C}\cdot\cdot\boldsymbol{D} = C_{ij}D_{ij}$ in Cartesian coordinates. The vector $\boldsymbol{g}$ is known as the "director production density" or "intrinsic director body force." It is an additional constitutive quantity beyond the requirements of the ordinary BOLTZMANN continuum with the unit $\mathrm{N/m^2}$ and subject to the constraint

$$
\left[\boldsymbol{\sigma} + (\nabla\otimes\boldsymbol{d})^{\mathrm{T}}\cdot\boldsymbol{\pi} - \boldsymbol{d}\otimes\boldsymbol{g}\right]_\times = \boldsymbol{0},
\tag{2}
$$

where the GIBBSian cross is defined by $(\boldsymbol{a}\otimes\boldsymbol{b})_\times = \boldsymbol{a}\times\boldsymbol{b}$. Finally, u is the specific internal energy, and $\boldsymbol{q}$ is the heat flux. Note that there is a connection between Eq. (1) and the general balance of spin, *e.g.*, [4, pg. 12], or [19, pp. 250]. This has been analyzed in [12, Sec. 10.1].

Before we can turn to the flow problem, constitutive equations for the (non-symmetric) stress tensor, $\boldsymbol{\sigma}$, the director stress tensor, $\boldsymbol{\pi}$, the director force vector, $\boldsymbol{g}$, and for the heat flux, $\boldsymbol{q}$, are required. It is assumed that the non-equilibrium parts of the constitutive functions depend linearly on the objective quantities $\boldsymbol{m} = \dot{\boldsymbol{d}} - \boldsymbol{W}\cdot\boldsymbol{d}$, $\boldsymbol{D}$, and ∇T, where $\boldsymbol{D}$ and $\boldsymbol{W}$ are the symmetric and skew symmetric part of the velocity gradient $\boldsymbol{L} = \boldsymbol{v}\otimes\nabla$, respectively. Then the usual techniques of rational thermodynamics can be used to derive the following explicit but lengthy non-linear constitutive relations (see [9] or [12] for details):

$$
\begin{aligned}
\boldsymbol{\sigma} &= -p\boldsymbol{1} + \boldsymbol{\sigma}^{\text{el.}} + \boldsymbol{\sigma}^{\text{visc.}}, \quad \boldsymbol{\sigma}^{\text{el.}} = -\frac{\partial(\rho_0 f)}{\partial \boldsymbol{R}^{\text{T}}}\cdot\boldsymbol{R},\\
\boldsymbol{\pi} &= \boldsymbol{\beta}\otimes\boldsymbol{d} + \frac{\partial(\rho_0 f)}{\partial \boldsymbol{R}^{\text{T}}}, \quad \boldsymbol{g} = \gamma\boldsymbol{d} - \boldsymbol{R}\cdot\boldsymbol{\beta} - \frac{\partial(\rho_0 f)}{\partial \boldsymbol{d}} + \boldsymbol{g}',\\
2\rho_0 f &= k_0(T) + k_{22}\boldsymbol{R}\cdot\cdot\,\boldsymbol{R} + (k_{11} - k_{22} - k_{24})(\nabla\cdot\boldsymbol{d})^2\\
&\qquad + (k_{33} - k_{22})(\boldsymbol{R}^{\text{T}}\cdot\boldsymbol{R})\cdot\cdot\,(\boldsymbol{d}\otimes\boldsymbol{d}) + k_{24}\boldsymbol{R}\cdot\cdot\,\boldsymbol{R}^{\text{T}},\\
\boldsymbol{\sigma}^{\text{visc.}} &= \mu_1(\boldsymbol{d}\cdot\boldsymbol{D}\cdot\boldsymbol{d})\boldsymbol{d}\otimes\boldsymbol{d} + \mu_2\boldsymbol{d}\otimes\boldsymbol{m} + \mu_3\boldsymbol{m}\otimes\boldsymbol{d}\\
&\qquad + \mu_4\boldsymbol{D} + \mu_5\boldsymbol{d}\otimes\boldsymbol{d}\cdot\boldsymbol{D} + \mu_6\boldsymbol{d}\cdot\boldsymbol{D}\otimes\boldsymbol{d},\\
\boldsymbol{g}' &= \lambda_1\boldsymbol{m} + \lambda_2\boldsymbol{d}\cdot\boldsymbol{D}, \quad \lambda_1 = \mu_2 - \mu_3, \quad \lambda_2 = \mu_5 - \mu_6,\\
\boldsymbol{q} &= -\left[\kappa_{\|}\boldsymbol{d}\otimes\boldsymbol{d} + \kappa_{\perp}(\boldsymbol{1} - \boldsymbol{d}\otimes\boldsymbol{d})\right]\cdot(\nabla T),\\
u &= f - T\frac{\partial f}{\partial T} = -T^2\frac{\partial}{\partial T}\left(\frac{f}{T}\right) = f - \frac{T}{2\rho_0}\frac{\partial k_0}{\partial T} = \hat{u}(T,\boldsymbol{d},\boldsymbol{R}),
\end{aligned}
\tag{3}
$$

where $\boldsymbol{R} = (\nabla\otimes\boldsymbol{d})^{\text{T}}$ is the director gradient and the abbreviations $\nabla_{\boldsymbol{d}} = \frac{\partial(\cdot)}{\partial d_i}\boldsymbol{e}_i$ as well as $\nabla_{\boldsymbol{R}^{\text{T}}} = \frac{\partial(\cdot)}{\partial d_{i,j}}\boldsymbol{e}_j\otimes\boldsymbol{e}_i$ are used to represent derivatives with respect to the director field and the director gradient, respectively. The symbols $\boldsymbol{\sigma}'$ and $\boldsymbol{g}'$ refer to dissipative terms of the stress tensor and the director force vector, respectively. Note that there is no dissipative term present in the director stress tensor, $\boldsymbol{\pi}$. The derivatives of the free energy are given by

$$
\begin{aligned}
\frac{\partial\rho_0 f}{\partial\boldsymbol{d}} &= (k_{33} - k_{22})\boldsymbol{d}\cdot\boldsymbol{R}^{\text{T}}\cdot\boldsymbol{R},\\
\frac{\partial\rho_0 f}{\partial\boldsymbol{R}^{\text{T}}} &= k_{22}\boldsymbol{R}^{\text{T}} + (k_{11} - k_{22} - k_{24})(\nabla\cdot\boldsymbol{d})\boldsymbol{1} + (k_{33} - k_{22})\boldsymbol{d}\otimes\boldsymbol{R}\cdot\boldsymbol{d} + k_{24}\boldsymbol{R}.
\end{aligned}
\tag{4}
$$

The ERICKSEN–LESLIE theory contains six different shear viscosities, μ_i, and several elastic stiffnesses, k_{ij}, which were originally introduced in [17]. This shows very clearly that the liquid crystals are at the borderline between fluids and solids. The stiffnesses occur in the elastic part of the free energy density, $\rho_0 f$, and they are related to deformation energy during splaying, bending, and twisting of a gradually changing director field, similarly as a gradient of displacement gives rise to the well known elastic energy in linear elastic HOOKEan solids (see [16, pp. 16]). The additional coefficients λ_1 and λ_2 are expressed in terms of the viscosities μ_i in order to satisfy the constraint in Eq. (2) identically, see [9]. The quantities $\boldsymbol{\beta}$ and γ are arbitrary parameters, which are used to ensure additional constraints, *e.g.*, the inextensibility of the director. The arbitrary parameters γ and $\boldsymbol{\beta}$ play a similar role as the pressure in an incompressible fluid, see also the discussion in [12, pp. 366]. Besides many constraints on the elastic and viscous parameters shown in, *e.g.*, [9,16], a famous constraint arises from the ONSAGER relations and was found by PARODI, [13],

$$\mu_2 + \mu_3 = \mu_6 - \mu_5. \tag{5}$$

3 Problem Statement and Simplified Field Equations

The two-dimensional problem of a lid driven cavity in Fig. 1 is analyzed. For the velocity, no-slip conditions are employed such that the top plate velocity $\boldsymbol{v}_0$ is directly transferred to the liquid crystal and the velocity at the other rigid walls vanishes. For the director, homeotropic strong anchoring is employed at the horizontal walls, *i.e.*, the director is perpendicular to the top and bottom plate. In order to employ consistent boundary conditions, it is assumed that the director points in vertical direction at the left and right wall such that it is vertical at all surfaces, *i.e.*, $\boldsymbol{d} = \boldsymbol{e}_y$. For simplicity, DIRICHLET boundary conditions are used for the temperature and a constant reservoir temperature, T_{res}, is imposed. Hence, the following boundary values are employed for the cavity:

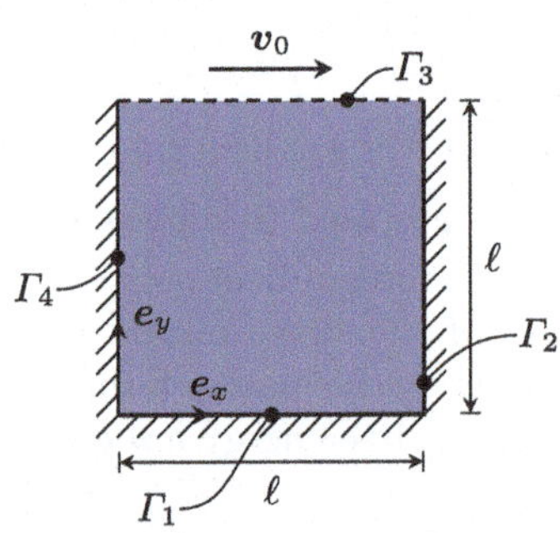

Fig. 1. Setting of a quadratic cavity.

$$\begin{cases} T(\boldsymbol{x}) = T_{\text{res}}, & \boldsymbol{d}(\boldsymbol{x}) = \boldsymbol{e}_y, & \boldsymbol{v}(\boldsymbol{x}) = v_{\text{ref}}\boldsymbol{e}_x, & \boldsymbol{x} \in \Gamma_3, \\ T(\boldsymbol{x}) = T_{\text{res}}, & \boldsymbol{d}(\boldsymbol{x}) = \boldsymbol{e}_y, & \boldsymbol{v}(\boldsymbol{x}) = \boldsymbol{0}, & \boldsymbol{x} \in (\Gamma_1 \cup \Gamma_2 \cup \Gamma_4), \end{cases} \tag{6}$$

From Eq. (3) it can be seen that $u = \hat{u}(T, \boldsymbol{d}, \boldsymbol{R})$ and the substantial time derivative of the internal energy reads

$$\dot{u} = \frac{\partial u}{\partial T}\dot{T} + \frac{\partial u}{\partial \boldsymbol{d}} \cdot \dot{\boldsymbol{d}} + \frac{\partial u}{\partial \boldsymbol{R}} \cdot\cdot\, \dot{\boldsymbol{R}} = c_v \dot{T} + \frac{\partial f}{\partial \boldsymbol{d}} \cdot \dot{\boldsymbol{d}} + \frac{\partial f}{\partial \boldsymbol{R}} \cdot\cdot\, \dot{\boldsymbol{R}}.$$

Therefore, the following set of equations is obtained,

$$
\begin{aligned}
\rho_0 \boldsymbol{L}\cdot\boldsymbol{v} = \nabla\cdot\boldsymbol{\sigma}, \quad (\boldsymbol{d}\cdot\boldsymbol{d}-1)^2 = 0, \quad 0 = \nabla\cdot\boldsymbol{v}, \\
\mathbf{0} = \nabla\cdot(\boldsymbol{\pi} - \rho_0^{\mathrm{d}}\boldsymbol{v}\otimes\boldsymbol{R}\cdot\boldsymbol{v}) + \boldsymbol{g}, \\
\rho_0 c_v (T\nabla)\cdot\boldsymbol{v} = -\nabla\cdot\boldsymbol{q} + \boldsymbol{\sigma}\cdot\cdot\,\boldsymbol{L}^{\mathrm{T}} + \boldsymbol{\pi}\cdot\cdot\,(\nabla\otimes(\boldsymbol{R}\cdot\boldsymbol{v})) \\
- \left(\left[\rho_0\frac{\partial f}{\partial \boldsymbol{d}} + \boldsymbol{g}\right]\cdot\boldsymbol{R} + \rho_0\frac{\partial f}{\partial \boldsymbol{R}}\cdot\cdot\,(\boldsymbol{R}\otimes\nabla)\right)\cdot\boldsymbol{v},
\end{aligned}
\tag{7}
$$

where the incompressibility constraint was used several times. In order to obtain a simplified version of the system of equations, it is assumed that k_{24} vanishes and all other elastic constants are equal, *i.e.*, $k_{11} = k_{22} = k_{33} =: k$. Following [1] and [10], most of viscosities are put to zero: $\mu_1 = \mu_2 = \mu_5 = 0$. Then, it follows from PARODI's relation (5) that $\mu_3 = \mu_6 =: \xi$. For stationary processes, the constitutive relations simplify to read:

$$
\begin{aligned}
\boldsymbol{\sigma} = -p\mathbf{1} - k\boldsymbol{R}^{\mathrm{T}}\cdot\boldsymbol{R} + \xi\big(\boldsymbol{R}\cdot\boldsymbol{v} + \boldsymbol{L}^{\mathrm{T}}\cdot\boldsymbol{d}\big)\otimes\boldsymbol{d} + \mu_4\boldsymbol{D}, \quad \boldsymbol{\pi} = k\boldsymbol{R}^{\mathrm{T}}, \\
\boldsymbol{g} = \gamma\boldsymbol{d} - \xi\big(\boldsymbol{R}\cdot\boldsymbol{v} + \boldsymbol{L}^{\mathrm{T}}\cdot\boldsymbol{d}\big), \quad \boldsymbol{q} = -\big[\kappa_\perp\mathbf{1} + (\kappa_\parallel - \kappa_\perp)\boldsymbol{d}\otimes\boldsymbol{d}\big]\cdot(\nabla T),
\end{aligned}
\tag{8}
$$

where $\boldsymbol{\beta}$ is neglected. For the numerical analysis a normalization is introduced. The primary normalizations $\boldsymbol{x} = \ell_{\mathrm{ref}}\tilde{\boldsymbol{x}}$ and $\boldsymbol{v} = v_{\mathrm{ref}}\tilde{\boldsymbol{v}}$ are employed, where the reference velocity is the top plate velocity $v_{\mathrm{ref}} = v_0$. The following normalizations are used

$$
p = \rho_0 v_{\mathrm{ref}}^2\tilde{p}, \quad \gamma = \frac{v_{\mathrm{ref}}\xi}{\ell_{\mathrm{ref}}}\tilde{\gamma}, \quad T - T_{\mathrm{res}} = \frac{\mu v_{\mathrm{ref}}^2}{\kappa_\parallel}\tilde{T}, \quad \boldsymbol{q} = \frac{\kappa_\parallel T_{\mathrm{ref}}}{\ell_{\mathrm{ref}}}\tilde{\boldsymbol{q}}. \tag{9}
$$

Then the dimensionless system of equations reads:

$$
\begin{aligned}
\tilde{\boldsymbol{L}}\cdot\tilde{\boldsymbol{v}} + \tilde{\nabla}\tilde{p} = -\frac{1}{Er\,Re}\big([\tilde{\Delta}\boldsymbol{d}]\cdot\tilde{\boldsymbol{R}} + \tilde{\boldsymbol{R}}\cdot\cdot\,[\tilde{\boldsymbol{R}}\otimes\tilde{\nabla}]\big) + \frac{M}{Re}\tilde{\boldsymbol{R}}\cdot\big(\tilde{\boldsymbol{R}}\cdot\tilde{\boldsymbol{v}} + \tilde{\boldsymbol{L}}^{\mathrm{T}}\cdot\boldsymbol{d}\big) \\
+ \frac{M}{Re}\big([\tilde{\nabla}\cdot\tilde{\boldsymbol{R}}]\cdot\tilde{\boldsymbol{v}} + \tilde{\boldsymbol{R}}\cdot\cdot\,\tilde{\boldsymbol{L}}^{\mathrm{T}} + [\tilde{\Delta}\tilde{\boldsymbol{v}}]\cdot\boldsymbol{d}\big)\boldsymbol{d} + \frac{1}{4Re}\tilde{\Delta}\tilde{\boldsymbol{v}}, \\
\tilde{\nabla}\cdot(\tilde{\boldsymbol{v}}\otimes\tilde{\boldsymbol{R}}\cdot\tilde{\boldsymbol{v}}) = \frac{1}{Re\,Er\,L^2}\tilde{\Delta}\boldsymbol{d} + \frac{M}{Re\,L^2}\tilde{\gamma}\boldsymbol{d} - \frac{M}{Re\,L^2}\big(\tilde{\boldsymbol{R}}\cdot\tilde{\boldsymbol{v}} + \tilde{\boldsymbol{L}}^{\mathrm{T}}\cdot\boldsymbol{d}\big), \\
Pr(\tilde{T}\tilde{\nabla})\cdot\tilde{\boldsymbol{v}} = -\frac{1}{Re}\tilde{\nabla}\cdot\tilde{\boldsymbol{q}} - \frac{1}{Er\,Re}(\tilde{\boldsymbol{R}}^{\mathrm{T}}\cdot\tilde{\boldsymbol{R}})\cdot\cdot\,\tilde{\boldsymbol{L}}^{\mathrm{T}} - \frac{M}{Re}\boldsymbol{d}\cdot\tilde{\boldsymbol{L}}\cdot\tilde{\boldsymbol{R}}\cdot\tilde{\boldsymbol{v}} \\
- \frac{M}{Re}\boldsymbol{d}\cdot\tilde{\boldsymbol{L}}\cdot\tilde{\boldsymbol{L}}^{\mathrm{T}}\cdot\boldsymbol{d} + \frac{2}{Re}\tilde{\boldsymbol{D}}\cdot\cdot\,\tilde{\boldsymbol{D}} + \frac{1}{Er\,Re}\tilde{\boldsymbol{R}}^{\mathrm{T}}\cdot\cdot\,(\tilde{\nabla}\otimes(\tilde{\boldsymbol{R}}\cdot\tilde{\boldsymbol{v}})) \\
- \left(\frac{M}{Re}\big[\boldsymbol{d} - (\tilde{\boldsymbol{R}}\cdot\tilde{\boldsymbol{v}} + \tilde{\boldsymbol{L}}^{\mathrm{T}}\cdot\boldsymbol{d})\big]\cdot\tilde{\boldsymbol{R}} + \frac{1}{Er\,Re}\tilde{\boldsymbol{R}}\cdot\cdot\,(\tilde{\boldsymbol{R}}\otimes\tilde{\nabla})\right)\cdot\tilde{\boldsymbol{v}},
\end{aligned}
\tag{10}
$$

with the dimensionless numbers:

$$
M = \frac{\xi}{\mu}, \quad Re = \frac{\rho_0 v_{\mathrm{ref}}\ell_{\mathrm{ref}}}{\mu}, \quad Er = \frac{\mu v_{\mathrm{ref}}\ell_{\mathrm{ref}}}{k}, \quad L = \frac{d_0}{\ell_{\mathrm{ref}}}, \quad Pr = \frac{\mu c_v}{\kappa_\parallel}, \quad \kappa = \frac{\kappa_\perp}{\kappa_\parallel}. \tag{11}
$$

Note that the fourth LESLIE viscosity and the classical shear viscosity are connected via $\mu_4 = 2\mu$.

4 Results

For obtaining numerical solutions of the dimensionless partial differential equations the finite element method is employed using the software package FEniCS. A spatial discretization based on the stable P_2-P_1 TAYLOR–HOOD element for velocity and pressure is employed, see [17]. Analogously, the director field and the field γ are also discretized by using a P_2-P_1 element, and for the temperature a P_1 element.

In order to assess the quality of the numerical results, a comparison to an analytical solution related to the considered problem is advantageous. In [15] an analytical solution for the COUETTE flow with a specific parameter set was derived. However, this parameter set is not consistent with the simplifications of the last section. Due to the lack of an analytical solution, a convergence analysis w.r.t. mesh refinement is performed. In particular, the volume averages of the incompressibility and of the unit sphere constraint are analyzed:

$$e^{v} = \frac{\int_V \left|\tilde{\nabla} \cdot \tilde{\boldsymbol{v}}\right| \, \mathrm{d}V}{\int_V \mathrm{d}V}, \qquad e^{d} = \frac{\int_V |\boldsymbol{d} \cdot \boldsymbol{d} - 1| \, \mathrm{d}V}{\int_V \mathrm{d}V}. \tag{12}$$

These averages are shown in Fig. 2 for an increasing grid size n, where the total number of grid points is n^2. One can see that the constraints are fulfilled with sufficient accuracy.

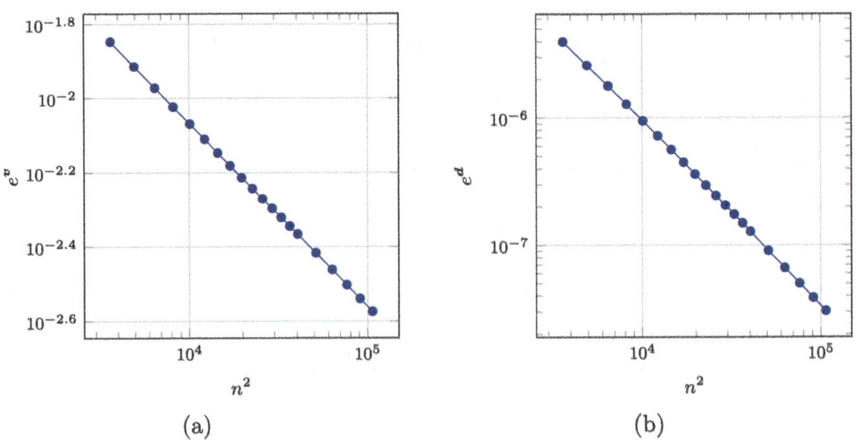

Fig. 2. Divergence of the velocity field (left) and the norm of the director field (right).

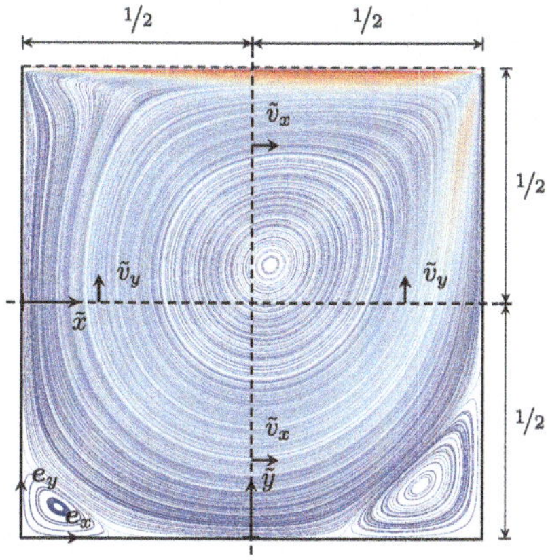

(a) Streamline plot with dashed lines for profiles.

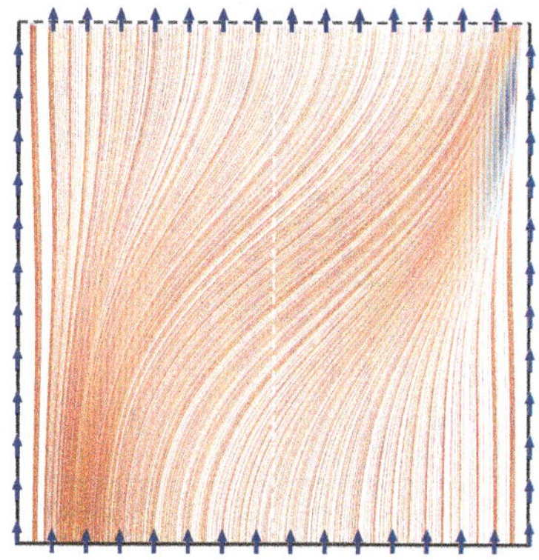

(b) Streamline plot with dashed lines for profiles.

Fig. 3. Cavity line plot scheme. In (a) and (b) stream tracer plots for the velocity and the director field are shown, respectively. Note that the blue indications of the director shown at the boundary of the cavity could also point downwards. The results are obtained for the base parameter set. Colors indicate the magnitude of the fields.

For visualization, the components of the velocity field are plotted over two lines through the cavity, *cf.* Fig. 3. For the director, the unit length is exploited. Since only a planar problem is considered, the vector can be represented by one angle,

$$\boldsymbol{d} = \sin\vartheta \boldsymbol{e}_x + \cos\vartheta \boldsymbol{e}_y,$$

and the scalar field $\vartheta = \hat{\vartheta}(\tilde{x}, \tilde{y})$ is used for visualization instead of the whole vector.

4.1 Parameter Study

The parameter study is performed starting from a base state for the dimensionless numbers in Eq. (11):

$$M = 0.2, \quad Re = 200, \quad Er = 80, \quad L = 10^{-4} \quad Pr = 2, \quad \kappa = 0.589, \tag{13}$$

which is motivated by the parameters given in [6] for the material MBBA at room temperature. Then, every dimensionless number is varied w.r.t. the base state. In order to save computational costs, a parameter continuation is performed, *i.e.*, the solution of the previous parameter set is used as an initial guess for the next simulation.

The first parameter to be varied is the REYNOLDS number in Fig. 4. From Fig. 3a one can see that a nearly circular main vortex is developed near the center in the upper right quadrant of the cavity. Therefore, in the line plots of Fig. 4, zeros in the velocities indicate the position of the main vortex. With an

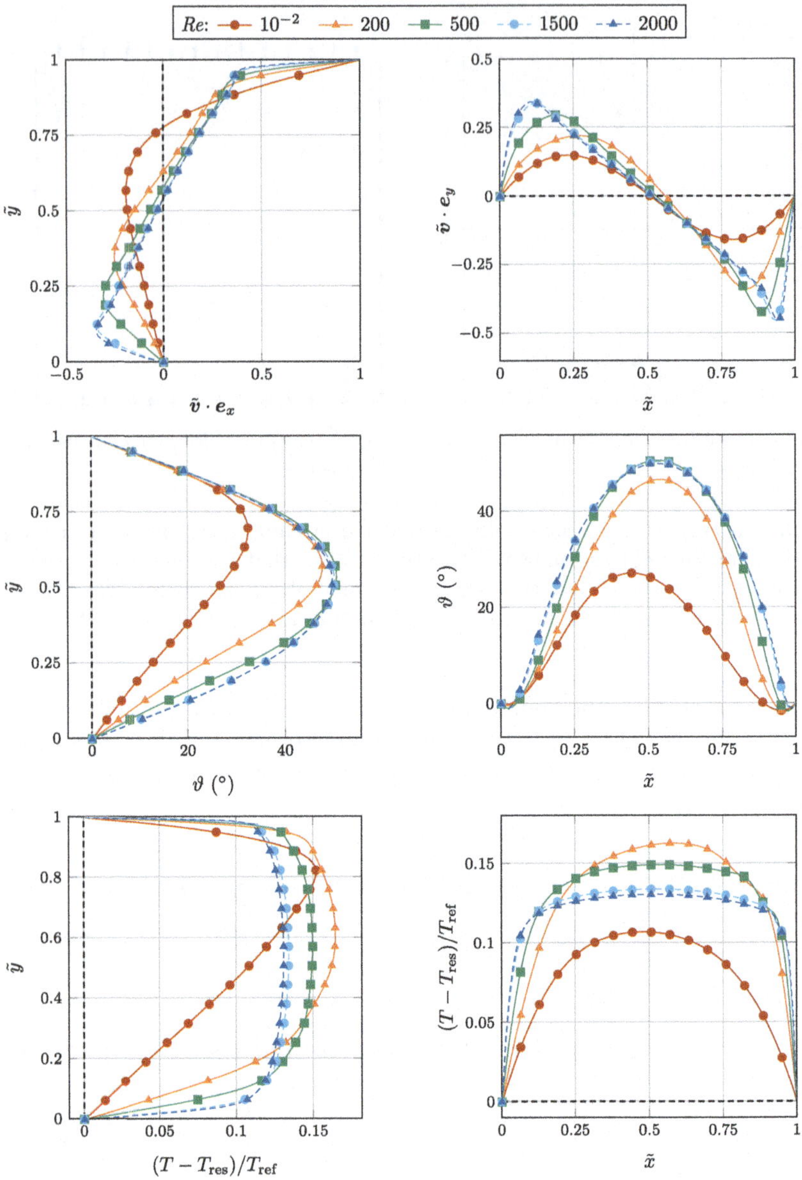

Fig. 4. Velocity, director angle and temperature profiles for a variation of the Reynolds number.

increasing Reynolds number the vortex starts to move downwards. This is also reflected in the director angle profile: With an increasing Reynolds number the maximal tilt increases and the maximal angle moves slightly towards the center of the cavity. At first glance, the temperature rise behaves as expected. As

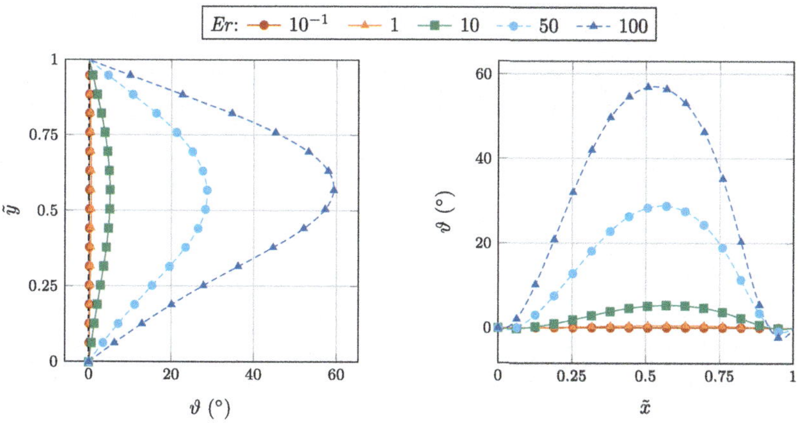

Fig. 5. Director angle profiles for a variation of the Ericksen number.

expected the temperature profiles broaden with increasing Reynolds number. However, the maximal temperature does not increase monotonically. The highest local temperature rise is obtained for $Re = 200$ w.r.t. the analyzed Reynolds numbers. Beyond this value the temperature forms plateau shaped profiles rather than parabolic ones. A similar effect for the Couette flow is also reported in [15].

Starting from the base state in Eq. (13), the variation of the Ericksen number strongly affects the director field and leaves the velocity and temperature field almost unaffected. Therefore, in Fig. 5 only the variations in the director angle are presented. The Ericksen number is the ratio of viscosity and elasticity. For small values of this number, elastic effects dominate and almost no flow alignment is visible. When the Ericksen number increases, the influence of viscous effects increases and the director field starts to tilt, so that the angle ϑ becomes larger. It is intuitively clear that the maximal rotation of the director is located at the center point of the main vortex of the velocity field. Because the velocity field is almost unaffected by the Ericksen number, the location of this main vortex (and, therefore, the position of the maximum of the angle function) is not affected.

In Fig. 6 the viscosity ratio M is varied over several orders of magnitude. Variations of M below values of 10^{-3} have no influence on all three fields. Small values of M correspond to the case of a dominating classical shear viscosity. As M increases, the influence of the non-classical viscosity increases. After increasing the viscosity ratio beyond the order of magnitude $\mathcal{O}(M) = 10^{-3}$, the velocity profiles change, but only slightly. In contrast to that, the director changes from almost no tilt to almost 50°. For the temperature (not shown here), the same sudden but slight change (as with the velocity) were observed. It can be noted that after the critical order of magnitude, no further changes in the fields can be observed.

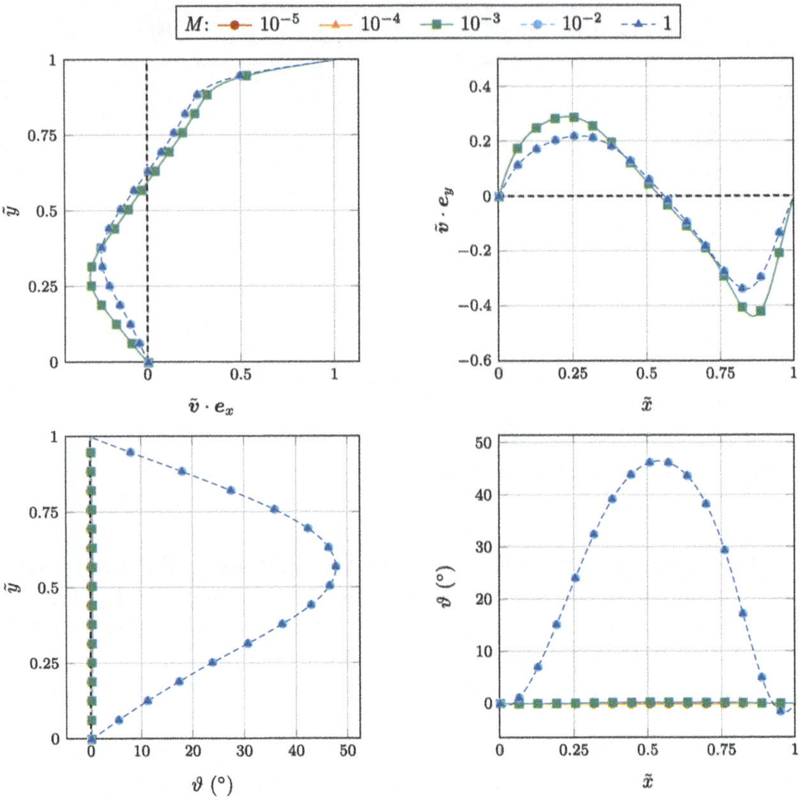

Fig. 6. Velocity, director angle and temperature profiles for a variation of the viscosity ratio.

The variations of the Prandtl number and the heat conductivity ratio are shown in Fig. 7. Therein, only the temperature profiles are shown since Pr and κ appear only in the energy balance which decouples from the mechanical equations and is solved separately. The effect of an increasing heat conductivity ratio on the temperature rise is strongly coupled to the director field. From Fig. 3b one can see that the director is almost vertical in the lower right quadrant. With decreasing $\kappa = \kappa_\perp/\kappa_\parallel$, the heat conduction in the direction of the lower right corner gains more influence and a boundary layer develops.

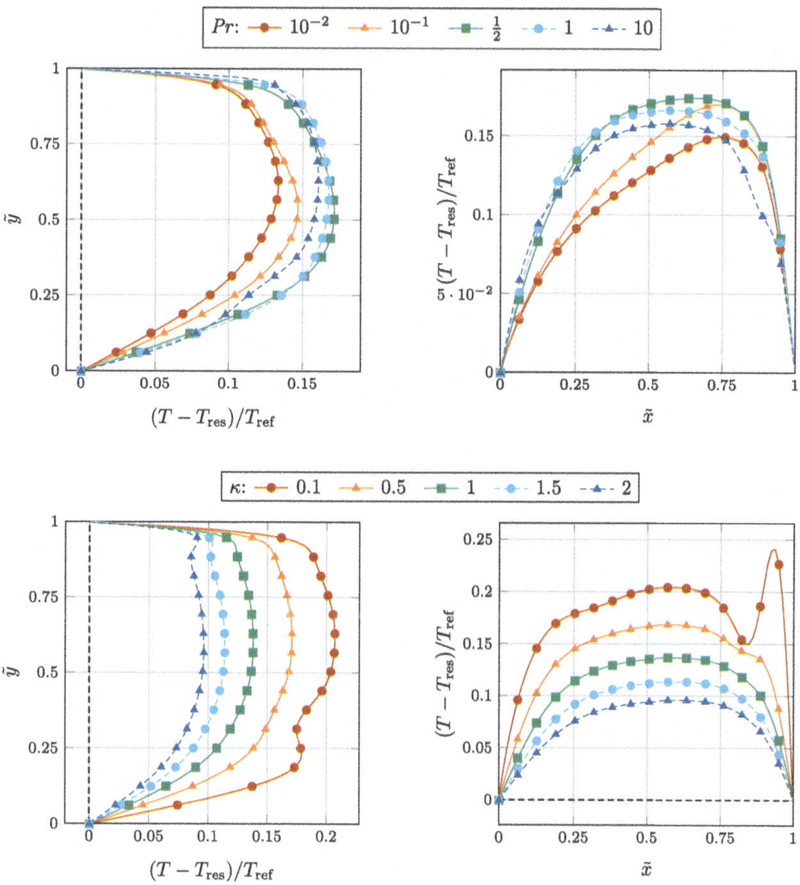

Fig. 7. Temperature profiles for variations of the PRANDTL number and the heat conductivity ratio.

5 Conclusion and Outlook

In this work, the governing equations of the ERICKSEN–LESLIE theory for the description of incompressible and rigid nematic liquid crystals were stated for non-isothermal conditions. They were simplified to the case of a two-dimensional lid-driven cavity under steady-state conditions. Furthermore, customary restrictions on the parameter set were employed. It was found that the unit sphere constraint for the director, *i.e.*, $\|\boldsymbol{d}\| = 1$, can be fulfilled by utilizing a field γ which was obtained in [9] as a byproduct of the methods of rational thermodynamics. This is favorable compared the penalty approach which is often employed in the literature.

The simplified equations are then reformulated in a dimensionless form and subsequently analyzed numerically. The numerical solutions were obtained for different sets of parameters. These sets were obtained by variation of one parameter from a base set. From this parameter study it was found that the REYNOLDS number has the biggest influence on all three analyzed fields, *i.e.*, velocity, director field and temperature rise due to viscous dissipation. In contrast to the REYNOLDS number, the ERICKSEN number has almost no influence on the velocity and temperature. Similarly, the employed viscosity ratio influences mostly the director field. For the simplified situation analyzed here, the energy balance decouples from the other balances and can be solved separately. Therefore, the thermal parameters solely influence the temperature rise.

The results obtained from the simplified equations with the restricted parameter set can be used as initial guesses for simulations with a non restricted parameter set. Then, it could be analyzed if the effects remain, *e.g.*, if the influence of the viscosity ratio M only depends on its magnitude and if other ratios such as μ_1/μ_4 have similar effects.

References

1. Becker, R., Feng, X., Prohl, A.: Finite element approximations of the Ericksen-Leslie model for nematic liquid crystal flow. SIAM J. Numer. Anal. **46**(4), 1704–1731 (2008)
2. Cruz, P.A., et al.: Numerical solution of the Ericksen-Leslie dynamic equations for two-dimensional nematic liquid crystal flows. J. Comput. Phys. **247**, 109–136 (2013). https://doi.org/10.1016/j.jcp.2013.03.061
3. Ericksen, J.L.: Inequalities in liquid crystal theory. Phys. Fluids **9**(6), 1205–1207 (1966)
4. Eringen, A.C.: Microcontinuum Field Theories- II Fluent Media. Springer Verlag, New York (2001)
5. Feireisl, E., et al.: A new approach to non-isothermal models for nematic liquid crystals. Arch. Ration. Mech. Anal. **205**(2), 651–672 (2012). https://doi.org/10.1007/s00205-012-0517-4
6. Finlayson, B.A.: Numerical computations for the flow of liquid crystals. In: Liquid Crystals and Ordered Fluids, 211–223. Springer, Boston (1974). https://doi.org/10.1007/978-1-4684-2727-1_20.
7. Frank, F.C.: I. Liquid crystals. On the theory of liquid crystals. Discuss. Faraday Soc. **25**, 19–28 (1958). https://doi.org/10.1039/DF9582500019
8. Hieber, M., Prüss, J.: Thermodynamical consistent modeling and analysis of nematic liquid crystal flows. In: Mathematical Fluid Dynamics, Present and Future. Springer, pp. 433–459 (2016)
9. Leslie, F.M.: Some constitutive equations for liquid crystals. Arch. Ration. Mech. Anal. **28**(4), 265–283 (1968). https://doi.org/10.1007/BF00251810
10. Li, J., Xin, Z.: Global existence of weak solutions to the non-isothermal nematic liquid crystals in 2D. Acta Mathematica Scientia **36**(4), 973–1014 (2016). https://doi.org/10.1016/s0252-9602(16)30054-6
11. Lin, F.H.: Nonlinear theory of defects in nematic liquid crystals; phase transition and flow phenomena. Commun. Pure Appl. Math. **42**(6), 789–814 (1989)
12. Müller, I.: Thermodynamics. Pitman (1985)
13. Parodi, O.: Stress tensor for a nematic liquid crystal. J. de Physique **31**(7), 581–584 (1970). https://doi.org/10.1051/jphys:01970003107058100

14. Pelcovits, R.A., Liu, P.: Nematic topological defects in the presence of axisymmetric fluid flow. arXiv preprint arXiv:1209.4914 (2012)
15. Rickert, W., Vilchevskaya, E.N., Müller, W.H.: A note on Couette flow of nematic crystals according to the Ericksen–Leslie theory. ZAMM – J. Appl. Math. Mech. (2019, in press)
16. Stewart, I.W.: The Static and Dynamic Continuum Theory of Liquid Crystals: A Mathematical Introduction. Taylor & Francis (2004)
17. Taylor, C., Hood, P.: A numerical solution of the Navier-Stokes equations using the finite element technique. Comput. Fluids **1**(1), 73–100 (1973). https://doi.org/10.1016/0045-7930(73)90027-3
18. Yang, X., et al.: 2-D lid-driven cavity flow of nematic polymers: an unsteady sea of defects. Soft Matter **6**(6), 1138(2010). https://doi.org/10.1039/b908502e
19. Zhilin, P.A.: Векторы и тензоры второго ранга в трехмерном пространстве (Vectors and tensors of second rank in three-dimensional space). НЕСТОР (2001). (in Russian)

HEDE Model vs Inner Pressure Model in Calculating the Strength of Hydrogenated Metals

Julia Sedova[1,2(✉)], Vladimir Polyansky[1,2], and Ivan Popov[1]

[1] Peter the Great St. Petersburg Polytechnic University, Saint Petersburg, Russia
julka0309@mail.ru

[2] Institute for Problems in Mechanical Engineering Russian Academy of Sciences (IPME RAS), Saint Petersburg, Russia

Abstract. The degrading effect of hydrogen on high strength steels is well recognized, but this problem still does not have an exact solution. In the present paper, we analyze two models that describe the mechanism of hydrogen embrittlement: HEDE and inner pressure models. We considered the effects of pre-stress load level and initial hydrogen concentration on the initiation time of cracking instigated by hydrogen in the steel bar and on the critical load level. We realized two important facts: (1) the HEDE model has several significant drawbacks such as the distribution of hydrogen in the sample and the place of initiation of the crack, as well as the ratio describing the relationship of stress and hydrogen concentration; (2) the coefficient of diffusion expansion needs research for the possible study of many phenomena in nature.

Keywords: Hydrogen embrittlement · HEDE model · FE simulation

1 Introduction

It is well known that hydrogen, which causes embrittlement and destruction of steel, is a serious problem for the oil and gas, construction and transport industries. Susceptibility to hydrogen embrittlement especially increases with increasing material strength and makes significant adjustments in the application of high-strength steels. In a hydrogen environment, metals lose their strength, ductility, toughness, and fail at much lower loads than hydrogen-unsaturated materials. Hydrogen consumption of steel occurs mainly in the process of its smelting (due to the presence of water in charge materials, in ligature and other materials used in steel production) and because of its being in a corrosive environment due to electrochemical reduction of water.

To date, several hypotheses have been proposed to describe the mechanism of hydrogen embrittlement, but this process remains poorly understood. In this paper, two models are analyzed: the hydrogen enhanced decohesion model (HEDE) and the model proposed by the ANSYS software system for engineering finite element analysis.

D. A. Indeitsev and A. M. Krivtsov (Eds.): APM 2019, LNME, pp. 206–212, 2020.
https://doi.org/10.1007/978-3-030-49882-5_20

2 HEDE Model Review

According to the HEDE model, interstitial hydrogen expands the metal atomic lattice, thereby reducing the cohesive strength of the atoms. Then the energy barrier for cracking decreases, what leads to the grain boundary or cleavage-like decohesion. It is assumed that it will occur in places of stress concentration, where the stress intensity factor is particularly high. For example, in internal corners, holes, grooves, or on the tips of a crack or notch. According to HEDE, hydrogen lowers the critical value of the stress intensity factor in these areas below the local voltage level caused by the application of a load. In other words, the concept of a critical hydrogen concentration is postulated, at which the critical stress value necessary for the start of crack development is equal to the applied stress, because of which destruction occurs.

Quite often, for numerical prediction and analysis of hydrogen embrittlement, researchers use cohesive zone modeling (CZM), based on the FE modeling method stated by Serebrinsky et al. [1] and Olden et al. [2, 3]. Y.F. Wang et al. [4] used this model to describe the nucleation of a crack in the rods, performed by the finite element method in the ABAQUS software package. A three-step process of finite element modeling was proposed: structural stress analysis, hydrogen diffusion analysis, and cohesive stress analysis.

A rectangular area with a spherical inclusion in the center ($r = 200\,\mu\text{m}$) with a uniform distribution of hydrogen on it ($c = 1,1\,\text{ppm}$) was taken as a two-dimensional geometric model of the rod. Pre-stress load is $\sigma = 650\,\text{MPa}$. The critical value of the stress intensity factor for the PSB1080 steel grade in question, obtained from experiments, is $K_{ic} = 71.5\,\text{MPa}\sqrt{m}$.

In the structural part, stress analysis was performed using the von Mises yield criterion and the isotropic strain hardening law. Further, the processes of the diffusion part were solved using the first and second Fick laws. In the cohesive stress analysis, in addition to the ratios connecting the stress intensity factor and the critical separation with the energy required for complete separation of cohesive surfaces, the researchers used traction separation law proposed by Serebrinsky et al. [1] based on suggestion of Jiang and Carter [5]:

$$\gamma(\theta) = \left(1 - 1,0467\theta + 0,1687\theta^2\right)\gamma(0) \tag{1}$$

$$\theta = \frac{c}{c + \exp\left(-\Delta g_b^0/RT\right)} \tag{2}$$

$$\sigma(\theta) = \left(1 - 1,0467\theta + 0,1687\theta^2\right)\sigma(0) \tag{3}$$

Here the first is relation linking hydrogen and surface energy with $\gamma(\theta)$ and without $\gamma(0)$ hydrogen. In the second Δg_b^0 is difference of Gibbs free energy in the hydrogen-adsorbed and bulk standard states of metal. Based on these two and considering decreasing effect of hydrogen on fracture energy the main relation (3) is deduced.

Using the procedure described above, the researchers calculated the dependence of the estimated time of onset of cracking and the critical hydrogen concentration the radius of the inclusion, pre-stressed level and the initial hydrogen concentration.

This approach has several significant drawbacks. Firstly, in the work, crack extension is modeled starting from the center of the bar. It is argued that according to microstructure examination of destructed high-strength steel bars, the number and size of inclusions near the center were much larger. The changes in the microstructure can be associated not with hydrogen, but with the stresses that occur during charging. There are studies [6–8] showing that with standard technology hydrogen charging its concentration inside the metal does not differ from the background. The next inaccuracy is an equal distribution of hydrogen throughout the bar. It is experimentally established that this is not the case [6–10]. Only a thin surface layer of the material up to 150 μm is saturated with hydrogen. Furthermore, the validity of using the dependency (1)–(2), proposed by Serebrinsky, is in doubts. This law is empirical, based on a numerical experiment and has not fundamental research.

3 FE Simulations

3.1 FE Solution

Consider another mechanism - an equation embedded in the ANSYS engineering complex. As in the procedure of HEDE modeling, in this case the FE simulation consists of structural and diffusion part.

At the stage of the associated structural-diffusion analysis, the total deformation consisted of two components: the elastic part that is familiar to all of us, described by Hooke's law, and the diffusion part, represented as follows:

$$\{\varepsilon\} = [E]^{-1}\{\sigma\} + \{\beta\}\Delta C \tag{4}$$

Here ΔC is concentration change, $\{\beta\}$ – vector of coefficients of diffusion expansion, which is a kind of analogue of the coefficient of thermal expansion.

In general, the essence of modeling is comparing the current stress intensity factor with its critical value. It is claimed that as soon as critical is exceeded, the crack will be initiated, which will lead to the destruction of the sample. And exactly relation (4) determines the dependence of the stress intensity factor on the concentration of a substance in this model. It should be noted that in the present work it is assumed that the crack will be initiated on the outer metal layer. But the distribution of hydrogen is also taken equal.

A cylindrical steel rod with a hub was taken as a sample. The geometric and physical parameters of the model are presented in following Fig. 1 and Table 1.

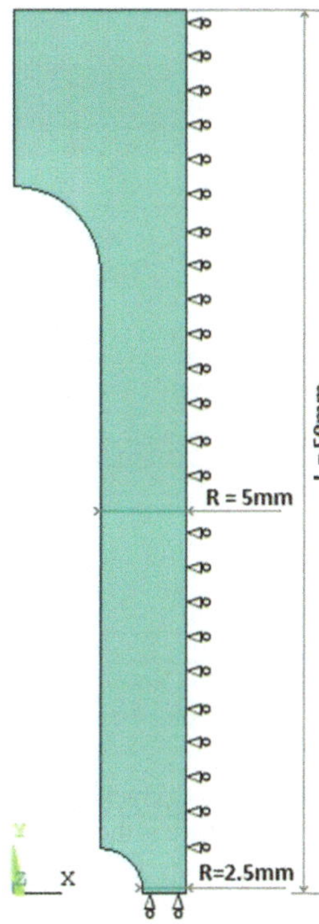

Fig. 1. The geometry and boundary conditions of the model.

Table 1. The physical conditions of the model.

Steel AISI 4135	
Young's modulus	$E = 2 * 10^5$ MPa
Poisson's ratio	$\nu = 0.32$
Density	$\rho = 7865\,\text{kg/m}^3$
Diffusion coefficient	$D = 4 * 10^{-11}\,\text{m}^2/\text{s}$
Coefficients of diffusion expansion	$\beta = 2 * 10^4$

3.2 The Results of FE Simulation

Following the procedure described above, the initiation time of cracking instigated by hydrogen in the bar was calculated for various values of pre-stress load level and initial substance concentration. The appropriate graphs are presented below.

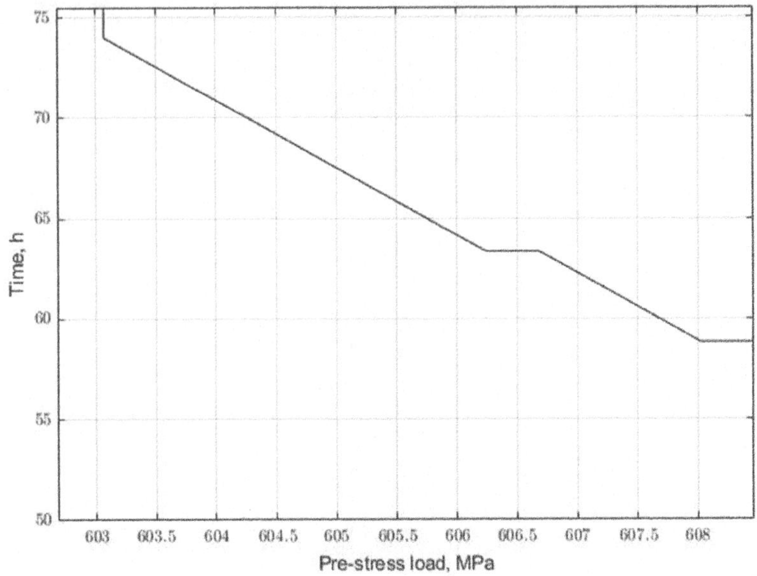

Fig. 2. The dependence of the initiation time of cracking on the pre-stress load level for the case initial hydrogen concentration $c = 1.1$ ppm (here is the load acting on the stress concentrator, that is, at the site of the initiation of the crack.).

Figure 2 shows the effect of pre-stress load on the initiation time of cracking in the bar. As the load on the stress concentrator increased from 603.5 to 608 MPa (which corresponds to the general load level from 195 to 215 MPa), the time has decreased from 74 to 59 h.

Figure 3 demonstrates that with increasing the initial hydrogen concentration, the initiation time of cracking decreases. Thus, for concentration level of 1.3 ppm the crack start time is 5 h, and for concentration of 0.9 ppm – around 98 h. It is interesting to note that for the initial hydrogen concentration of 0.8 ppm the initiation time of cracking did not manage to obtain (it exceeded hundreds of thousands of hours).

Additionally, the bars critical pre-stress load level was calculated for various values of initial hydrogen concentration (Fig. 4). The resulting dependency also shows that acceptable bars load level is inversely proportional to the initial hydrogen concentration, and with the increase of hydrogen in the metal, the load that the rod can withstand significantly decreases.

The important fact is that, coefficient of diffusion expansion, on which the whole model is based, has not yet been studied and, accordingly, there are no precise values for steel. So, its value was chosen empirically, until a tangible contribution to the total deformation of the model was obtained. During the study, it was found that for the local value of the stress intensity factor to exceed the critical, the value of the diffuse expansion coefficient must be of the order of $10^4 \div 10^5$, and it somewhat defies common sense.

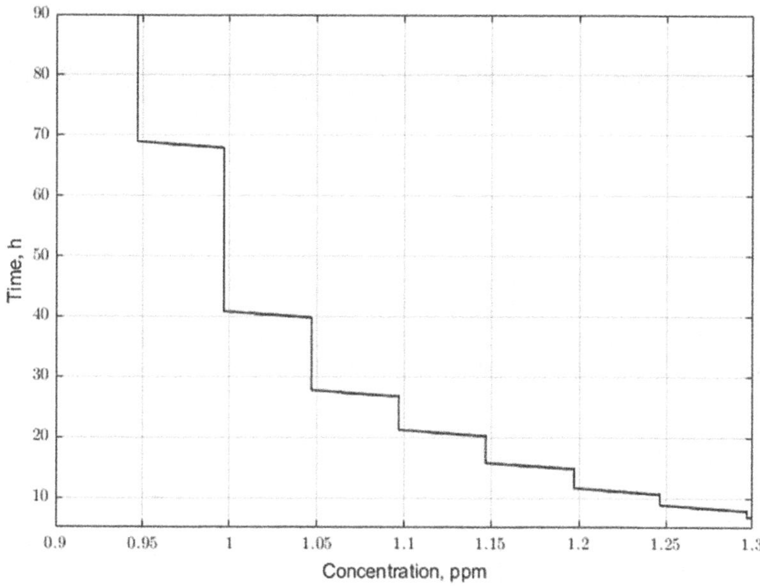

Fig. 3. The dependence of the initiation time of cracking on initial hydrogen concentration for the case pre-stress load level $\sigma = 210\,\text{MPa}$.

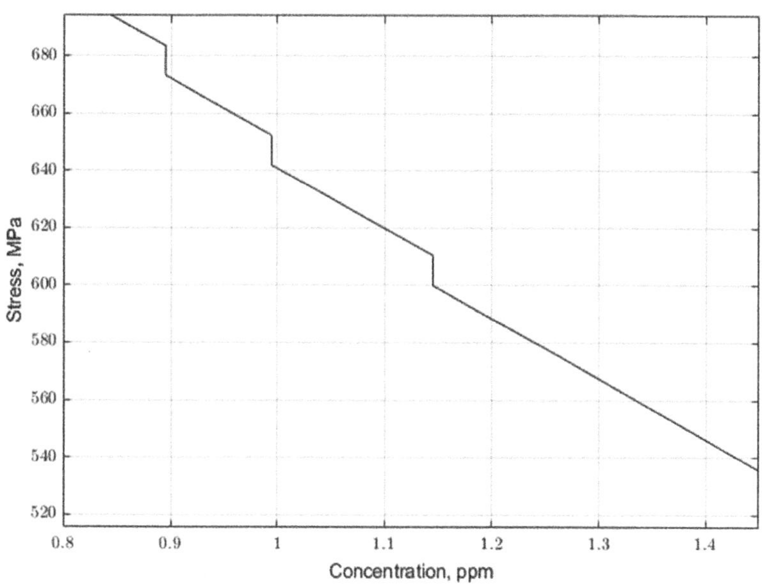

Fig. 4. The dependence of the pre-stress load level on initial hydrogen concentration for the case task calculation time $t = 100\,\text{h}$.

4 Conclusion

Thus, it was found that the current hypotheses describing the mechanism of hydrogen embrittlement really allow us to simulate the growth and development of cracks in samples, but they have significant drawbacks. For the first case, they are related to the fact that the relations used in the calculation of the process are empirical and have no physical justification. For the second model, we had to use parameters that have a nonphysical value. Therefore, further research into the phenomenon of hydrogen embrittlement is necessary.

Acknowledgments. This paper is based on research carried out with the financial support of the grant of the Russian Science Foundation (project no. 18-19-00160).

References

1. Serebrinsky, S., Carter, E.A., Ortiz, M.: A quantum-mechanically informed continuum model of hydrogen embrittlement. J. Mech. Phys. Solids **52**(10), 2403–2430 (2004)
2. Olden, V., Thaulow, C., Johnsen, R., Østby, E.: Cohesive zone modeling of hydrogen-induced stress cracking in 25% Cr duplex stainless steel. Scripta Materialia **57**(7), 615–618 (2007)
3. Olden, V., Thaulow, C., Johnsen, R., Østby, E., Berstad, T.: Application of hydrogen influenced cohesive laws in the prediction of hydrogen induced stress cracking in 25% Cr duplex stainless steel. Eng. Fract. Mech. **75**(8), 2333–2351 (2008)
4. Wang, Y.-F., Li, X.-F., Song, X.-L., Dou, D.-Y., Shen, L.-M., Gong, J.-M.: Failure analysis of pre-stressed high strength steel bars used in a wind turbine foundation: experimental and FE simulation. Mater. Corros. **67**(4) (2015)
5. Jiang, D.E., Carter, E.A.: First principles assessment of ideal fracture energies of materials with mobile impurities: implications for hydrogen embrittlement of metals. Acta Mater. **52** (16), 4801–4807 (2004)
6. Alekseeva, E.L., Belyaev, A.K., Zegzhda, A.S., Polyanskiy, A.M., Polyanskiy, V.A., Frolova, K.P., Yakovlev, Y.A: Boundary layer influence on the distribution of hydrogen concentrations during hydrogen-induced cracking test of steels. Diagn. Resour. Mech. Mater. Struct. **3**, 43–56 (2018)
7. Omura, T., Nakamura, J., Hirata, H., Jotoku, K., Ueyama, M., Osuki, T., Terunuma, M.: Effect of surface hydrogen concentration on hydrogen embrittlement properties of stainless steels and Ni based alloys. ISIJ Int. **56**(3), 405–412 (2016)
8. Yagodzinskyy, Y., Todoshchenko, O., Papula, S., Hänninen, H.: Hydrogen solubility and diffusion in austenitic stainless steels studied with thermal desorption spectroscopy. Steel Res. Int. **82**(1), 20–25 (2011)
9. Hadam, U., Zakroczymski, T.: Absorption of hydrogen in tensile strained iron and high-carbon steel studied by electrochemical permeation and desorption techniques. Int. J. Hydrogen Energy **34**(5), 2449–2459 (2009)
10. Martinsson, A., Sandström, R.: Hydrogen depth prole in phosphorus-doped, oxygen – free copper after cathodic charging. J. Mater. Sci. **47**(19), 6768–6776 (2012)

The Influence of Balcony Greening of High-Rise Buildings on Urban Wind and Thermal Environment: A Case of an Ideal City

Ying-Ming Su[1(✉)] and Chiao-Jou Hsieh[2]

[1] Department of Architecture, Master Program of Architecture and Urban Design, National Taipei University of Technology, No. 1, Sec.3, Zhong-xiao E. Rd., Taipei City, Taiwan (R.O.C.)
ymsu@ntut.edu.tw

[2] Department of Architecture, Master Program of Architecture and Urban Design, National Taipei University of Technology, 3F., No. 498, Yongji Rd., Taipei City 110, Taiwan (R.O.C.)
archi21th@gmail.com

Abstract. The balcony greening of high-rise buildings is one of the vertical greening strategies that have been tried in recent years due to the lack of green space in urban areas. In this study, the influence of balcony greening of high-rise building on wind speed and air temperature at different heights (1.75 m and 50 m) through two different variables including "balcony depth" (none, 3 m) and "green cover ratio" (0%, 50%, 100%) of 3 m-balcony are discussed. The key findings can be summarized as follows: 1. There are some differences in wind speeds and air temperatures between the buildings with 3 m-balcony and no balcony. 2. When the 3 m-balcony increased their green cover ratio (up to 50% or 100%), it was found to simultaneously achieve the effect of slowing down the urban wind speed and reducing the urban temperature.

Keywords: High-rise building · Balcony greening · Computational fluid dynamics (CFD)

1 Introduction

In recent years, with the rapid development of metropolitan areas around the world, in the case of limited urban land but dense population, the urban area buildings have a trend of high-rise and high-density development, thus the urban wind environment is shielded, which in turn aggravates the urban heat island effect. Lin et al. [1] have pointed out that urban ventilation paths have a mitigating effect on urban thermal accumulation, while plantings have the functions of guiding wind, cooling, and purifying air. At the same time, parks and large green spaces in urban areas also have a key influence on easing the urban heat island effect. In urban areas where land is limited and the area of green space and open space is insufficient, many cities have begun to develop a vertical greening model that uses roofs, balconies, and walls to increase greening space so as to achieve the purpose of energy conservation, carbon reduction and to help release the urban heat island effect as a countermeasure.

D. A. Indeitsev and A. M. Krivtsov (Eds.): APM 2019, LNME, pp. 213–221, 2020.
https://doi.org/10.1007/978-3-030-49882-5_21

Among the vertical greening model, as to the greening part of the balcony, due to the lack of space in the early built balconies (mostly within 2 m in depth), these balconies mainly use potted plants to plant herbs or shrubs, but in recent years, there have been a new trend to build balconies with over 3 m in depth and to plant trees (or shrubs) on these balconies that covered with solid soil, such as the Parkroyal on Pickering in Singapore and the vertical forests in Milan, Ruo-Shan Apartment I in Taiwan and so on. According to Zheng et al. [2], planting trees helps in improving thermal comfort; however, planting trees also has the effect of slowing wind speed. When we comprehensively promote the balcony greening (planting trees) of the high-rise buildings in the urban areas, what is the impact of the vertical greening of the balconies on the urban wind environment and thermal environment? It is worth further discussion.

2 Study Design and Verification

2.1 Basic Model and Environmental Setting

This study took an ideal city with high-rise buildings proposed by Hang et al. [3] as a theoretical model. The model consisted of 66 high-rise buildings (each featuring a volume measuring 30 m, 30 m, and 80 m in length, width, and height, respectively). The model was tested using computational fluid dynamics (CFD) and by performing wind tunnel tests, which confirmed that the deviations in wind speed data at measuring points V1 and V6 were within the acceptable range. A basic model is rebuilt in this study according to the above-mentioned ideal city model. As for the boundary conditions of the computational domain, this study referred to the methods employed by Hang et al. [3] (see Fig. 1).

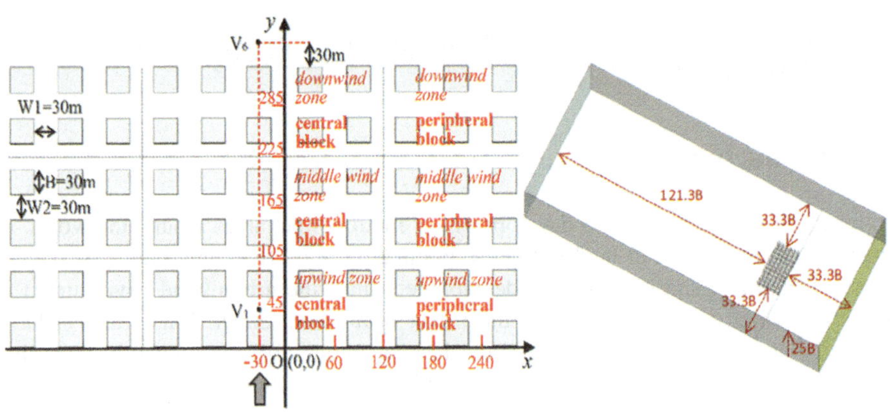

Fig. 1. Basic model and computational domain of the ideal city with high-rise buildings. Note. Adapted from Hang et al. [3]

Based on the abovementioned basic model, this study used two different variables of the "balcony depth" (none, 3 m) and "balcony green cover ratio" (0%, 50%, 100%), and then used ANSYS Fluent 18.0 software to perform CFD numerical simulations to investigate the impact of high-rise buildings' balconies and their green cover ratio on pedestrian wind field (height: 1.75 m) and urban wind field (height: 50 m). Among them, the study first divided the architectural group of the ideal city into "upwind zone", "middle zone" and "downwind zone" according to the distances from the main wind direction. According to the characteristics of urban regional development, the buildings are divided into "central blocks" and "peripheral blocks" (also see Fig. 1).

For the mesh independence test, this study referred to the process utilized by Hang et al. [3], testing three grid configuration types, i.e., coarse grids (number of grids: 6,588,060), medium grids (number of grids: 12,067,443), and fine grids (number of grids: 28,907,122). Wind speed simulations were performed for the three grid types using ANSYS Fluent v18, wherein the results were compared with the wind tunnel test data obtained from a theoretical model employed by Hang et al. [3]. The mesh quality assessment results are presented in Table 1.

Table 1. Comparison between the wind speed simulation data obtained by a basic model employed in this study and those obtained by a theoretical model used by Hang et al.

Measuring point	Theoretical model of Hang et al. (Wind tunnel data)	Basic model of this study (CFD results)					
		Coarse grid (grid: 6,588,060)		Medium grid (gnd: 12,067,443)		Fine grid (grid: 28,907,122)	
	Average wind speed	Average wind speed	Error	Average wind speed	Error	Average wind speed	Error
VI	1.72 m/s	2.25 m/s	30.53%	2.00 m/s	16.23%	1.55 m/s	−9.83%
V6	1.48 m/s	1.88 m/s	27.42%	1.68 m/s	13.65%	1.29 m/s	−12.97%
Total	–	–	28.98%	–	14.94%	–	−11.40%
Mesh quality assessment		Value	Result	Value	Result	Value	Result
Orthogonal quality (Min)		0.193	Acceptable	0.152	Acceptable	0.144	Acceptable
Skewness (Max)		0.854	Acceptable	0.848	Acceptable	0.856	Acceptable

Table 1 shows that the quality of the grids (i.e., coarse grids, medium grids, and fine grids) used by the basic model of this study was acceptable, and that the overall deviations generated by the medium and fine grids were within an acceptable range (i.e., a deviation of less than 20%). Because of time limitations and because fine grids could only improve grid quality and simulation result accuracy minimally, this study used medium grids for subsequent CFD simulations.

2.2 Designing the Study Variables

This study set balcony depth (i.e., none and 3 m) and balcony green cover ratio (i.e., 0%, 50%, and 100%) as the variables, producing four possible model types (i.e., none-0%, 3 m-0%, 3 m-50%, 3 m-100%). Because planting trees required a growth space of two floors, this study adopted split-level balconies, as shown in Table 2. By using the aforementioned four model types, this study investigated the effects of balcony depths and balcony green cover ratios on wind speeds and temperature in cities.

Table 2. Condition settings of the proposed model

Model (depth of balcony – Green coverage ratio of balcony)	**Schematic diagram**
None-0%	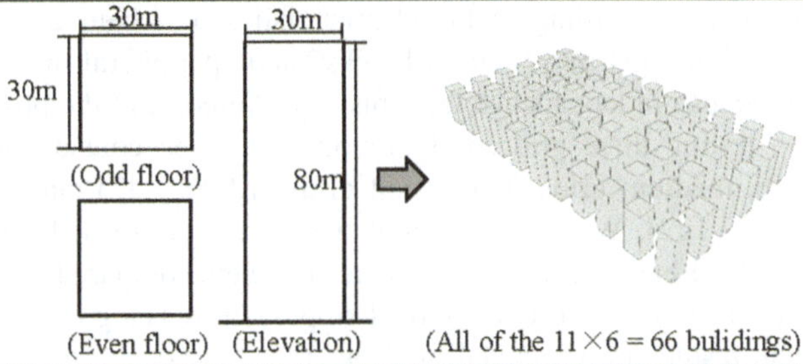
3m-0%	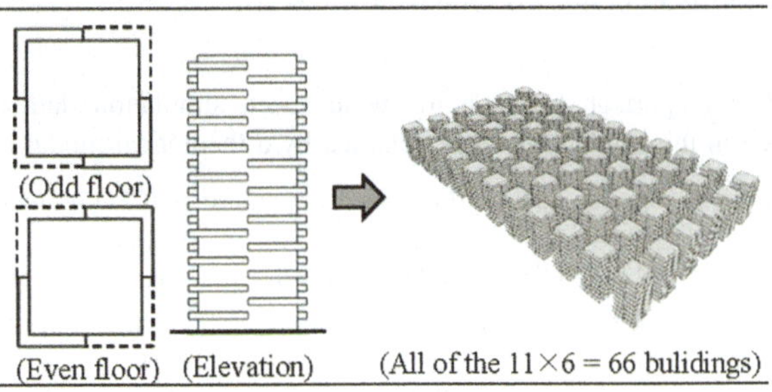
3m-50%	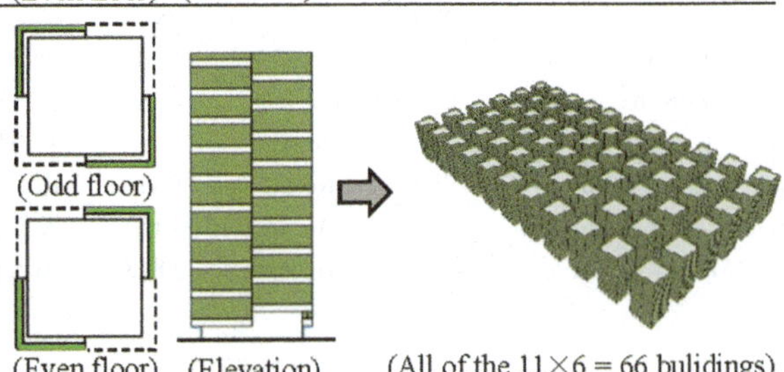
3m-100%	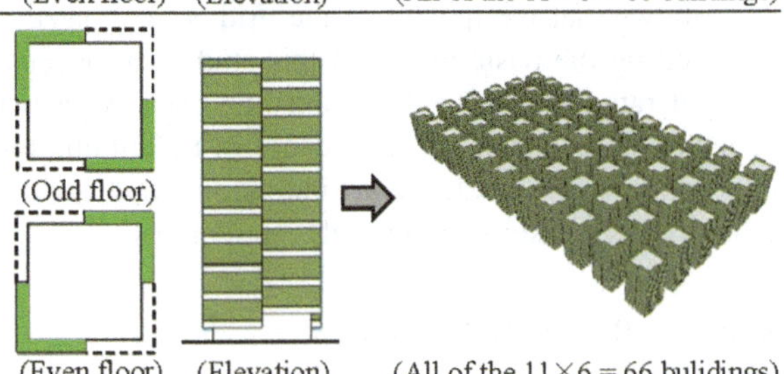

3 CFD Simulation Results and Analysis

This study is based on the analysis of wind speed and temperature simulation results of urban high-rise buildings by the balcony depth and green cover ratio of the 3 m-balcony. The main findings are as follows:

Adding a 3 m-balcony will reduce the wind speed of the X-axis street in the central block, but increase the wind speed of the Y-axis street slightly.

After adding a 3 m-balcony to the high-rise building, the wind speed of the Y-axis street (i.e., the street parallel to the wind direction) can be moderately increased at the pedestrian height (1.75 m). However, there is no significant difference upon the depth of balcony to the wind speed of the Y-axis street at the height of 50 m (see Fig. 2). Viewing the vertical wind speed profile, the 3 m-balcony slightly reduces the wind speed of the X-axis street in the central block; however, this phenomenon is not found in the peripheral block (see Fig. 3). Therefore, the addition of a 3 m- balcony is more helpful for increasing the wind speed of the Y-axis street (i.e. the main urban ventilation path), but the wind speed of the X-axis street (including the windward side and leeward side of the buildings) of the central block may be slightly reduced.

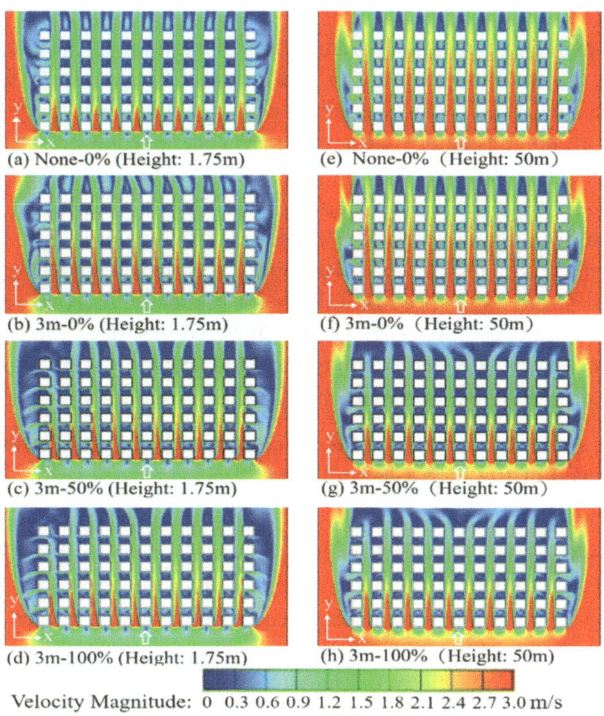

Fig. 2. Simulation results of the horizontal wind speed for different models of balcony greening

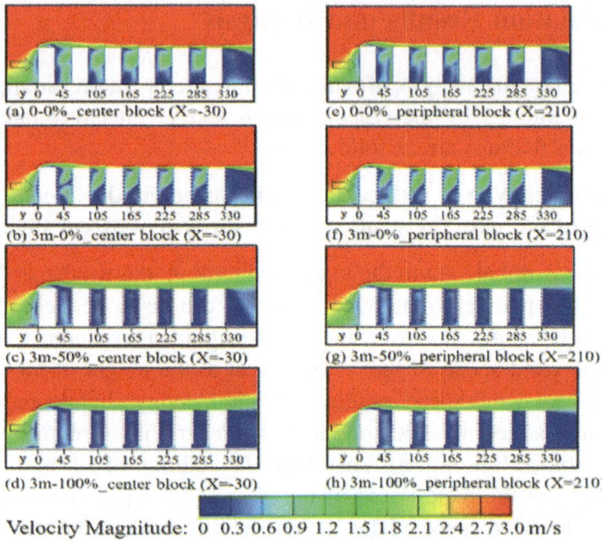

Fig. 3. Simulation results of the vertical wind speed profile for different models of balcony greening

The above-mentioned phenomenon may be related to the addition of 3 m-balcony, which reduced the street width to 24 m and increased the venture effect and in turn slightly increased the Y-axis wind speeds in pedestrian (height: 1.75 m) wind fields. At a height of 50 m, no considerable differences in wind speeds were observed between none-balcony and 3 m-balcony. This may be resulted from wind speeds increase with height. The inlet wind speed was weaker at 1.75 m in height, causing a shorter extended length of the wind speed of 3.0 m/s (or higher) in Y-axis streets with none-balcony. By contrast, because the inlet wind speed was stronger at a height of 50 m, the extended length of the wind speed of 3.0 m/s (or higher) in Y-axis streets with none-balcony was considerably longer. Even the street widths were diminished by 3 m-balcony, the changes in extended length of the wind speed of 3.0 m/s (or above) in the Y-axis street were negligible at the height of 50 m.

Increasing the green cover ratio of the balcony has the effect of slowing down the urban wind speed of the high-rise buildings.

When the balcony green cover ratio is 50% and 100%, either at the height of 1.75 m or 50 m, the extension length of the Y-axis street wind speed of 3.0 m/s or over is significantly shorter than that of the balcony green cover ratio of 0% (also see Fig. 2). Obviously, the higher the balcony green cover ratio, the slower the wind speed of the Y-axis street. According to the vertical wind speed profile, the wind speed of the balcony with a green cover ratio of 0% may reach more than 3 m/s at a height of 80 m or more in the central block or the peripheral block. Whereas while the balcony green cover ratio is 50% and 100%, when at the height between 80 m and 105 m, the wind speed is slowed down (even to 1.2 m/s) (also see Fig. 3). It shows that increasing the green cover ratio of the balcony may produce the effect of slowing down the wind speed due to the evapotranspiration of the plant within about 25 m above the top floor of the high-rise building. In summary, increasing the green cover ratio of the balcony does have the effect of moderately slowing the urban wind speed. The effect of slowing wind speed may be caused by planting trees, which is consistent with the findings of Zheng et al. [2].

Adding a 3 m-balcony will cause the temperature of the farther urban area away from the main wind direction to rise partially.

At the height of 1.75 m, the temperature of the 3 m-balcony on the leeward side of the Y-axis street and the leeward side of the overall building group are significantly higher than the temperature of none-balcony. As for the commonality between 3 m-balcony and none-balcony at the height of 1.75 m, there is a common trend in the X-axis street and Y-axis street that the temperatures are distributed from low to high as follows: upwind zone < middle wind zone < downwind zone. However, at the height of 50 m, there is no significant difference in temperature between 3 m-balcony and none-balcony (see Fig. 4). According to the vertical temperature profile, whether in the central block or in the peripheral block, it can be seen that in two row of farther buildings with 3 m-balcony that are away from the main wind direction, their temperatures in the windward side may be raised up to over 38 °C; however, there is no such phenomenon in the buildings with no balcony (see Fig. 5). From this, it can be

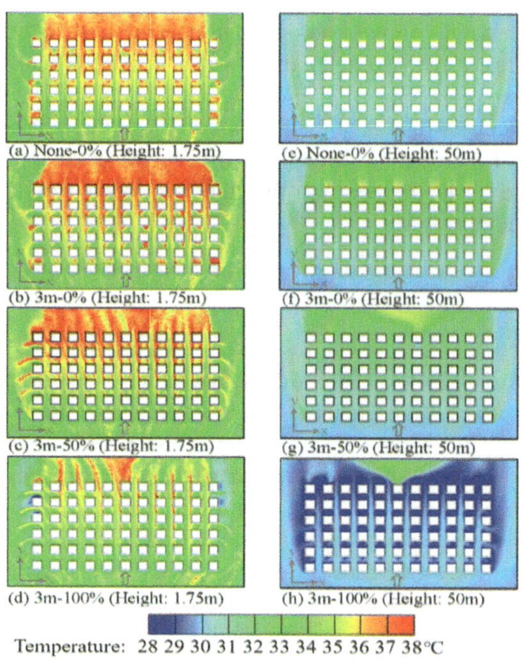

Fig. 4. Simulation results of the horizontal air temperature for different models of balcony greening

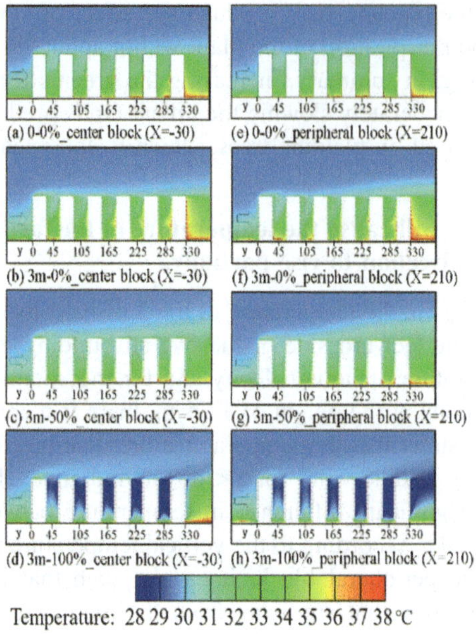

Fig. 5. Simulation results of the vertical air temperature profile for different models of balcony greening

seen that the addition of a 3 m-balcony does cause an increase in the temperature of the urban area that is farther away from the main wind direction. This phenomenon may be due to the obvious weakening of the wind speed in the downwind zone, and the addition of 3 m-balcony will lead to the increase in the amount of heat absorbed by the concrete, which will increase the accumulated heat. Meanwhile, there is not enough wind speed to take away the heat, so that the 3 m-balcony will result in the phenomenon of High temperature.

Adding the green cover ratio of the balcony can effectively reduce the urban temperature.

As for the difference in temperature between different balcony green cover ratios, whether at a height of 1.75 m or 50 m, the temperature of the original balcony which has a green cover ratio of 0% is higher, whereas when green cover ratios of the balcony increased to 50% and 100%, the temperature turns downward trend (also see Fig. 4). In addition, according to the vertical temperature profile, the vertical temperature profile of the X-axis street gradually decreases with the increase of the balcony green cover ratio in the central block or the peripheral block (also see Fig. 5). This demonstrates that an increase in balcony green cover ratio can decrease urban temperature; moreover, this result is consistent with the findings of higher green cover ratios can lead to lower temperature which is presented by Adams et al. [4].

4 Conclusions

Overall, the addition of a 3 m-balcony to urban high-rise buildings may result in a slight increase in the wind speed on the streets parallel to the wind direction (i.e., Y-axis street) and a slight decrease in the wind speed on the streets perpendicular to the wind direction (i.e., X-axis street), and may cause a local increase in the temperature of the farther urban area from the main wind direction. However, when the green cover ratio of 3 m-balcony was increased (up to 50% or 100%), it was found to simultaneously achieve the obvious effect of slowing down the urban wind speed and reducing the urban temperature. High-rise buildings may cause the problems of strong winds around high-rise buildings and deteriorating the urban heat island effect. This study found that adding the 3 m-balcony and increasing its green cover ratios (up to 50% or 100%) can moderately slow down the wind speed and lower the temperature, which is helpful for improving the comfort of the overall urban environment. Therefore, it is suggested that green cover ratio of the balcony should be at least a certain ratio (such as 50%, etc.), and it need to be included in the relevant building regulations, or to establish incentive mechanisms in the future for improving the comfort of the building environment, mitigating the urban heat island effect, and to help solve the problem of strong winds around high-rise buildings.

Acknowledgements. This work was financially supported by the Ministry of Science and Technology, Taiwan (R.O.C.) (Grant no. MOST 107-2221-E-027-020) and the "Research Center of Energy Conservation for New Generation of Residential, Commercial, and Industrial Sectors" from the "Featured Areas Research Center Program" within the framework of the "Higher Education Sprout Project" by the Ministry of Education (MOE), Taiwan (R.O.C.).

References

1. Lin, H.T., Lee, K.P., Chen, K.T., Lin, L.J., Kuo, H.C., Chen, T.C.: Analysis of the heat island effect in the four metropolitan areas of Taiwan. J. Arch. **31**, 51–73 (1999)
2. Zheng, S., Guldmann, J.M., Liu, Z., Zhao, L.: Influence of trees on the outdoor thermal environment in subtropical areas: an experimental study in Guangzhou, China. Sustain. Cities Soc. **42**, 482–497 (2018). https://doi.org/10.1016/j.scs.2018.07.025
3. Hang, J., Li, Y., Sandberg, M.: Experimental and numerical studies of flows through and within high-rise building arrays and their link to ventilation strategy. J. Wind Eng. Ind. Aerodyn. **99**, 1036–1055 (2011). https://doi.org/10.1016/j.jweia.2011.07.004
4. Adams, M.P., Smith, P.L.: A systematic approach to model the influence of the type and density of vegetation cover on urban heat using remote sensing. Landscape Urban Plan. **132**, 47–54 (2014). https://doi.org/10.1016/j.landurbplan.2014.08.008

Contour Convergence Regularities for Openings in Coal

Vitaly A. Trofimov(✉) and Yury A. Filippov

Institute of Comprehensive Exploitation of Mineral Resources Russian Academy of Sciences, Krukovsky tup., 4, Moscow, Russia
asas2001@mail.ru

Abstract. The article is devoted to the applied problem solution of a continuous medium inelastic deformation as applied to the description of the heaving of rocks in the workings. This is the ubiquitous manifestation of rock pressure in workings, both in coal and ore deposits. It is expressed in raising the soil, which value reaches significant values compared with the size of mining working and lead to the loss of its functional purpose. Heaving is a process stretched in time for weeks, months, or even years.

In the article, the construction of the heaving problem solution is carried out within the framework of the viscoplasticity theory. This approach can reflect the time dependence of the deformation process, since the equations of state are formulated in increments or strain rates. Creep strain is characterized by the strain rate, which is generally a function of the stresses, time, and possibly temperature.

The form of this function can be constructed for each particular material experimentally as a result of numerous experiments on samples with subsequent statistical processing of the results obtained. However, theoretical studies use a variety of simplified analytical dependencies, mostly of a power type.

The article discusses options for the numerical calculation of viscoplastic deformation both with a constant in time deformation rate, and with a decreasing rate, when the total deformation after a certain period of time reaches a constant level.

To preserve the excavation working capacity after a certain time, a "undermining" of the soil is produced, removing the layer of rock extruded into the mining working. It is known that such actions are performed 2–4 times, after which the heaving on this place completely stops. In addition, from the practice is known such a phenomenon as a gap of the soil; a situation where, in addition to raising the soil, a tension crack is formed in it, which runs along the workings along its center line. Within the framework of the developed approach with the use of a viscoplastic model, the main regularities of the course of such processes are considered.

Keywords: Coal seam · Contour convergence · Viscoplasticity · Undermining · Heaving

D. A. Indeitsev and A. M. Krivtsov (Eds.): APM 2019, LNME, pp. 222–232, 2020.
https://doi.org/10.1007/978-3-030-49882-5_22

1 Introduction

Heaving of rocks is ubiquitous manifestation of rock pressure in single mine workings, both in coal and ore deposits. It is expressed in the form of raising the soil, which can reach significant values comparable to the size of mine workings, leading to the loss of its functional purpose. In order to preserve the working capacity of the mine, after a certain time, the soil undermines, removing the layer of rock extruded into the mine. This is quite a costly undertaking, but inevitable for the functioning of a mine.

For many years of work of mining enterprises in conditions of soil heaving, numerous preventive measures have been developed and put into practice, designed to reduce or completely eliminate this harmful manifestation. Almost all of them are based on the generalization of numerous field observations inclusive some theoretical considerations. This may include, in particular, fixing the contour of the mine with the help of anchors, especially in the lower part of the mine, creating unloading slots, loosening and hardening the soil rocks directly under the mine, etc. They all showed their effectiveness in various mining conditions.

It should be noted that such a phenomenon could not pass by the attention of many theoreticians who tried to build conceptual, mechanical and mathematical model, to explain the mechanism of the deformation processes occurring in this process.

In the historical aspect, we note the works of K. Terzaghi, P.M. Tsymbarevich, V. D. Slesarev, who tried to describe the soil heaving within the framework of the statics of a granular medium. They could not be successful due to the fact that heaving is a process stretched in time for weeks, months, and even years. Theories of elasticity, plasticity, granular medium in their various versions and modifications do not include in their relations the time parameter, and therefore are unacceptable in this situation.

The concept for solving the problem under consideration within the framework of continuum mechanics, which is usually used in solving various problems of geomechanics, is a hereditary theory of creep. In this regard, it is necessary to mention a number of researchers, who achieved definite results in description of the development of temporal processes in a rock massif - among them prominent scientists in the field of continuum mechanics Zh.S. Erzhanov, M.I. Rozovsky, L.Ya. Parchevsky, V.T. Glushko, Yu.N. Rabotnov, G.L. Fisenko, K.V. Ruppenate, V.A. Lytkin, A.A. Ilyushin, A.Yu. Ishlinsky, L.S. Leybenzon, N.A. Tsytovich, A.N. Stavrogin and many others.

It should be noted that so far it has not been possible to construct an adequate description of the heaving process, as evidenced by the interest to this problem, which has not been weakened up to now. Many possible mechanisms of this phenomenon were built, among which two can be considered the most adequate: expansion of rocks as a result of their swelling due to water and squeezing of rocks into workings under the action of the reference pressure [1–6]. In addition, it is considered that the most acceptable theoretical approach is the hypothesis about the loss of the elastic-plastic stability of the rock massif in the vicinity of a single workings [7–9].

Nevertheless, we note that the deformation of the contour of workings is, as a rule, quite complex in nature and is not associated only with the soil heaving. The deviation of the sidewalls of the workings can also reach values in tens of centimeters, often leading to a complete overlap of its clearance. At the level of loading existing in the massif, this kind of deformation is difficult to attribute to loss of stability with such a complex shape of the deformed boundary. This indicates a different mechanism of deformation during heaving.

Let us dwell in more detail on the possibility of applying the hereditary theory of creep to describe the behavior of heaving rocks. Its main theoretical positions and the corresponding equations are well known and are widely used for solving problems of continuum mechanics, including some problems of geomechanics.

In the framework of this theory, the medium equation of state is represented by the Boltzmann-Volterra integral equation, which in the general case of non-linear material deformation has a form

$$\varepsilon(t) = \frac{1}{E_M}\left[\sigma(t) + \int_0^t K(t,\tau)f(\sigma)d\tau\right], \tag{1}$$

where t is time, EM is the instantaneous modulus of elasticity, K is the core of the integral equation, ε, σ is the strain and stress at a point, f(σ) is a certain stress function characterizing the strain non-linearity.

Along with the usual equations of continuum mechanics, i.e. equilibrium, deformation continuity, geometric equations and the corresponding boundary and initial conditions, they completely define the problem and in principle allow us to obtain a solution on the deformation of the rock massif in time at constant load, which occurs during heaving.

It should be noted that the form of the kernel K and the function f could be written out in an explicit form only for very simple rheological models, which, however, in many cases are sufficient to obtain an acceptable solution. In these cases, the integral in the above formula is taken in the final form and the final relations between the deformations and stresses, which explicitly include time, are obtained.

Finally, the solution of the problem is the dependences of the distribution of stresses, strains (strain rates) and displacements within the entire computational domain on time. Note that, despite the final form of all the relations in the problem, the acceptable result can be obtained only in a few cases. Nevertheless, the described approach is quite acceptable.

2 The Use of Viscoplasticity Theory

Let us consider a slightly different approach for constructing a solution of the heaving problem within the framework of the viscoplasticity theory. This approach may also reflect the time dependence of the deformation process, since the equations of state are formulated in increments or strain rates ε. We characterize the deformation during creep by the deformation rate ε. Obviously, this value in the general case depends on the effective stresses σ, time t and, possibly, temperature T. Bearing in mind the deformation of rocks, the last value can be ignored. In other words, the following relation holds

$$\dot{\varepsilon} = F(\sigma, t). \tag{2}$$

The form of this dependence can be constructed experimentally for each specific material as a result of numerous experiments on samples with subsequent statistical processing of the results obtained. Nevertheless, theoretical studies use a variety of simplified analytical expressions for (2). Bearing in mind the use of the finite elemental complex ANSYS for solving heaving problems, we note that various relations of the form (2) are built in ANSYS for solving visco-plastic problems. Some of them are shown in Table 1 in form they are used in ANSYS. C_i values are constants that determine the nature of material deformation. The choice of a certain heaving equation is based on a comparison of the calculation results for the deformation process with the values obtained by measuring the corresponding parameters in natural conditions.

Thus, from now on, for solving the model problem of soil heaving in the workings, Eq. 2 (Time hardening) was chosen as the simplest one, allowing one to take into account the deceleration of the heaving process with time. In this case, it was considered that $C_2 = C_4 = 0$. The choice of $C_2 = 0$ requires an explanation. It is obvious that in places where the stress is greater, the strain rate should also be greater. If we associate the increase in stresses with increasing depth, then at constant depth, the influence of the term σ^{C2} in the chosen ratio is equivalent to the influence of C_1 due to

Table 1. Various expressions for strain rate

Creep model	Name	Equation	
1	Strain Hardening	$\dot{\varepsilon}_{cr} = C_1\sigma^{C2}\varepsilon_{cr}^{C3}e^{C_4/T}$	$C_1 > 0$
2	Time Hardening	$\dot{\varepsilon}_{cr} = C_1\sigma^{C2}t^{C3}e^{-C4/T}$	$C_1 > 0$
3	Generalized Exponential	$\dot{\varepsilon}_{cr} = C_1\sigma^{C2}re^{-rt}$, $r = C_5\sigma^{C3}e^{-C4/T}$	$C_1 > 0$, $C_5 > 0$
4	Generalized Graham	$\dot{\varepsilon}_{cr} = C_1\sigma^{C2}(t^{C_3} + C_4t^{C_5} + C_6t^{C_7})e^{-C_{8/T}}$	$C_1 > 0$
5	Generalized Blackburn	$\dot{\varepsilon}_{cr} = f(1 - e^{-rt}) + gt$, $f = C_1e^{C_2\sigma}, r = C_3\left(\frac{\sigma}{C_4}\right)^{C_5}, g = C_6e^{C_7\sigma}$	$C_1 > 0$, $C_3 > 0$, $C_6 > 0$
6	Modified Time Hardening	$\dot{\varepsilon}_{cr} = C_1\sigma^{C_2}t^{C_3+1}e^{-\frac{C_4}{T}}/(C_3 + 1)$	$C_1 > 0$
7	Modified Strain Hardening	$\dot{\varepsilon}_{cr} = \left\{C_1\sigma^{C_2}[(C_3 + 1)\varepsilon_{cr}]^{C_3}\right\}^{1/(C_3+1)}e^{-C_4/T}$	$C_1 > 0$
8	Generalized Garofalo	$\dot{\varepsilon}_{cr} = C_1[\sinh(C_2\sigma)]^{C_3}e^{-C_4/T}$	$C_1 > 0$
9	Exponential form	$\dot{\varepsilon}_{cr} = C_1e^{\sigma/C_2}e^{-C_3/T}$	$C_1 > 0$
10	Norton	$\dot{\varepsilon}_{cr} = C_1\sigma^{C_2}e^{-C_3/T}$	$C_1 > 0$
11	Combined Time Hardening	$\frac{\dot{\varepsilon}_{cr} = C_1\sigma^{C_2}t^{C_3+1}e^{-\frac{C_4}{T}}}{C_3 + 1} + C_5\sigma^{C_6}te^{-C_7/T}$	$C_1 > 0$, $C_5 > 0$

σ = const. Thus, it can be simply ignored in the presence of C_1. If we connect the increased stresses with the places of their concentration near the workings, then their influence zones are small and confined to the corner (or similar) points of the contour. It is possible that there is significant creep locally there, but in situ it may simply not be noticed without making special observations.

Thus, relation (2) was taken as

$$\dot{\varepsilon} = C_1 t^{C_3}. \tag{3}$$

Here, the constant C_1 has the dimension of the strain rate and is equal to it at t = 1. At the same time, C_3 is a dimensionless quantity. Both of them are not material constants, but depend on the used time unit. Initially it was thought that $C_3 = 0$, i.e. deformation with a constant strain rate was considered.

Figure 1 shows the main elements of the formulation of soil heaving problem. By virtue of the symmetry of the problem with respect to the vertical line passing through the center of the opening, the computational area was considered only for a half of research domain. This vertical line was considered as a line of symmetry and the conditions of symmetry were set on it, i.e. zero horizontal offsets. On the outer vertical and on the upper horizontal boundaries, normal stresses were set, which magnitude was equal to 12.5 MPa, which corresponds to the depth of the output in 500 m. At the same time, the opening surface was free from stress.

Note that the creep properties in the considered variant have only soil rocks. It was considered that $C_1 = 4 * 10^{-8}$. This value of the strain rate was chosen from the consideration that the maximum rising of the soil for 1 month was ~20 ÷ 30 cm.

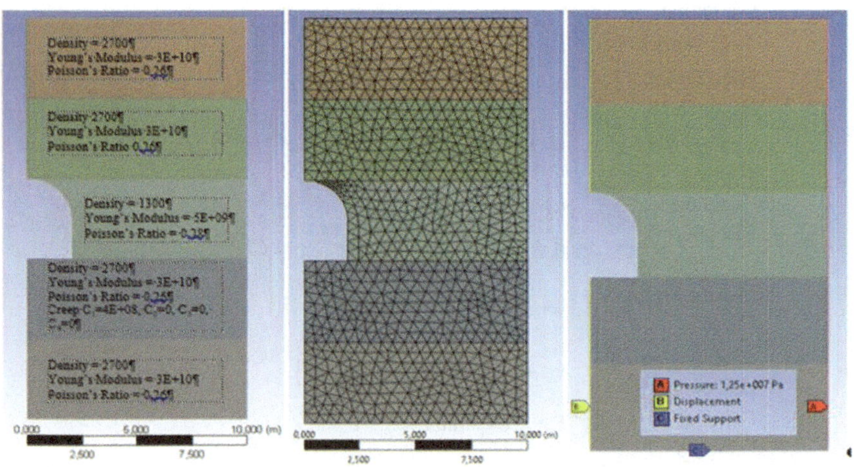

Fig. 1. Geometric model, with a finite element mesh, loads and boundary conditions, physical and mechanical properties

Figures 2, 3, 4, 5, 6 and 7 show some calculation results.

Figure 2 shows the total displacement vectors at the points of the computational domain at the time point of 30 days after the opening has been created. They show the values (the length of the arrows and the color) and the direction of the displacements for the points of the array. It is seen that the rocks of the soil have received the greatest displacement inside the opening, which can be interpreted as heaving of the soil. Further, the amount of heaving will be characterized by the greatest displacement, which is achieved at the average ("reference") point of the opening soil, and which in case under consideration is equal ~ 0.21 m. From the figure, one can see the trajectories of moving viscoplastic material, which are directed from the area of the viscoplastic layer into the opening.

It is of interest to develop heaving in time. Figure 3 shows the offset of the "reference" point in time. Note that this is a direct line over the entire time interval until the end of the calculation. This is due to the fact that the strain rate was adopted as a constant. In this case, the value of the final displacement is directly proportional to value C_1.

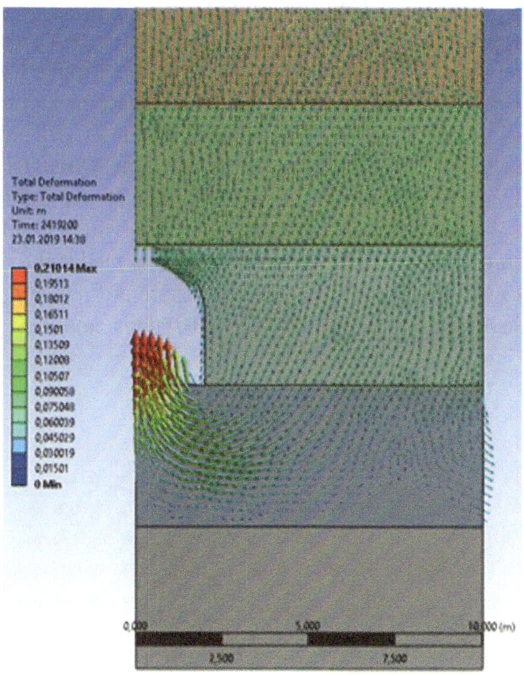

Fig. 2. Soil displacement vectors

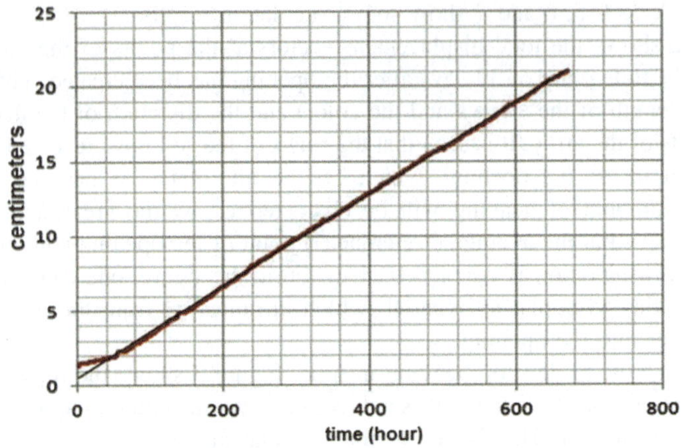

Fig. 3. The offset of the "reference" point in time

We note the following circumstance. Practice shows that in real conditions, there is usually no such straightness. Over time, the growth rate of the "reference" point decreases and the corresponding curve bends downwards, what is obvious in accordance with a general physical point of view. Thus, the assumption $C_3 = 0$ for large time intervals turns out to be untenable. For a more adequate description of the heaving process, it is necessary to determine two constants (C_1 and C_3) from the condition that the maximum soil rise in 30 days, as before, is ~ 0.2 m. Obviously, the value of C_1, previously defined, in this case is unacceptable. Actually, the values of these parameters themselves in this situation do not matter, since the study conducted is not tied to any particular mining situation. The calculations performed only show that it is possible in principle to solve the inverse problem if there are relevant experimental data. Gouging quickly develops in the first few (up to 10) days, and then the rate of increase gradually drop.

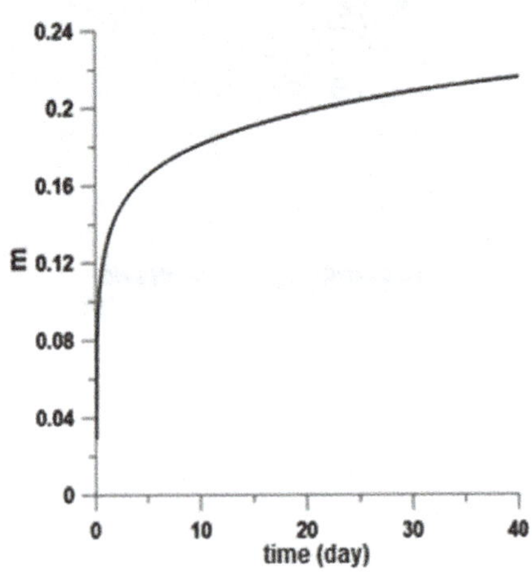

Fig. 4. Soil raising at a "reference" point over time

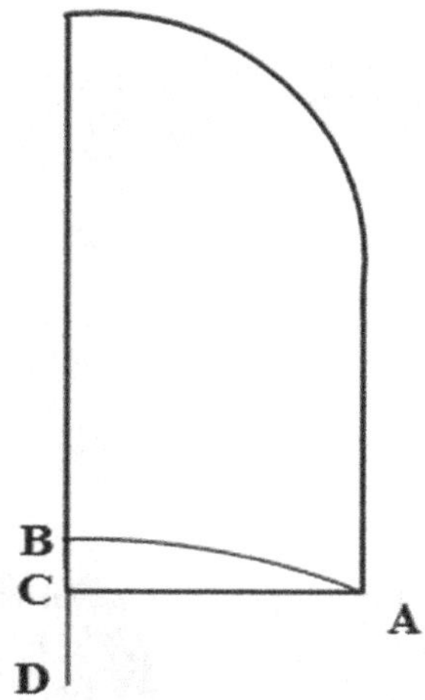

Fig. 5. Pattern of soil deformation in workings

From practice it is known such a phenomenon as a break in the soil; a situation where, in addition to raising the soil along its centerline, a crack with diverging shores passes along the opening. Obviously, this is an expanding tensile crack.

This situation is also easily modeled within the framework of the stated approach. An analysis of the stresses in the soil rocks shows that their horizontal component is tensile, despite the compressive initial stresses in massif. This is due to the fact that rising soil rocks form a "arch" (Fig. 5). The length of the generatrix of the "arch" AB is significantly longer than the corresponding distance AC at the soil level, i.e. line AC stretches to AB. This means that, even taking into account the compression initial stresses, the occurrence of tensile horizontal stresses, leading to the formation of tensile cracks, is possible.

If conditions are set on the BCD line with low tensile strength, then a discontinuity in the calculations will be revealed, as shown in Fig. 6.

Returning to the development of contour convergence, consider the situation when coal, like soil rocks, also has creep properties. Figure 7 shows the total displacements of the rock mass in this case, where in addition to the heaving of soil rocks, there is a significant deformation of the lateral walls of the opening. For seeing imagery, the scale of the displacements of the points in rock massive in the figure is increased by 10 times. It is also controlled by C_1 and C_3, but different from previous, namely for coal. By choosing the appropriate values of the C_1 and C_3 parameters for each of the rocks in the vicinity of the mine, a model can be built to evaluate the time development of the deformation processes in the massif for a long period of time.

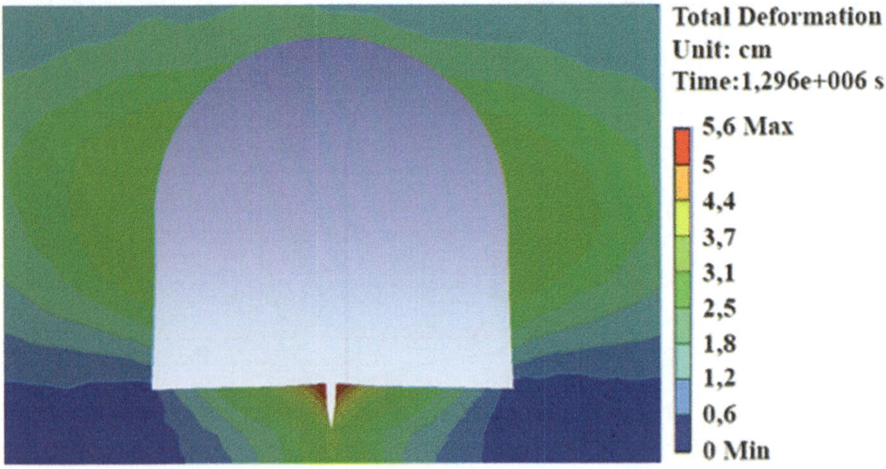

Fig. 6. The distribution of horizontal stresses around the opening 30 days after its excavation

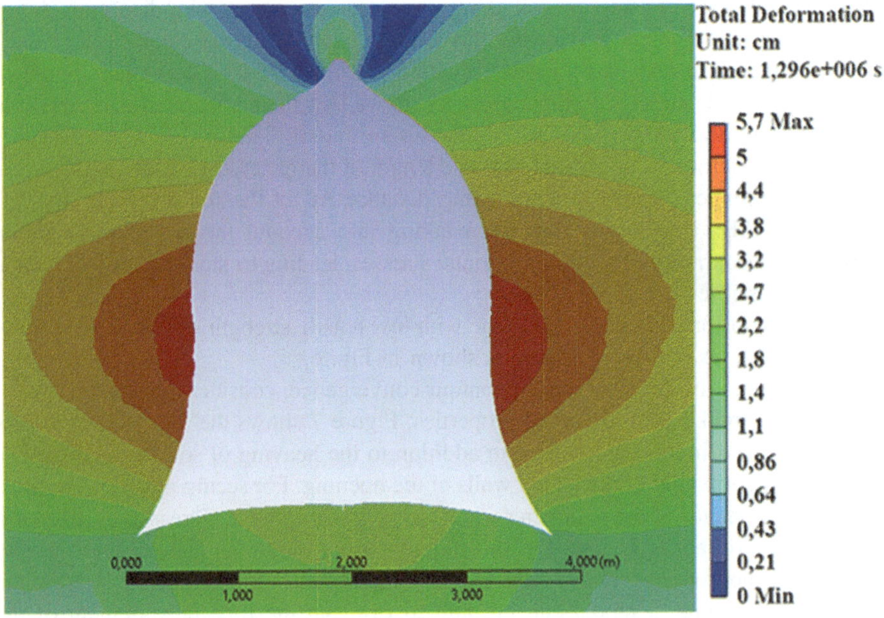

Fig. 7. Deformation of the contour of production in the presence of creeping rocks in the soil and sides of the opening

And now consider the basic laws of the processes occurring during the undermining of the soil in order to eliminate heaving. It is known that such actions are performed 2–4 times, after which the heaving process at this place completely stops.

Let us proceed from the heaving curve obtained earlier and shown in Fig. 4. It is also schematically shown in Fig. 8 (line 0abc), where z is the vertical distance from the initial position of the soil at the beginning of heaving to the current position at time t. When the rising reaches the maximum technologically permissible value h_n, which is determined by the conditions for the preservation of the possibility of functioning of the mine, a bulging rock mass is removed. This layer of soil is removed and the soil passes back to the state shown in Fig. 8 by point t_1, i.e. to the initial level, $z = 0$.

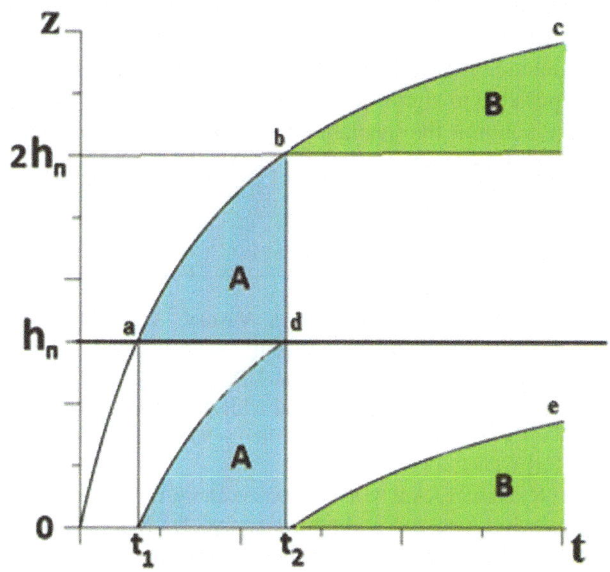

Fig. 8. Pattern of consecutive soil "underminings"

The subsequent rising will occur along the curve t_1d, which is obtained from the ab by transferring it down by h_n. Thus, it is considered that the speed of the process of heaving is not influenced by the "undermining" and it continues, following the general regularity of 0abc, but each time at a new level. When the heaving reaches again the value of h_n, i.e. point d everything repeats. It should be noted that the time intervals between "undermining" gradually increase and already after the second (in this case), the t_2e curve (part of the curve of bc demolished down by $2h_n$) can no longer cross the limit line $z = h_n$. Rising at this step is already small and in the subsequent "demolitions" will not be required. Thus, the shift of the "reference" point because of a series of "undermining" will occur in accordance with the curve $0at_1dt_2e$ for given technological rising h_n.

3 Conclusions

In the mountain practice, there are often situations when the rocks over time experience large deformations comparable to the size of the excavated workings, which can lead to a complete loss of their functional purposes. In this case, the displacement of the contour of the workings can be of the order of a meter in a relatively short period of time - 1 ÷ 2 months. These values vary greatly and depend primarily on the properties of the rocks in which the excavations are completed. Largely, coal, salt rocks and other so-called "pulling" rocks are subject to this.

The use of the theory of viscoplasticity makes it possible in many respects to solve the problems associated with taking into account large displacements within the framework of continuum mechanics and to obtain final results for various mining engineering situations. Directly specifying the dependence of the strain rate on the determining parameters—stresses, time, temperature—in the form of a particular function, makes it possible to significantly simplify the solution. This approach allows us to include in the solution the available experimental data for the selection of the parameters of the functions used.

References

1. Kazanin, O.I., Sidorenko, A.A., Ilinec A.A., Vasilev, V.F.: Chislennye issledovaniya pucheniya pochvy shtrekov pri primenenii razgruzochnyh shchelej na shahte TALDINSKAYA-ZAPADNAYA 2. Izvestiya TulGU. Nauki o Zemle Vyp **3,** 179–187 (2018). (Numerical studies of soil drift heaving with the use of unloading slots at the mine TALDINSKAYA-WESTERN 2) CONTOUR CONVERGENCE REGULARITIES FOR OPENINGS IN COAL 11
2. Kovalenko, V.V., Ryazancev, A.P.: Obosnovanie parametrov sposoba borby s pucheniem porod pochvy v usloviyah ugol'nyh shaht: monografiya. D.: NGU, 119 (2013). (Justification of the parameters of the method of combating the heaving of soil rocks in coal mines: monograph)
3. Buslov, A.S., Kalachyova, N.: Reologiya vyazkoplasticheskih opolznej v prirodnom sostoyanii i pri stabilizacii ih svayami. Vestnik MGSU **11**, 4554 (2012). (Rheology of viscous-plastic landslides in the natural state and with their stabilization by piles)
4. Vyalov, S.S.: Reologicheskie osnovy mekhaniki gruntov. M.:Vyssh. shk., 447 (1978). (Rheological basis of soil mechanics)
5. Ogibalov, P.M., Mirzadzhanzade, A.H.: Nestacionarnye dvizheniya vyazkoplastichnyh sred. M.: Izd-vo MGU, 373 (1977). (Non-stationary motions of viscoplastic media)
6. Anand, L., Kothari, M.: A computational procedure for rate independent crystal plasticity. J. Mech. Phys. Solids **44**(4), 525–558 (1996)
7. Kuzmin, S.V., Salvasser, I.A., Meshkov, S.A.: Mekhanizm razvitiya pucheniya porod pochvy i sposoby bor'by s nim. GIAB **3**, 120–126 (2014). (The mechanism of development of soil heaving and ways to combat it)
8. Smirnov, A.V.: Obobshchenie naturnyh issledovanij processa deformirovaniya porodnogo massiva v okrestnosti protyazhennyh vyrabotok. Izvestiya vuzov. Gornyj zhurnal **5**, 75–80 (2015). (Generalization of field studies of the rock massif deformation process in the vicinity of extended workings)
9. Cherdancev, N.V.: Reshenie zadachi o puchenii porodnogo sloya pochvy vyrabotki, projdennoj po ugol'nomu plastu. Izvestiya vuzov. Gornyj zhurnal **8,** 32–39 (2016). (The solution of the problem of heaving of the soil in excavation, passed through the coal seam)

On Finite Element Analysis in Generalized Mechanics

Hua Yang(✉), Bilen Emek Abali, and Wolfgang H. Müller

Institute of Mechanics, Technische Universität Berlin,
Einsteinufer 5, 10587 Berlin, Germany
hua.yang@campus.tu-berlin.de

Abstract. Different numerical implementations have been proposed in the literature for computation in generalized mechanics. A computational benchmark problem is beneficial to highlight the differences or even validate an approach. We briefly present the strain gradient elasticity theory and its weak form. A relatively simple analytic solution in strain gradient elasticity theory is shown. The closed-form solutions for a plate under simple shearing in plane strain are investigated for different boundary conditions. Moreover, IsoGeometric Analysis (IGA) within the finite element method is used. By employing open source packages developed under the FEniCS project, we develop a general framework and use the analytical solution to verify the numerical implementation. Comparison of the computation to the closed-form solutions shows that the numerical implementation is accurate and reliable.

Keywords: Strain gradient elasticity · Finite element method · Isogeometric analysis

1 Introduction

Additive manufacturing techniques enable constructing designs with substructures of different length scales resulting in tailored macroscopic deformation behaviors. Because the material response depends on the substructure, such structures are called *metamaterials*. Pantographic structures, which have been largely studied by dell'Isola et al. (2015, 2016, 2019); De Angelo et al. (2019); Barchiesi et al. (2018, 2019a), are a special kind of metamaterials with a substructure composed of two orthogonal arrays of fibers connected by pivots. Similar structures cause unexpected deformation patterns (Leismann and Mahnken 2015; Kochmann and Bertoldi 2017) and experimental analysis is a valuable tool to comprehend the behavior (Turco et al. 2017; Ganzosch et al. 2018; Barchiesi et al. 2019b; Juritza et al. 2019). A direct finite element simulation involving details of the microscopic substructures can also be employed to inspect the deformation response, being able to capture the deformation behavior effected by the substructure made of polyamide shown in Yang et al. (2018); Yang and Müller (2019) or in textiles with twinning as in Böhlke et al. (2007); Placidi

D. A. Indeitsev and A. M. Krivtsov (Eds.): APM 2019, LNME, pp. 233–245, 2020.
https://doi.org/10.1007/978-3-030-49882-5_23

et al. (2016); Laudato and Barchiesi (2019). Nevertheless, such direct computation is quite time-consuming due to the enormous number of degrees of freedom and the computational burden is still a challenge even for modern computers. Hence, different types of reduced order models are suggested, for instance by using Kirchhoff space rods (Greco and Cuomo 2016), Kirchhoff–Love shells presented in Greco et al. (2018) as well as plane-curved beams displayed in Cazzani et al. (2016). The reduced models have limitations because they are designed and tested for a particular substructure under only a few loading scenarios. A more general approach is known as homogenization, which has been an active research field for many years (Forest et al. 1999; Kochmann and Venturini 2013; Arabnejad and Pasini 2013; Rahali et al. 2015; Barboura and Li 2018; Ganghoffer et al. 2018; Solyaev et al. 2019). After a successful homogenization the metamaterial is modeled directly on the macroscale by using generalized continuum mechanics, which started with the works of Toupin (1962); Mindlin (1965); Eringen (1968) and have been under investigation, especially in the last decade (Auffray et al. 2009; Altenbach and Eremeyev 2009; Askes and Aifantis 2011; Polizzotto 2013a, 2013b; Auffray et al. 2015; Ivanova and Vilchevskaya 2016; Müller and Vilchevskaya 2017; Abdoul-Anziz and Seppecher 2018; Barchiesi and Khakalo 2019).

Various finite elements implementations have been proposed in the literature for the solution of boundary value problems involving the strain-gradient effect (Jeong et al. 2009; Cuomo et al. 2014; Abali et al. 2015; Niiranen et al. 2016; Reiher et al. 2016). In this paper, we present a numerical implementation of strain gradient elasticity based on IsoGeometric Analysis (IGA), as studied in details in Fischer et al. (2011); Rudraraju et al. (2014); Hughes et al. (2005); Khakalo and Niiranen (2017); Makvandi et al. (2018). The geometric model is discretized within patches (like the elements in the finite element method). Within the patches the continuity of unknowns is chosen as high as necessary by using NURBS shape functions. As the CAD model topology is represented by NURBS, too, one of the advantages of IGA is that the geometric model is built exactly by means of the shape functions generating the unknowns. In the context of strain gradient elasticity this method is expected to provide accurate results, because the displacement is the unknown with a necessarily higher order continuity. In this paper the basic formulation of strain gradient elasticity is summarized and the weak form of strain gradient elasticity is presented. A simple analytic solution of a strain gradient elasticity problem and its numerical solutions using IGA are shown and compared. The IGA implementation is based on making use of the FEniCS project and was developed by Kamensky and Bazilevs (2019). FEniCS is an open-source computing platform for solving partial differential equations (PDEs), it enables symbolic differentiation for linearization, see Abali (2017) for more engineering applications solved by using FEniCS.

2 Strain Gradient Elasticity

Conventional continuum mechanics theories assume that the stress in a material point is a function of strain of the same point. This local assumption has long been proved to be frequently adequate. However, when the wavelength of a deformation field is comparable to the micro-structural length scale of the material, the assumption must be questioned, because the behavior of material at a point is influenced by the deformation of the neighboring point. In other words, in this case not only the strain $\varepsilon_{ij} = (u_{i,j} + u_{j,i})/2$ but also the strain gradient $\varepsilon_{ij,k} = (u_{i,jk} + u_{j,ik})/2$ should be taken into account. Herein $\boldsymbol{u}$ is the displacement field and a comma means a partial differentiation in space. All fields are expressed in Cartesian coordinates.

In this section, the concepts of strain gradient elasticity will be briefly repeated. We use the theory of classical analytical mechanics and define a Lagrange density $\mathfrak{L}$, which depends on the primitive variables ϕ_A and its derivatives. For details of the formulations and justification, we refer to Abali et al. (2015, 2017). By neglecting the kinetic energy, body forces, and boundary terms acting on edges according to least action principles, for a domain Ω, we have the following integral form:

$$\begin{aligned} \int_\Omega \Big(\frac{\partial \mathfrak{L}}{\partial \phi_A}\delta\phi_A + \frac{\partial \mathfrak{L}}{\partial \phi_{A,i}}\delta\phi_{A,i} \\ + \frac{\partial \mathfrak{L}}{\partial \phi_{A,ij}}\delta\phi_{A,ij}\Big) dV + \int_{\partial\Omega} \Big(\frac{\partial W_s}{\partial \phi_A}\delta\phi_A + \frac{\partial W_s}{\partial \phi_{A,i}}\delta\phi_{A,i}\Big) dA = 0, \end{aligned} \tag{1}$$

where W_s refers to the external work done on the surface or edge; dV and dA denote infinitesimal volume and surface elements, respectively. In strain gradient elasticity theory, the primitive variable is the displacement $\phi_A = u_i$; for nonpolar homogeneous materials, we propose the following Lagrange density:

$$\mathfrak{L} = -w. \tag{2}$$

w is the stored deformation energy density and depends on the first and second gradient of the primitive variable, or equivalently,

$$w = w(\varepsilon_{ij}, \varepsilon_{ij,k}). \tag{3}$$

By inserting the Lagrange density Eq. (1) becomes

$$\int_\Omega \Big(-\frac{\partial w}{\partial u_{i,j}}\delta u_{i,j} - \frac{\partial w}{\partial u_{i,jk}}\delta u_{i,jk}\Big) dV + \int_{\partial\Omega} \Big(\frac{\partial W_s}{\partial u_i}\delta u_i + \frac{\partial W_s}{\partial u_{i,j}}\delta u_{i,j}\Big) dA = 0. \tag{4}$$

Equation (4) is also known as a weak form. After applying integration by parts and GAUSS'S theorem, we obtain

$$\begin{aligned} &\int_{\Omega} \Big(-\frac{\partial w}{\partial u_{i,j}} + \big(\frac{\partial w}{\partial u_{i,jk}}\big)_{,k} \Big)_{,j} \delta u_i dV \\ &+ \int_{\partial\Omega} \left(n_j \Big(\frac{\partial w}{\partial u_{i,j}} - \big(\frac{\partial w}{\partial u_{i,jk}}\big)_{,k} \Big) - \frac{\partial W_s}{\partial u_i} \right) \delta u_i dA \\ &+ \int_{\partial\Omega} \Big(n_k \frac{\partial w}{\partial u_{i,jk}} - \frac{\partial W_s}{\partial u_{i,j}} \Big) \delta u_{i,j} dA = 0. \end{aligned} \tag{5}$$

We assume that W_s on NEUMANN boundaries is given by the expression:

$$W_s = t_i u_i + m_{ij} u_{i,j}, \tag{6}$$

where t_i and m_{ij} should fulfill the following conditions:

$$t_i = \frac{\partial W_s}{\partial u_i} = n_j \Big(\frac{\partial w}{\partial u_{i,j}} - \big(\frac{\partial w}{\partial u_{i,jk}}\big)_{,k} \Big), \qquad m_{ij} = \frac{\partial W_s}{\partial u_{i,j}} = n_k \frac{\partial w}{\partial u_{i,jk}}. \tag{7}$$

These are also known as traction and double traction, respectively. From Eq. (5), we can also obtain the governing equations, which read

$$\Big(\frac{\partial w}{\partial u_{i,j}} - \big(\frac{\partial w}{\partial u_{i,jk}}\big)_{,k} \Big)_{,j} = 0. \tag{8}$$

3 Analytical Solution

We present an analytical solution in the framework of strain gradient elasticity which will be used for verifying the numerical implementations in the following. The strain energy density $w = w(\varepsilon_{ij}, \varepsilon_{ij,k})$ for centro-symmetric and isotropic materials are given by

$$w(\varepsilon_{ij}, \varepsilon_{ij,k}) = \frac{1}{2} \varepsilon_{ij} C_{ijkl} \varepsilon_{kl} + \frac{1}{2} \varepsilon_{ij,k} D_{ijklmn} \varepsilon_{lm,n}. \tag{9}$$

The fourth-order classical stiffness tensor C_{ijkl} and the sixth-order strain gradient stiffness tensor D_{ijklmn} read

$$\begin{aligned} C_{ijkl} =& c_1 \delta_{ij} \delta_{kl} + c_2 (\delta_{ik} \delta_{jl} + \delta_{il} \delta_{jk}), \\ D_{ijklmn} =& c_3 (\delta_{ij} \delta_{kl} \delta_{mn} + \delta_{in} \delta_{jk} \delta_{lm} + \delta_{ij} \delta_{km} \delta_{ln} + \delta_{ik} \delta_{jn} \delta_{lm}) \\ &+ c_4 \delta_{ij} \delta_{kn} \delta_{ml} \\ &+ c_5 (\delta_{ik} \delta_{jl} \delta_{mn} + \delta_{im} \delta_{jk} \delta_{ln} + \delta_{ik} \delta_{jm} \delta_{ln} + \delta_{il} \delta_{jk} \delta_{mn}) \\ &+ c_6 (\delta_{il} \delta_{jm} \delta_{kn} + \delta_{im} \delta_{jl} \delta_{kn}) \\ &+ c_7 (\delta_{il} \delta_{jn} \delta_{mk} + \delta_{im} \delta_{jn} \delta_{lk} + \delta_{in} \delta_{jl} \delta_{km} + \delta_{in} \delta_{jm} \delta_{kl}) \end{aligned} \tag{10}$$

where c_1, c_2 are two LAMÉ constants, and c_3, c_4, c_5, c_6, c_7 are five additional material parameters. After inserting Eq. (9) into Eq. (8) and applying chain rule we find

$$\begin{aligned} \frac{\partial w}{\partial u_{i,j}} &= \frac{\partial w}{\partial \varepsilon_{kl}} \frac{\partial \varepsilon_{kl}}{\partial u_{i,j}} = C_{ijkl}\varepsilon_{kl}, \\ \frac{\partial w}{\partial u_{i,jk}} &= D_{ijklmn}\varepsilon_{lm,n}. \end{aligned} \tag{11}$$

By using Eq. (10) and Eq. (11) we obtain

$$\frac{\partial w}{\partial u_{i,j}} = c_1 \delta_{ij}\varepsilon_{kk} + 2c_2\varepsilon_{ij}, \tag{12}$$

$$\begin{aligned} \frac{\partial w}{\partial u_{i,jk}} =& c_3(\delta_{ij}\varepsilon_{km,m} + \delta_{jk}\varepsilon_{mm,i} + \delta_{ij}\varepsilon_{nk,n} + \delta_{ik}\varepsilon_{nk,n}) + c_4\delta_{ij}\varepsilon_{ll,k} \\ &+ c_5(\delta_{ik}\varepsilon_{jn,n} + \delta_{jk}\varepsilon_{li,l} + \delta_{ik}\varepsilon_{nj,n} + \delta_{jk}\varepsilon_{im,m}) \\ &+ c_6(\varepsilon_{ij,k} + \varepsilon_{ji,k}) + c_7(\varepsilon_{ik,j} + \varepsilon_{ki,j} + \varepsilon_{jk,i} + \varepsilon_{kj,i}). \end{aligned} \tag{13}$$

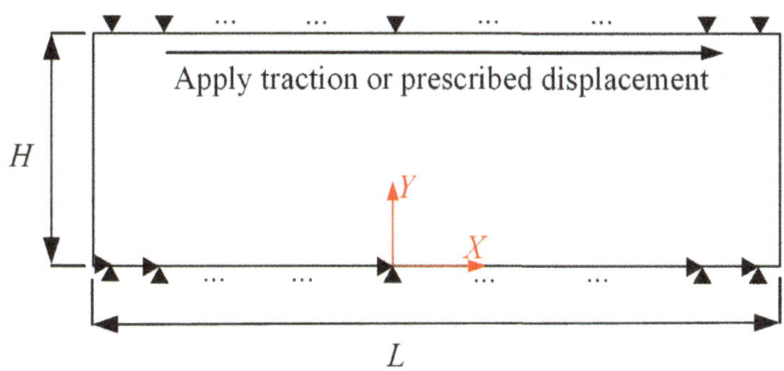

Fig. 1. Schematic of a plate under shear.

Consider the problem of simple shearing in plane strain of a plate of thickness H, extending infinitely in the lateral direction, analogously to Zervos et al. (2009). The infinite plate will be modeled with a finite length, L. In order to eliminate boundary effects, the length is chosen suitably large, $L = 100H$. The bottom boundary of the model is fixed in both directions (X and Y), and the lateral boundaries are free of constraints and traction or double traction. Now we use a semi-inverse ansatz and reduce to a one-dimensional problem with $u_x = u(y)$ such that only $u_{x,y} = u'$ and $u_{x,yy} = u''$ are non-zero. In this one-dimensional case, the non-zero quantities can be obtained (Fig. 1)

$$\frac{\partial w}{\partial u_{x,y}} = c_2 u', \tag{14}$$

$$\frac{\partial w}{\partial u_{x,yy}} = (c_5 + c_6 + c_7)u''. \tag{15}$$

Substituting Eq. (14) and Eq. (15) into Eq. (8), we obtain the governing equation:

$$c_2 u'' - (c_5 + c_6 + c_7)u'''' = 0. \tag{16}$$

The general solution of the fourth order differential equation is

$$u_x(y) = K_1 + K_2 y + K_3 \sinh(y/l) + K_4 \cosh(y/l), \tag{17}$$

where $l = \sqrt{(c_5 + c_6 + c_7)/c_2}$ and K_1, K_2, K_3, K_4 are four integration constants.

In order to determine the four integration constants, four boundary conditions should be applied. Note that the boundary conditions should satisfy Eq. (5). One can prescribe two *classical* boundary conditions for $u(0)$ and $u(H)$, and two *non-classical* conditions by describing $u'(0)$ and $u'(H)$. Actually, in view of Eq. (5), in the case of the classical boundary conditions, either displacements u_i or the traction t_i should be prescribed. In the case of higher order boundary conditions, the gradient of displacement $u_{i,j}$ or double traction m_{ij} should be prescribed as shown in Tsepoura et al. (2002). In what follows, two specific cases under different boundary conditions will be investigated.

3.1 Shear Test, Prescribed Displacement

Consider the following boundary conditions:

$$u_x(0) = 0, \qquad u_x(H) = u_p, \qquad m_{xy}(0) = 0, \qquad u'(H) = 0. \tag{18}$$

On its bottom the plate is fixed by setting $u(0) = 0$. A displacement u_p is prescribed on the top of the plate along the x direction. As discussed above, two more higher-order boundary conditions are needed to close the system of equations. We intentionally specify the double traction $m_{xy}(0) = 0$ as well as displacement gradient $u'(H) = 0$. Using Eq. (17), Eq. (18) and by taking Eq. (7), Eq. (14), Eq. (15) into account, we obtain

$$\begin{aligned}
u_x(0) &= K_1 + K_4 = 0, \\
u_x(H) &= K_1 + K_2 H + K_3 \sinh\left(\frac{H}{l}\right) + K_4 \cosh\left(\frac{H}{l}\right) = u_p, \\
m_{xy}(0) &= (c_5 + c_6 + c_7)\frac{K_4}{l^2} = 0, \\
u'(H) &= K_2 H + \frac{K_3}{l}\cosh\left(\frac{H}{l}\right) + \frac{K_4}{l}\sinh\left(\frac{H}{l}\right) = 0.
\end{aligned} \tag{19}$$

After solving the four equations, the four integration constants read

$$\begin{aligned}
K_1 &= 0, \qquad K_2 = \frac{u_p \cosh(H/l)}{H\cosh(H/l) - l\sinh(H/l)}, \\
K_3 &= -\frac{u_p l}{H\cosh(H/l) - l\sinh(H/l)}, \qquad K_4 = 0.
\end{aligned} \tag{20}$$

Setting $H = 0.1\,\mathrm{mm}$, $L = 10.0\,\mathrm{mm}$, prescribed displacement, $u_p = 0.01\,\mathrm{mm}$, YOUNG's modulus, $E = 400\,\mathrm{MPa}$, POISSON's ratio, $\nu = 0.49$, with additional parameters, $c_5 + c_6 + c_7 = 0.2\,\mathrm{N}$, the analytical ($u^{\mathrm{a}}$) and numerical solution (u^{n}) are studied along the Y axis. It is shown in Fig. 2. A good agreement between the analytic and numerical solutions indicates that the numerical solutions is accurate and reliable.

In order to interpret the role of the additional parameters, different values of $c_5 + c_6 + c_7$ are specified and the results of displacements along Y axis are shown in Fig. 3. It is observed that with increasing parameters the strain gradient effect becomes more dominant.

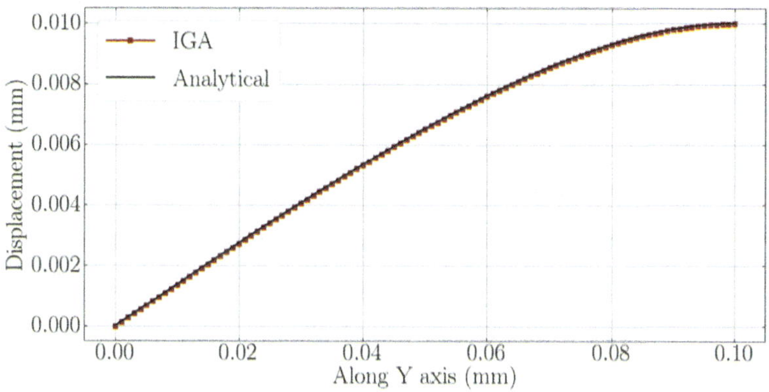

Fig. 2. Comparisons of displacements along Y axis of the plate between the analytic and IGA solutions.

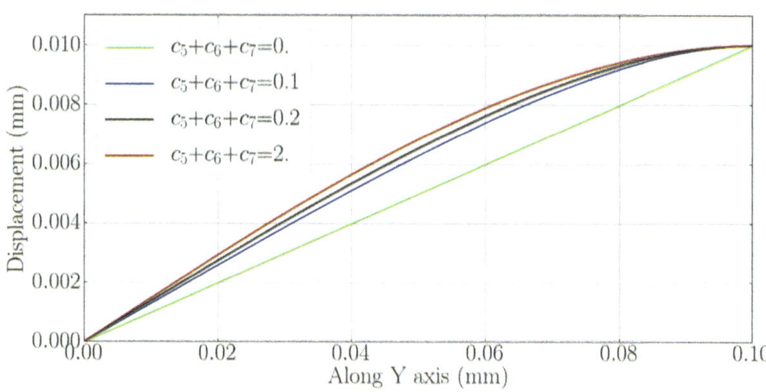

Fig. 3. Comparisons of displacements for different values of the additional material parameters in case of prescribed displacement.

The numerical solution is also evaluated here by calculating the so-called L^2 norm of error between the numerical and analytical results, given by

$$e_{L^2} = \left[\int ||u^{\mathrm{n}} - u^{\mathrm{a}}|| dV \right]^{\frac{1}{2}}. \tag{21}$$

As shown in Table 1 with relatively small L^2 errors for different cases of $c_5 + c_6 + c_7$, the numerical solutions are all convergent and have enough accuracy.

Table 1. The L^2 errors calculated for different value of $c_5 + c_6 + c_7$ in the case of prescribed displacement boundary condition.

$c_5 + c_6 + c_7$	0 N	0.1 N	0.2 N	2.0 N
L^2 error	2.28×10^{-5}	1.29×10^{-6}	1.54×10^{-6}	5.74×10^{-6}

3.2 Shear Test Steered by Traction

In order to validate the numerical implementation further, another set of boundary conditions is considered,

$$u_x(0) = 0, \qquad t_x(H) = \hat{t}, \qquad u'(0) = 0, \qquad m_{xy}(H) = 0. \tag{22}$$

The bottom of the plate is fixed by setting $u(0) = 0$. A traction $t_x(H) = \hat{t}$ is applied on the top of the plate along x direction. Moreover, the higher-order boundary conditions $u'(0) = 0$ and $m_{xy}(H) = 0$ are applied. Likewise, by using Eq. (17) and Eq. (18) and by taking Eqs. (7), (14), (15) into account we obtain

$$\begin{aligned} u_x(0) &= K_1 + K_4 = 0, \\ t_x(H) &= c_2\left(K_2 + \frac{K_3}{l}\cosh\left(\frac{H}{l}\right) + \frac{K_4}{l}\sinh\left(\frac{H}{l}\right)\right) \\ &\quad - (c_5 + c_6 + c_7)\left(\frac{K_3}{l^3}\cosh\left(\frac{H}{l}\right) + \frac{K_4}{l^3}\sinh\left(\frac{H}{l}\right)\right) = \hat{t}, \\ u'(0) &= K_2 + \frac{K_3}{l} = 0, \\ m_{xy}(H) &= (c_5 + c_6 + c_7)\left(\frac{K_3}{l^2}\sinh\left(\frac{H}{l}\right) + \frac{K_4}{l^2}\cosh\left(\frac{H}{l}\right)\right) = 0. \end{aligned} \tag{23}$$

After solving the equations, the four integration constants read

$$\begin{aligned} K_1 &= l^3\hat{t}\sinh\left(\frac{H}{l}\right)\xi, & K_2 &= -l^2\hat{t}\cosh\left(\frac{H}{l}\right)\xi, \\ K_3 &= l^3\hat{t}\cosh\left(\frac{H}{l}\right)\xi, & K_4 &= -l^3\hat{t}\sinh\left(\frac{H}{l}\right)\xi, \end{aligned} \tag{24}$$

with

$$\xi = \Big(c_2 \cosh\left(\frac{H}{l}\right) l^2 - \cosh^2\left(\frac{H}{l}\right)\big(c_2 l^2 - (c_5 + c_6 + c_7)\big) + \sinh^2\left(\frac{H}{l}\right)\big(c_2 l^2 - (c_5 + c_6 + c_7)\big)\Big)^{-1}. \tag{25}$$

Setting $H = 0.1\,\text{mm}$, $L = 10.0\,\text{mm}$, with an applied traction $\hat{t} = 1\,\text{MPa}$, YOUNG's modulus $E = 400\,\text{MPa}$, POISSON's ratio $\nu = 0.49$, and using additional parameters $c_5 + c_6 + c_7 = 0.3\,\text{N}$, a good agreement between analytical and numerical solution is acquired as shown in Fig. 4. Different values of $c_5 + c_6 + c_7$ are chosen, and the results of displacements along the Y axis are presented in Fig. 5, which clearly indicates a strain gradient effect for increasing additional material parameters $c_5 + c_6 + c_7$. As shown in Table 2 with relatively small L^2 norm of errors for different cases of $c_5 + c_6 + c_7$, the accuracy of numerical solutions is guaranteed.

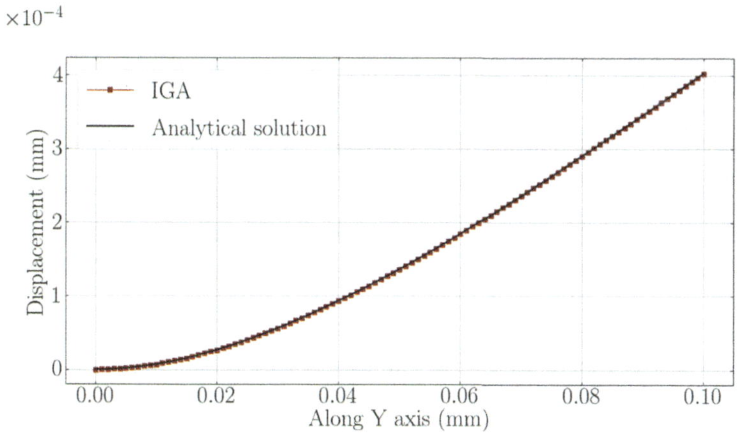

Fig. 4. Comparisons of displacements along Y axis for the analytic and IGA solutions.

Table 2. The L^2 errors calculated for different value of $c_5 + c_6 + c_7$ in case of applied traction.

$c_5 + c_6 + c_7$	0 N	0.15 N	0.3 N	3.0 N
L^2 error	1.94×10^{-6}	3.33×10^{-7}	3.60×10^{-7}	1.62×10^{-7}

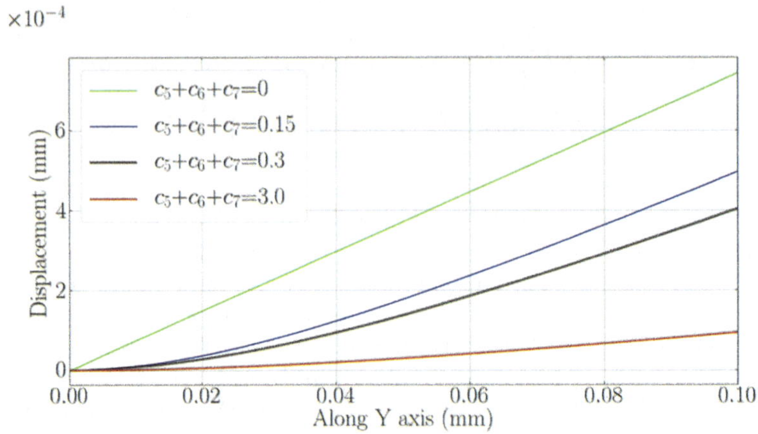

Fig. 5. Comparison of displacements for different values of the additional material parameters in case of applied traction.

4 Discussion and Conclusion

In this paper the basic formulation of strain gradient elasticity and the weak form was revisited. An analytic solution for a simple case of strain gradient elasticity was presented in order to validate the numerical implementation of IGA for two cases of different boundary conditions and different specifications of additional parameters. All the numerical implementations are based on FEniCS projects which provides a novel tool also for isogeometric analysis. Computational results demonstrate the accuracy of the numerical implementation. We refer to Placidi et al. (2015, 2017) for closed-form solutions of more sophisticated cases under different boundary conditions, to be used for assessing the merits and shortcomings of the implementation in terms of accuracy, simplicity and computational efficiency. Such a study is left to future work.

References

Abali, B.E.: Computational Reality, Solving Nonlinear and Coupled Problems in Continuum Mechanics. Advanced Structured Materials, vol. 55. Springer, Singapore (2017)

Abali, B.E., Müller, W.H., Eremeyev, V.A.: Strain gradient elasticity with geometric nonlinearities and its computational evaluation. Mech. Adv. Mater. Mod. Process. **1**(1), 1–11 (2015)

Abali, B.E., Müller, W.H., dell'Isola, F.: Theory and computation of higher gradient elasticity theories based on action principles. Arch. Appl. Mech. **87**(9), 1495–1510 (2017)

Abdoul-Anziz, H., Seppecher, P.: Strain gradient and generalized continua obtained by homogenizing frame lattices. Math. Mech. Complex Syst. **6**(3), 213–250 (2018)

Altenbach, H., Eremeyev, V.: On the linear theory of micropolar plates. ZAMM-J. Appl. Math. Mech./Z. für Angew. Math. Mechanik **89**(4), 242–256 (2009)

Arabnejad, S., Pasini, D.: Mechanical properties of lattice materials via asymptotic homogenization and comparison with alternative homogenization methods. Int. J. Mech. Sci. **77**, 249–262 (2013)
Askes, H., Aifantis, E.C.: Gradient elasticity in statics and dynamics: an overview of formulations, length scale identification procedures, finite element implementations and new results. Int. J. Solids Struct. **48**(13), 1962–1990 (2011)
Auffray, N., Bouchet, R., Brechet, Y.: Derivation of anisotropic matrix for bi-dimensional strain-gradient elasticity behavior. Int. J. Solids Struct. **46**(2), 440–454 (2009)
Auffray, N., dell'Isola, F., Eremeyev, V.A., Madeo, A., Rosi, G.: Analytical continuum mechanics à la Hamilton-Piola least action principle for second gradient continua and capillary fluids. Math. Mech. Solids **20**(4), 375–417 (2015)
Barboura, S., Li, J.: Establishment of strain gradient constitutive relations by using asymptotic analysis and the finite element method for complex periodic microstructures. Int. J. Solids Struct. **136**, 60–76 (2018)
Barchiesi, E., Khakalo, S.: Variational asymptotic homogenization of beam-like square lattice structures. Math. Mech. Solids **24**, 3295–3318 (2019). https://doi.org/10.1177/1081286519843155
Barchiesi, E., Laudato, M., Di Cosmo, F.: Wave dispersion in non-linear pantographic beams. Mech. Res. Commun. **94**, 128–132 (2018)
Barchiesi, E., Eugster, S.R., Placidi, L., dell'Isola, F.: Pantographic beam: a complete second gradient 1D-continuum in plane. Z. für Angew. Math. Phys. **70**(5), 135 (2019a)
Barchiesi, E., Ganzosch, G., Liebold, C., Placidi, L., Grygoruk, R., Müller, W.H.: Out-of-plane buckling of pantographic fabrics in displacement-controlled shear tests: experimental results and model validation. Continuum Mech. Thermodyn. **31**(1), 33–45 (2019b)
Böhlke, T., Glüge, R., Klöden, B., Skrotzki, W., Bertram, A.: Finite element simulation of texture evolution and swift effect in nial under torsion. Modell. Simul. Mater. Sci. Eng. **15**(6), 619 (2007)
Cazzani, A., Malagù, M., Turco, E.: Isogeometric analysis of plane-curved beams. Math. Mech. Solids **21**(5), 562–577 (2016)
Cuomo, M., Contrafatto, L., Greco, L.: A variational model based on isogeometric interpolation for the analysis of cracked bodies. Int. J. Eng. Sci. **80**, 173–188 (2014)
De Angelo, M., Barchiesi, E., Giorgio, I., Abali, B.E.: Numerical identification of constitutive parameters in reduced-order bi-dimensional models for pantographic structures: application to out-of-plane buckling. Arch. Appl. Mech. **89**, 1333–1358 (2019)
dell'Isola, F., Lekszycki, T., Pawlikowski, M., Grygoruk, R., Greco, L.: Designing a light fabric metamaterial being highly macroscopically tough under directional extension: first experimental evidence. Z. für Angew. Math. Phys. **66**(6), 3473–3498 (2015)
dell'Isola, F., Giorgio, I., Pawlikowski, M., Rizzi, N.: Large deformations of planar extensible beams and pantographic lattices: heuristic homogenization, experimental and numerical examples of equilibrium. Proc. R. Soc. A: Math. Phys. Eng. Sci. **472**(2185), 20150,790 (2016)
dell'Isola, F., Seppecher, P., Spagnuolo, M., Barchiesi, E., Hild, F., Lekszycki, T., Giorgio, I., Placidi, L., Andreaus, U., Cuomo, M., Eugster, S.R., Pfaff, A., Hoschke, K., Langkemper, R., Turco, E., Sarikaya, R., Misra, A., De Angelo, M., D'Annibale, F., Bouterf, A., Pinelli, X., Misra, A., Desmorat, B., Pawlikowski, M., Dupuy, C., Scerrato, D., Peyre, P., Laudato, M., Manzari, L., Göransson, P., Hesch, C., Hesch, S., Franciosi, P., Dirrenberger, J., Maurin, F., Vangelatos, Z., Grigoropoulos, C., Melissinaki, V., Farsari, M., Müller, W., Abali, B.E., Liebold, C., Ganzosch, G., Harrison,

P., Drobnicki, R., Igumnov, L., Alzahrani, F., Hayat, T.: Advances in pantographic structures: design, manufacturing, models, experiments and image analyses. Continuum Mech. Thermodyn. **31**(4), 1231–1282 (2019)

Eringen, A.C.: Mechanics of micromorphic continua. In: Mechanics of Generalized Continua, pp. 18–35. Springer (1968)

Fischer, P., Klassen, M., Mergheim, J., Steinmann, P., Müller, R.: Isogeometric analysis of 2D gradient elasticity. Comput. Mech. **47**(3), 325–334 (2011)

Forest, S., Dendievel, R., Canova, G.R.: Estimating the overall properties of heterogeneous cosserat materials. Modell. Simul. Mater. Sci. Eng. **7**(5), 829 (1999)

Ganghoffer, J., Goda, I., Novotny, A., Rahouadj, R., Sokolowski, J.: Homogenized couple stress model of optimal auxetic microstructures computed by topology optimization. ZAMM-J. Appl. Math. Mech./Z. für Angew. Math. Mechanik **98**(5), 696–717 (2018)

Ganzosch, G., Hoschke, K., Lekszycki, T., Giorgio, I., Turco, E., Müller, W.: 3D-measurements of 3D-deformations of pantographic structures. Tech. Mechanik **38**(3), 233–245 (2018)

Greco, L., Cuomo, M.: An isogeometric implicit G1 mixed finite element for Kirchhoff space rods. Comput. Methods Appl. Mech. Eng. **298**, 325–349 (2016)

Greco, L., Cuomo, M., Contrafatto, L.: A reconstructed local B formulation for isogeometric Kirchhoff-Love shells. Comput. Methods Appl. Mech. Eng. **332**, 462–487 (2018)

Hughes, T.J., Cottrell, J.A., Bazilevs, Y.: Isogeometric analysis: CAD, finite elements, NURBS, exact geometry and mesh refinement. Comput. Methods Appl. Mech. Eng. **194**(39–41), 4135–4195 (2005)

Ivanova, E.A., Vilchevskaya, E.N.: Micropolar continuum in spatial description. Continuum Mech. Thermodyn. **28**(6), 1759–1780 (2016)

Jeong, J., Ramézani, H., Münch, I., Neff, P.: A numerical study for linear isotropic cosserat elasticity with conformally invariant curvature. ZAMM-J. Appl. Math. Mech./Z. für Angew. Math. Mechanik: Appl. Math. Mech. **89**(7), 552–569 (2009)

Juritza, A., Yang, H., Ganzosch, G.: Qualitative investigations of experiments performed on 3D-FDM-printed pantographic structures made out of PLA. In: New Achievements in Continuum Mechanics and Thermodynamics, pp. 197–209. Springer (2019)

Kamensky, D., Bazilevs, Y.: tIGAr: automating isogeometric analysis with FEniCS. Comput. Methods Appl. Mech. Eng. **344**, 477–498 (2019)

Khakalo, S., Niiranen, J.: Isogeometric analysis of higher-order gradient elasticity by user elements of a commercial finite element software. Comput. Aided Des. **82**, 154–169 (2017)

Kochmann, D.M., Bertoldi, K.: Exploiting microstructural instabilities in solids and structures: from metamaterials to structural transitions. Appl. Mech. Rev. **69**(5), 050,801 (2017)

Kochmann, D.M., Venturini, G.N.: Homogenized mechanical properties of auxetic composite materials in finite-strain elasticity. Smart Mater. Struct. **22**(8), 084,004 (2013)

Laudato, M., Barchiesi, E.: Non-linear dynamics of pantographic fabrics: modelling and numerical study. In: Wave Dynamics, Mechanics and Physics of Microstructured Metamaterials, pp. 241–254. Springer (2019)

Leismann, T., Mahnken, R.: Comparison of hyperelastic micromorphic, micropolar and microstrain continua. Int. J. Non-Linear Mech. **77**, 115–127 (2015)

Makvandi, R., Reiher, J.C., Bertram, A., Juhre, D.: Isogeometric analysis of first and second strain gradient elasticity. Comput. Mech. **61**(3), 351–363 (2018)

Mindlin, R.D.: Second gradient of strain and surface-tension in linear elasticity. Int. J. Solids Struct. **1**(4), 417–438 (1965)

Müller, W.H., Vilchevskaya, E.N.: Micropolar theory from the viewpoint of mesoscopic and mixture theories. Phys. Mesomech. **20**(3), 263–279 (2017)

Niiranen, J., Khakalo, S., Balobanov, V., Niemi, A.H.: Variational formulation and isogeometric analysis for fourth-order boundary value problems of gradient-elastic bar and plane strain/stress problems. Comput. Methods Appl. Mech. Eng. **308**, 182–211 (2016)

Placidi, L., Andreaus, U., Della Corte, A., Lekszycki, T.: Gedanken experiments for the determination of two-dimensional linear second gradient elasticity coefficients. Z. für Angew. Math. Phys. **66**(6), 3699–3725 (2015)

Placidi, L., Greco, L., Bucci, S., Turco, E., Rizzi, N.L.: A second gradient formulation for a 2D fabric sheet with inextensible fibres. Z. für Angew. Math. Phys. **67**(5), 114 (2016)

Placidi, L., Andreaus, U., Giorgio, I.: Identification of two-dimensional pantographic structure via a linear D4 orthotropic second gradient elastic model. J. Eng. Math. **103**(1), 1–21 (2017)

Polizzotto, C.: A second strain gradient elasticity theory with second velocity gradient inertia-part i: constitutive equations and quasi-static behavior. Int. J. Solids Struct. **50**(24), 3749–3765 (2013a)

Polizzotto, C.: A second strain gradient elasticity theory with second velocity gradient inertia-part ii: dynamic behavior. Int. J. Solids Struct. **50**(24), 3766–3777 (2013b)

Rahali, Y., Giorgio, I., Ganghoffer, J., dell'Isola, F.: Homogenization à la Piola produces second gradient continuum models for linear pantographic lattices. Int. J. Eng. Sci. **97**, 148–172 (2015)

Reiher, J.C., Giorgio, I., Bertram, A.: Finite-element analysis of polyhedra under point and line forces in second-strain gradient elasticity. J. Eng. Mech. **143**(2), 04016,112 (2016)

Rudraraju, S., Van der Ven, A., Garikipati, K.: Three-dimensional isogeometric solutions to general boundary value problems of Toupin's gradient elasticity theory at finite strains. Comput. Methods Appl. Mech. Eng. **278**, 705–728 (2014)

Solyaev, Y., Lurie, S., Barchiesi, E., Placidi, L.: On the dependence of standard and gradient elastic material constants on a field of defects. Math. Mech. Solids **25**, 35–45 (2019). https://doi.org/10.1177/1081286519861827

Toupin, R.A.: Elastic materials with couple-stresses. Arch. Ration. Mech. Anal. **11**(1), 385–414 (1962)

Tsepoura, K., Papargyri-Beskou, S., Polyzos, D., Beskos, D.: Static and dynamic analysis of a gradient-elastic bar in tension. Arch. Appl. Mech. **72**(6–7), 483–497 (2002)

Turco, E., Golaszewski, M., Giorgio, I., D'Annibale, F.: Pantographic lattices with non-orthogonal fibres: experiments and their numerical simulations. Compos. B Eng. **118**, 1–14 (2017)

Yang, H., Müller, W.H.: Computation and experimental comparison of the deformation behavior of pantographic structures with different micro-geometry under shear and torsion. J. Theor. Appl. Mech. **57**, 421–434 (2019)

Yang, H., Ganzosch, G., Giorgio, I., Abali, B.E.: Material characterization and computations of a polymeric metamaterial with a pantographic substructure. Z. für Angew. Math. Phys. **69**(4), 105 (2018)

Zervos, A., Papanicolopulos, S.A., Vardoulakis, I.: Two finite-element discretizations for gradient elasticity. J. Eng. Mech. **135**(3), 203–213 (2009)

Model of the Effect of Low Natural Concentrations of Hydrogen on Cylindrical Steel Samples

Aleksandr Zegzhda[1(✉)] and Vladimir Polyansky[2]

[1] Peter the Great St. Petersburg Polytechnic University, Saint Petersburg, Russia
alex.zegzhda@gmail.com

[2] Institute for Problems in Mechanical Engineering Russian Academy of Sciences (IPME RAS), Saint Petersburg, Russia

Abstract. Small amount of hydrogen concentration impact appears significantly in material fatigue and cracks propagation. To simulate a significant deterioration of the physicomechanical properties of the model, similar to what is happening in the experiments, a model with varying bilinear elastic-plastic properties is proposed, corresponding to the saturation of hydrogen after its accumulation in the actual structure for several years. The finite element method used to simulate the effect of degradation of the mechanical characteristics of the entire sample caused by hydrogen saturation. The simulation carried out for the case of cylindrical corset samples, for which there are extensive experimental data. Analysis of the calculation results shows that the behavior of the material under the influence of hydrogen occurs not only in the zone of elastic deformations. It is noticed that the degradation of the mechanical characteristics of the metal under the influence of hydrogen leads to a significant change in deformation and strength. The material model used for the entire sample allows one to describe experimentally observed effects.

Keywords: Hydrogen embrittlement · Tension model · FE simulation

1 Introduction

Hydrogen effect on high strength steel is well known problem due to the degradation of mechanical properties caused by hydrogen accumulation.

The influence of small natural concentrations of hydrogen on material properties is very considerable. It becomes observable at the level of about a single atom per 50000 atoms of matrix of the matter. On one hand, this influence of hydrogen cannot be ignored since, in many cases, it is the only explanation of changes in material properties in time and under external loading. On the other hand, this influence is difficult to describe due to smallness of this factor and variety of consequences. For example, under change in the hydrogen concentrations by fifty times at the level of one atom per 1000 matrix atoms the mechanical properties of the material can change from brittleness to superplasticity.

An important role is played by the bonding energy of hydrogen in material. It is known that, inside the materials, hydrogen is found in traps with different bonding

D. A. Indeitsev and A. M. Krivtsov (Eds.): APM 2019, LNME, pp. 246–255, 2020.
https://doi.org/10.1007/978-3-030-49882-5_24

energies. In steels the total hydrogen content is 0.1–40 ppm, while it is only hydrogen with a low binding energy that affects the strength, i.e. diffusively mobile hydrogen. Quite the contrary, the hydrogen with low bonding energy tends to zones of tensile stresses (Gorsky's effect [1]).

Accumulation of hydrogen in the destruction zone occurs both by the input from outside and by redistribution of natural hydrogen inside the material.

The well-known models of hydrogen influence for the materials properties are based on the experimental dependences [2]. These models describe only degradation of the material properties during hydrogen saturation. Moreover, complex models, which explain uneven degradation of material properties, were created [3]. The well-established point of view that the presence of hydrogen leads to the uniform deterioration of the material properties this example points out that the hydrogen presence is much more sophisticated. It affects the zones of the stress concentrations, especially those of tensile stresses, and causes a certain chain reaction in these zones. Increasing tensile stresses leads to higher concentration of hydrogen, which in turn, increases the stresses etc. This results in local collapse of the structure.

Important part of such kind of feedback, during chain reaction, is that hydrogen, which absorbed from the environment, are concentrated in the thin boundary layer of metal. It were detected experimentally [4–6] and were confirmed by mathematical modelling [7–9]. From a mechanical point of view, the problem arises of the influence of a multitude of small measure—a thin surface layer on the strength of the entire metal part or test specimen subjected to hydrogen absorption.

This task is very important for practice, as all modern methods of testing metals for the hydrogen embrittlement are based on the man-made charging of samples with hydrogen, described in [4–6]. These methods are standardized by ISO and ASME. Tens of thousands of tests per month are conducted in industry and the results of these tests are important for the further use of finished products made from these metals.

If we describe the degradation due to effect only in a thin surface layer, no critical change in the properties of the metal sample is observed, according to the data of [10]. The application of the moment theory also does not allow to explain the phenomenon of the influence of the surface layer on the entire sample, according [11]. The use of generally accepted HELP hydrogen embrittlement models, as [12–15] and HEDE [16], suggests the pre-definition of initiating destruction of a macrocrack, which can be considered an artificial technique, since the conditions of destruction depend on its parameters, according [15, 16].

Thus, there is a phenomenon of the influence of a thin, hydrogen-saturated surface layer of a metal sample, on strength, according [7]. This phenomenon is widely meets in practice, and there is no model describing the degradation of the properties of the entire metal due to the surface layer.

To simulate a significant degradation in the physicomechanical properties of a metal, similar to what is happening in experiments, we used a bi-continuum model of a material containing hydrogen [17].

2 Tension Model of Sample with Hydrogen

The objective of this study was to simulate the effect of a small, natural concentration of hydrogen on the stress-strain state of cylindrical steel samples of a metal with a groove.

The basis of our approach is the separation of all internal hydrogen in a metal into "bound" and diffusely-mobile. The bi-continuum model describes mobile hydrogen as an ideal fluid inside the diffusion channels, and bound hydrogen, as a metal-weakening element of the rheological model [17].

The bi-continuum model is non-linear. Linearization will not allow to describe the process of destruction, as in this case there may be significant local deformations. This non-linearity does not allow direct use of commercial finite-element packages for the calculation.

We conducted the following adaptation of the bi-continuum model for use in finite element calculations.

The bi-continuum model describes the hydrogen redistribution in zones of tensile stresses. We can develop a finite element model for real structures if we consider the outside hydrogen saturation.

The stress-strain curve for steel is changing due to the hydrogen saturation demonstrates the relaxation of the yield stress in steel with time in the presence of hydrogen cf. [18–20]. The yield stress decreases which leads to the exponential weakening of the material strength as time progresses, [17].

Hydrogen saturation time depends strongly on the temperature and varies in the range of ten minutes at the temperatures about 900 °C up to several days at the temperature near 20 °C (Fig. 1).

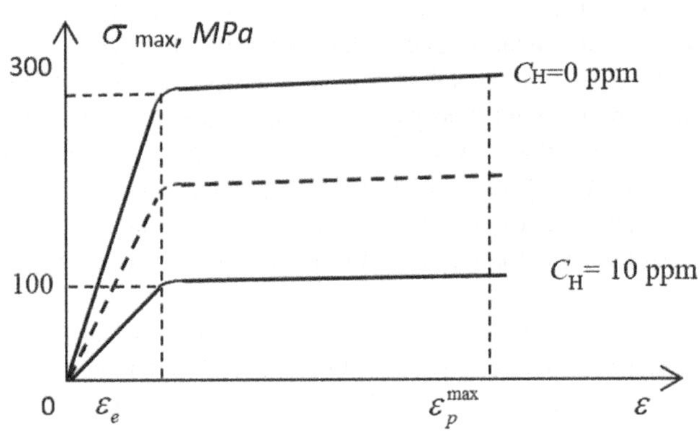

Fig. 1. The stress-strain curve for steel with different hydrogen concentration C_H

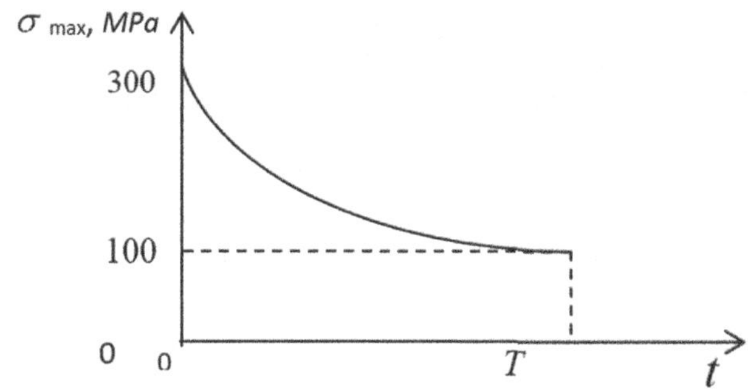

Fig. 2. The relaxation of the yield stress in steel with time in the presence of hydrogen

The algorithm of the calculation code is given in Fig. 3.

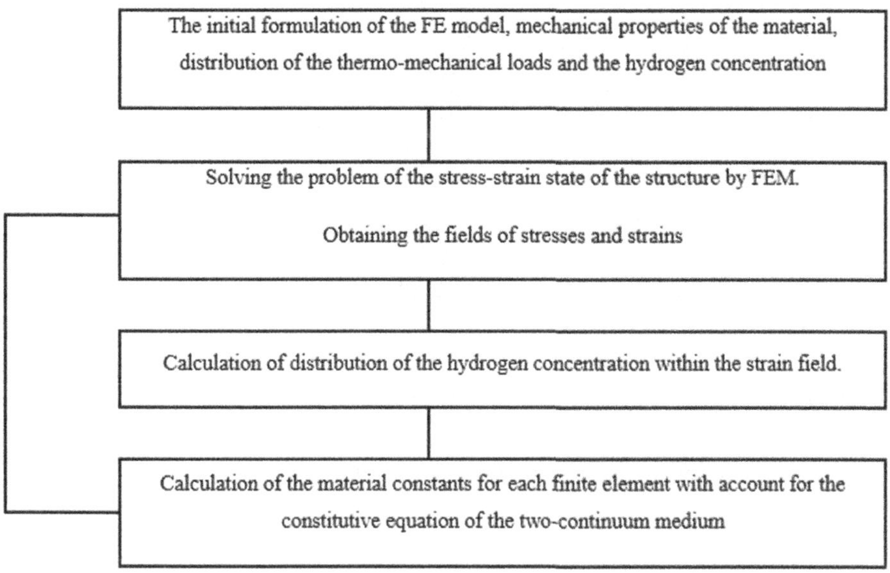

Fig. 3. Algorithm of the FEM code

The calculation is considered to be completed when two sequential iterations yield close fields of stresses and strains. This fact corresponds to total relaxation of stresses under the hydrogen redistribution. The relaxation time depends on many factors and can take up to several tens of years.

The proposed model follow algorithm with several iterations: in the first step, the calculation is made without considering the influence of hydrogen. The stress-strain state will be calculated and observed. In the next step, based on calculation result of distribution of the hydrogen concentration within the strain field [3] for steal 20, material properties will be modified to simulate hydrogen saturation. At the last step it will represent 10 ppm saturation. This concentration and following material properties are set for the entire model of the sample and corresponds to the bi-continuum model, which describes the hydrogen redistribution in zones of tensile stresses. The stress-strain state will be observed, which describe hydrogen redistribution in real structures after few years.

Figure 4 shows a finite element model with an applied load.

2D model are meshed by 2D Solid 162 axisymmetric elements. The Fig. 4 are show 3D revolving a section in 360° with specific mesh in stress concentrations area.

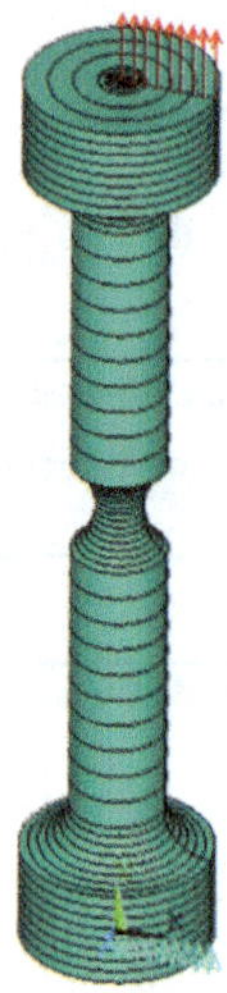

Fig. 4. Longitudinal tension of the plate by the force of 3.15 kN. The finite element mesh, the fixing (the lowest edge) and the force (the upper edge)

Boundary conditions (loads and constraints) are:

$y = 0$: $U_y = 0$,

$y = L$: $F_y = 3.15$ kN,

$x = R$: $n \cdot \boldsymbol{\sigma} = 0$.

The lowest edge of section is fixed. The force is applied to the upper edge witn constant value 3.15 kN. Material properties are based on data from cf. [3] about steel properties and hydrogen concentration and they are shown in the following Table 1.

Table 1. Material properties.

Parameter	Value
E	207 GPa
ν	0.29
σ_{max}	300 MPa
ρ	7850 kg/m3

3 Results and Discussion

Hydrogen saturation effect on mechanical properties degradation in the maximum of the stresses and the elastic strain were analyzed.

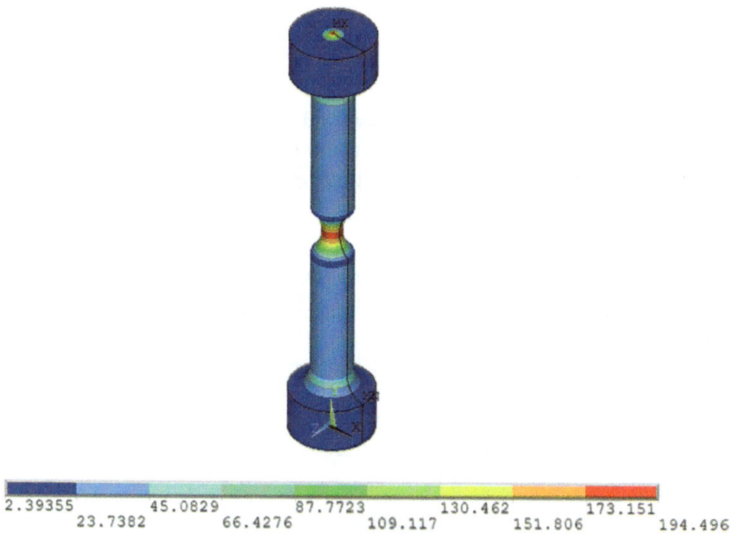

Fig. 5. Longitudinal tension of the plate by the force of 3.15 kN. The maximum von Mises stress **before** the hydrogen saturation is 195 MPa.

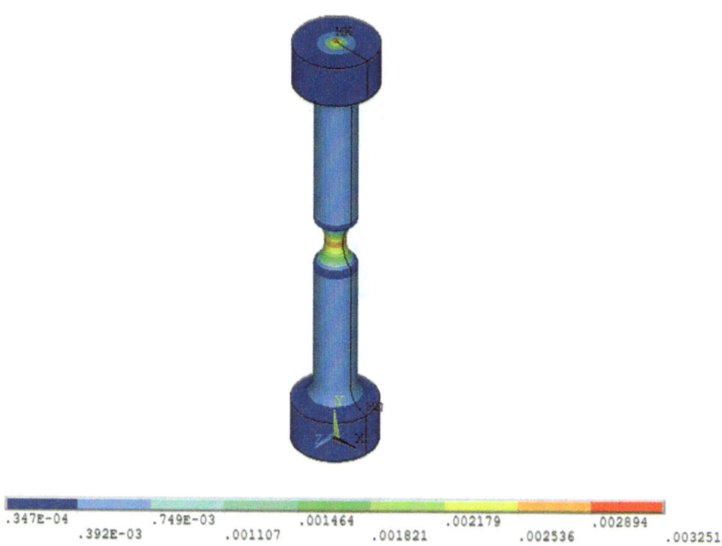

Fig. 6. Longitudinal tension of the plate by the force of 3.15 MN. The maximum elastic strain before the hydrogen saturation is 0.32%

Figures 5 and 6 shows the von Mises stresses and the elastic strain resulting from the calculation of the initial step. The initial calculation is done without the influence of hydrogen.

After determining the initial strain the iterative process runs according to the scheme shown in Fig. 3. At the first step, the following material parameters were used: Young's modulus 207 GPa, yield stress 300 MPa. These steel properties correspond condition without hydrogen ($\sim$0 ppm). The maximum stress calculation result is 195 MPa and strain is 0.32% by the force of 3.15 kN. According to the next algorithm step, the hydrogen concentration and material properties change were calculated based on the stress-strain result. It was conducted in [3] and result is shown in Fig. 2. Therefore, the last iteration material properties are Young's modulus 69 GPa, yield strength 145 MPa as it corresponds to the limit condition, when two sequential iterations yield close fields of stresses and strains. The last iteration material properties equal hydrogen saturation of 10 ppm. They are defined for the entire samples, which allow us to study the stress-strain state in the zone of maximum stresses. This process simulates the gradual redistribution of hydrogen inside the metal, which can occur in real structures in a few years.

The final picture of the stresses and strains (after hydrogen saturation) is shown in Figs. 7 and 8.

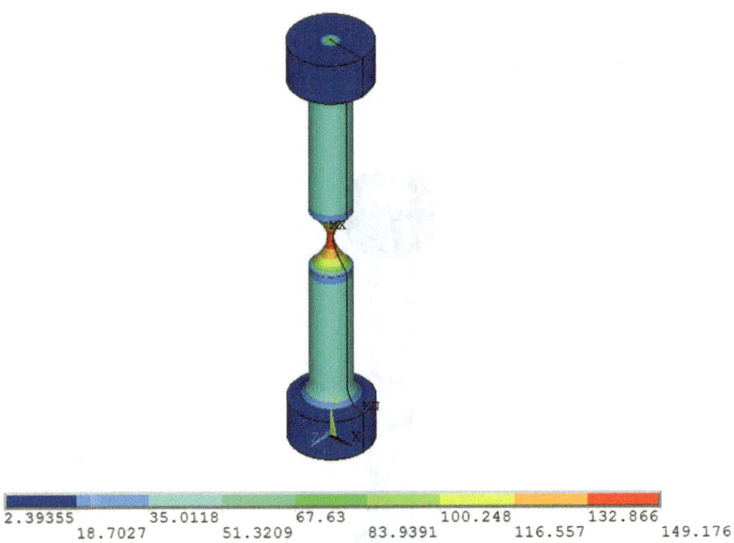

Fig. 7. Longitudinal tension of the plate of the force of 3.15 kN. The maximum von Mises stress **after** hydrogen saturation is 149 Mpa.

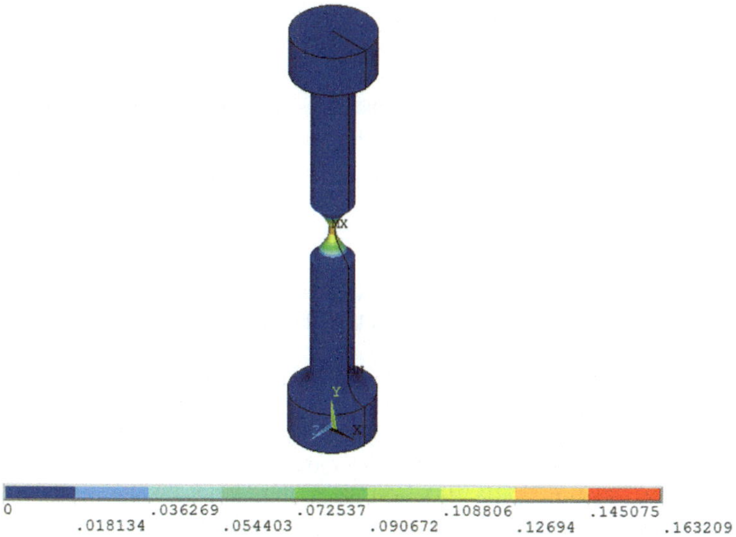

Fig. 8. Longitudinal tension of the plate of the force of 3.15 kN. The maximum strain **after** hydrogen saturation is 16.3%.

Analysis of the results of the calculation shows that the effect of hydrogen on the deformation of the material is in contrast to the yielding. In the zones of tensile stress, the hydrogen weakens the material and is stored at the same time which, in turn, reduces the elasticity and yield stress. This results in the additional deformations in this area and leads to greater saturation of the zones of tensile stress by the hydrogen, which can lead to the local destruction.

4 Conclusion

A finite element model was constructed, showing that one of the leading weakening factors of the structure is the accumulation of hydrogen concentrations from the external environment under uniaxial loading of steel samples.

The analysis of the effect of hydrogen saturation on the stress-strain state of a cylindrical sample of steel has been performed. Moreover, this particular example shows that the influence of small "natural concentrations" of hydrogen on the structural health is very considerable. In contrast to the well-established point of view that the presence of hydrogen leads to the uniform deterioration of the material properties this example points out that the hydrogen presence is much more sophisticated. It affects the zones of the stress concentrations, especially those of tensile stresses, and causes a certain chain reaction in these zones. Increasing tensile stresses leads to higher concentration of hydrogen, which in turn, increases the stresses etc. This results in local collapse of the structure. These collapses can have the consequences which are hardly observable (because of small size) but they can be much more dangerous.

The model used for the entire sample allows one to describe the experimentally observed effects.

Acknowledgments. The financial support of the Russian Foundation for Basic Research grants 17-08-00783-a and 18-08-00201-a, 18-19-00160 are acknowledged.

References

1. Gorsky, W.S.: Theorie der elastischen Nachwirkung in ungeordneten Mischkristallen (elastische Nachwirkung zweiter Art). Physikalische Zeitschrift der Sowjetunion **8**, 457–471 (1935)
2. Ahn, D.C., Sofronis, P., Dodds Jr., R.: Modeling of hydrogen-assisted ductile crack propagation in metals and alloys. Int. J. Fract. **145**, 135–157 (2007)
3. Vladimir, A.P., Belyaev Theoretical and Experimental Study of Influence of Hydrogen in Materials under Thermo-Mechanical Loading (2011)
4. Martinsson, A., Sandström, R.: Hydrogen depth prole in phosphorus-doped, oxygen-free copper after cathodic charging. J. Mater. Sci. **47**(19), 6768–6776 (2012)
5. Wu, R., Ahlström, J., Magnusson, H., Frisk, K., Martinsson, A.: Charging, degassing and distribution of hydrogen in cast iron. Swerea KIMAB, pp. 1–41 (2015)
6. Belyaev, A.K., Mansyrev, D.E., Polyanskiy, V.A., Polyanskiy, A.M., Tretyakov, D.A., Yakovlev, Y.A.: Diagnostics, resource and mechanics of materials and structures (4), 32–43 (2017)
7. Omura, T., Nakamura, J., Hirata, H., Jotoku, K., Ueyama, M., Osuki, T., Terunuma, M.: Effect of surface hydrogen concentration on hydrogen embrittlement properties of stainless steels and ni based alloys. ISIJ Int. **56**(3), 405–412 (2016)
8. Yagodzinskyy, Y., Todoshchenko, O., Papula, S., Hnninen, H.: Hydrogen solubility and diffusion in austenitic stainless steels studied with thermal desorption spectroscopy. Steel Res. Int. **82**(1), 20–25 (2011)
9. Hadam, U., Zakroczymski, T.: Absorption of hydrogen in tensile strained iron and high-carbon steel studied by electrochemical permeation and desorption techniques. Int. J. Hydrogen Energy **34**(5), 2449–2459 (2009)
10. Polyanskiy, V.A., Yakovlev, Y.A., Zegzhda, A.S.: Model of the effect of low concentrations of diffusion - mobile hydrogen on the cracks propagation. Adv. Prob. Mech. **2018**, 203–209 (2018)
11. Frolova, K., Vilchevskaya, E., Polyanskiy, V., Alekseeva, E.: Modelling of a hydrogen saturated layer within the micropolar approach. In: Abali, B., Altenbach, H., dell'Isola, F., Eremeyev, V., Öchsner, A. (eds.) New Achievements in Continuum Mechanics and Thermodynamics. Advanced Structured Materials, vol. 108, pp. 73–84. Springer, Cham (2019)
12. Birnbaum, H., Sofronis, P.: Hydrogen-enhanced localized plasticityamechanism for hydrogenrelated fracture. Mater. Sci. Eng. A **176**(12), 191–202 (1994)
13. Delafosse, D., Magnin, T.: Interfaces in stress corrosion cracking: a case study in duplex stainless steels. In: Solid State Phenomena, vol. 59, pp. 221–250. Trans Tech Publications (1998)
14. Delafosse, D., Magnin, T.: Hydrogen induced plasticity in stress corrosion cracking of engineering systems. Eng. Fract. Mech. **68**(6), 693–729 (2001)
15. Sofronis, P., Liang, Y., Aravas, N.: Hydrogen induced shear localization of the plastic flow in metals and alloys. Eur. J. Mech. A/Solids **20**(6), 857–872 (2001)

16. Varias, A., Massih, A.: Simulation of hydrogen embrittlement in zirconium alloys under stress, and temperature gradients. J. Nucl. Mater. **279**(23), 273–285 (2000)
17. Belyaev, A.K. Polyanskiy, A.M., Polyanskiy, V.A., Yakovlev, Y.A.: An approach to modeling structural materials with low hydrogen concentration. In: Altenbach, H., Belyaev, A., Eremeyev, V., Krivtsov, A., Porubov, A. (eds.) Dynamical Processes in Generalized Continua and Structures. Advanced Structured Materials, vol. 103, pp. 63–87. Springer, Cham
18. Oriani, R.A., Josephic, P.H.: The effects of hydrogen on the room-temperature creep of spheroidized 1040-steel. In: Perspectives in Hydrogen in Metals, pp. 609–614. Pergamon (1986)
19. Lunarska, E.: Effect of hydrogen on relaxation phenomena in pure iron. Scr. Metall. **11**(4), 283–287 (1977)
20. Lunarska, E., Wokulski, Z.: Effect of hydrogen charging on stress-strain curves for iron whiskers. In: Perspectives in Hydrogen in Metals, pp. 591–597. Pergamon (1986)

Calculation of the Activation Energies of Hydrogen in Titanium Manufactured with 3D Printing Technology by Means of a Multichannel Diffusion Model

P. A. Zumberov[1,2(✉)], Yu. A. Yakovlev[1], and V. A. Polyanskiy[1,2]

[1] Institute for Problems in Mechanical Engineering of the Russian Academy of Sciences, St. Petersburg, Russia
peterz.98@yandex.ru

[2] Peter the Great St. Petersburg Polytechnic University, St. Petersburg, Russia

Abstract. The most popular method for calculating activation energies of hydrogen is the method of thermo-desorption spectra (TDS). The disadvantage of this method is that it does not consider diffusion inside the metal. Essentially, only surface sorption and desorption can be adequately studied using this method. All this leads to large variations of experimentally evaluated activation energies for the same materials.

The present paper proposes a method for calculation of the activation energies of hydrogen based on experimental data. Measurements were carried out using the industrial mass-spectrometric hydrogen analyzer AV-1, which uses the hot vacuum extraction method. The authors of the paper realize a gradual heating in a vacuum to measure the activation energies of hydrogen in titanium. The mathematical model is based on the model of multichannel diffusion of hydrogen. Calculation of hydrogen flows and their time integrals, which characterize the integral experimental results on every temperature step, were performed using a Fortran program. The parameters of the sample and the experiment conditions were entered into the program. As a result, the activation energies of hydrogen diffusion were calculated. Using these activation energies, a graph of discrete thermo-diffusion spectrum was plotted, which was then compared to the graph obtained during the experiment.

Keywords: Hydrogen · Activation energy · Hydrogen embrittlement

1 Introduction

High concentration of hydrogen in metals can cause various problems. Hydrogen can cause bubbles in rolled metal products, problems during welding, microcracking in monocrystalline parts and hydrogen embrittlement.

Titanium has gained widespread use in various fields: mechanical engineering, aerospace, chemical, petroleum, marine and other industries. The reasons for its popularity are high corrosion resistance, low density to ultimate tensile strength ratio and

D. A. Indeitsev and A. M. Krivtsov (Eds.): APM 2019, LNME, pp. 256–263, 2020.
https://doi.org/10.1007/978-3-030-49882-5_25

cracking resistance. However, titanium is highly susceptible to hydrogen embrittlement due to its ability to form hydrides.

A high corrosion resistance of titanium caused by its durable oxide layer, which forms on the surface of titanium when it is exposed to the air. This oxide layer is resistant to hydrogen, however, hydrogen can infuse into the material, when titanium alloys are exposed to high temperatures, for example, during welding.

Additive technologies are increasingly used in recent years. They allow to reduce the costs and time required to create complex parts, which otherwise require many different manufacturing processes and the use of machines.

Additive manufacturing is a process of creating three-dimensional objects by building the components layer-by-layer. One of the most popular types of additive manufacturing is Selective Laser Melting (SLM). A part is created on a platform, which can move along a vertical axis. A metal powder is evenly spread in a thin layer. After that, a high-power laser melts (welds) the powder and it forms a solid cross-section of the part. Then the platform moves down and the process repeats for the next layer.

Comparison of the experimental data shows that the concentration of hydrogen in titanium powder is significantly higher than in titanium obtained by metallurgical method.

This hydrogen can remain in the material after the powder was welded together to form a part during SLM.

Other studies show fatigue cracks forming in pores in additive processed parts. The porosity is explained by the absorption of gasses during the welding. A few of the tested subjects failed at very early stages of testing cf. [1].

All this indicate the need for methods of accurate measurement of the concentration of hydrogen and prediction of the hydrogen effect.

Hydrogen in metals is located in various traps of different nature cf. [2]. The activation energy of the trap characterizes the interaction between hydrogen and the metal and the mobility of hydrogen inside the material. This value is important because hydrogen can concentrate locally and cause failures cf. [3–5].

One of the most widespread method for calculating these energies is the method of thermo-desorption spectra or TDS, which is based on Kissinger model cf. [6]. However, this method can produce vastly different results for the same materials [7, 8]. This is due to the fact that it can only account for sorption and desorption on the surface of a material [9]. Despite this, almost all studies nowadays are still using this method.

These shortcomings of TDS have led us to use a new method for determining binding energies for compact titanium samples, which takes into account the diffusion of hydrogen in its model.

2 Method Description

2.1 Measurements

The measurements were carried out using the industrial hydrogen analyzer AV-1. Its schematic is shown in Fig. 1.

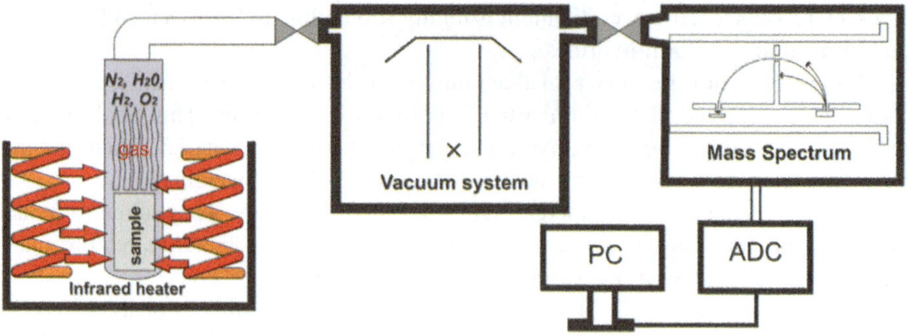

Fig. 1. The schematic of hydrogen analyzer

The apparatus consists of an extraction system, vacuum system, a mass spectrometer, an analog-to-digital converter and a PC. The extraction system consists of a quartz glass extractor and an infrared heater. The extractor is made from three chambers. The first one is used for storing multiple samples before the analysis. The second (the hot part) one is used to carry out the analysis. The third one is used for storing samples after the analysis. This construction allows one to analyze several samples without breaking the vacuum and to heat up the samples much quicker than usual and carry out the analysis at a constant temperature, because the hot part of the extractor can be pre-heated up to a certain temperature. The vacuum system can pump a vacuum up to 10^{-5} Pa.

Usually, the amount of hydrogen in a sample is estimated by heating the sample at a constant rate and then evaluating the amount of hydrogen based on the temperature dependence of the flow of hydrogen. However, with the increase in temperature, background flows of hydrogen from the extraction system grow exponentially, thus making increasing the uncertainty cf. [10, 11].

When an extraction system is maintained at a certain constant temperature, after some time (usually about an hour) the background flows settle to a constant level. This effect can be used to reliably subtract them from the total flow of hydrogen to calculate the hydrogen flow from a sample. To use this effect to get a more precise measurement a procedure of gradual heating was implemented. The algorithm of this procedure is shown in Fig. 2. After all the samples are placed in the extraction system and the vacuum is pumped, the extraction temperature is set lower, than usually needed to extract all the hydrogen from a sample. When the background flows settle down, a sample is dropped into the hot part of the extractor, where it quickly heats up to the extraction temperature. When the sample heats up, the analysis is carried out by the standard procedure of recording the extraction curve and calculating the amount of hydrogen released at this temperature. After the flow of hydrogen from a sample stops, the sample is moved back into the cold part of the extractor. Then, the empty extractor is heated up to the next temperature step and the cycle continues.

As a result of this procedure, a discrete thermo-desorption spectra (DTDS) is obtained. DTDS represents a temperature dependence of the amount of the extracted hydrogen. The advantage of a discrete thermo-desorption spectra over a conventional thermo-desorption spectra is the fact that the background flows are subtracted and it only represents the hydrogen extracted from a sample itself. There is a similar method called "isotheral spectra". However, in this method the temperature changes in the extraction system, which doesn't allow to account for the background flows as accurately, as in case of the DTDS method.

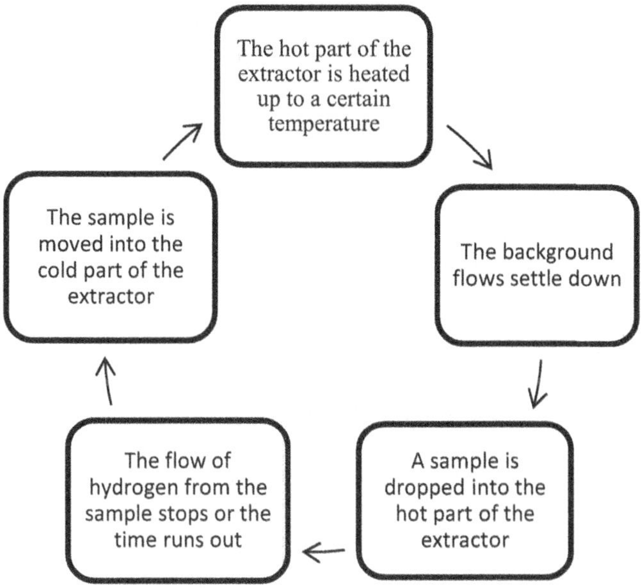

Fig. 2. The algorithm of step heating

2.2 Methods of Evaluation of the Activation Energies

Analysis of the Experiment Data

The sample was made out of 99.99% titanium. The hydrogen charging was done in 0,1 N NaOH solution with the current density of 1250 mA/cm^2 during 30 min.

An extraction curve was recorded on each temperature step. The evaluation of the amount of hydrogen extracted at each temperature step was performed by integration of the area between the extraction curve and the background flows' level in hydrogen analyzer software. After that, the DTDS was plotted.

Modelling

The multichannel diffusion model is used [9]. We assume that in a compact titanium sample hydrogen diffuses through several channels simultaneously. The equations for time-dependent hydrogen diffusion in the channel with number i are as follows:

$$\Delta C_i = \frac{1}{D_{0i}} \frac{\partial C_i}{\partial t}, \tag{1}$$

$$C_i\big|_s = 0, \quad C_i\big|_{t=0} = C_{i0}, \quad i = 1, \ldots, l,$$

where C_{0i} is the initial concentration of hydrogen in a diffusion channel; D_{0i} is a diffusion constant in a diffusion channel; t is time.

According to the Arrhenius law, the dependence of the diffusion coefficient D_i on the absolute temperature of the sample T has the form

$$D_i = D_{0i} \exp\left(-\frac{u_i}{kT}\right),$$

where u_i is the activation energy, k is the Boltzman constant, D_{0i} is the diffusion constant.

The equation for the heating of a sample is

$$\frac{dT}{dt} = \frac{\sigma S}{c\rho V} \cdot 0.2\left(T_0^4 - T^4\right), \tag{2}$$

where T is the absolute temperature of the sample; T_0 is the absolute temperature of the heater, S is the surface area of the sample, C is the specific heat capacity; ρ is the density of the sample; V is its volume, 0.2 is the infrared absorption coefficient for titanium.

From these equations we get the equation for the flow of hydrogen from a sample:

$$q(t) = \frac{16}{\pi^2}\left(\frac{1}{a^2} + \frac{1}{b^2} + \frac{1}{l^2}\right) \sum\nolimits_i C_{0i} \times D_{0i} \exp\left(-\frac{u_i}{kT}\right) f_i(t, u_i, D_{0i}) \tag{3}$$

where q is flow of hydrogen in the vacuum extraction system; a, b, l are the dimensions of a sample; u_i – activation energy, $f_i = f_i(t, u_i, D_{0i})$ is the solution of the Eq. (2) with the initial condition T = 295 °K and Eqs. (4) for each diffusion channel:

$$\begin{cases} \dot{f}_i + D_{0i} \cdot \exp\left(-\frac{u_i}{kT}\right)\left(\frac{\pi^2}{a^2} + \frac{\pi^2}{b^2} + \frac{\pi^2}{l^2}\right) f_i = 0 \\ f_i(0, u_i, D_{0i}) = 1 \end{cases} \tag{4}$$

A Fortran program estimates the extraction curves for each temperature step using the constants C_{0i}, u_i, D_{0i}. These constants are determined by fitting the experimental time dependences of the hydrogen flow from the sample on time at each temperature step of DTDS (extraction curves). The number of diffusion channels is determined by the number of peaks in the extraction curves. The known parameters of a sample were inputted into the program. It is important that all constants are set equal for all temperature steps. The whole process of DTDS is integrated with one initial condition, as a process of stepwise change in temperature of the extraction system over time. For each temperature step the model amount of the extracted hydrogen is calculated using integration with an explicit fourth-order Runge-Kutta method. The quality of the fitting of the diffusion parameters is determined by the coincidence of the amount of hydrogen extracted at each heating step from the temperature of the step (experimental and model).

The activation energies and diffusion constants were determined as the result of fitting the model extraction curves to real ones.

3 Results

As a result, activation energies and DTDS points were calculated (see Fig. 3). The absolute amount of hydrogen is given in mass concentrations and the temperature is given in degrees Celsius for the most useful representation.

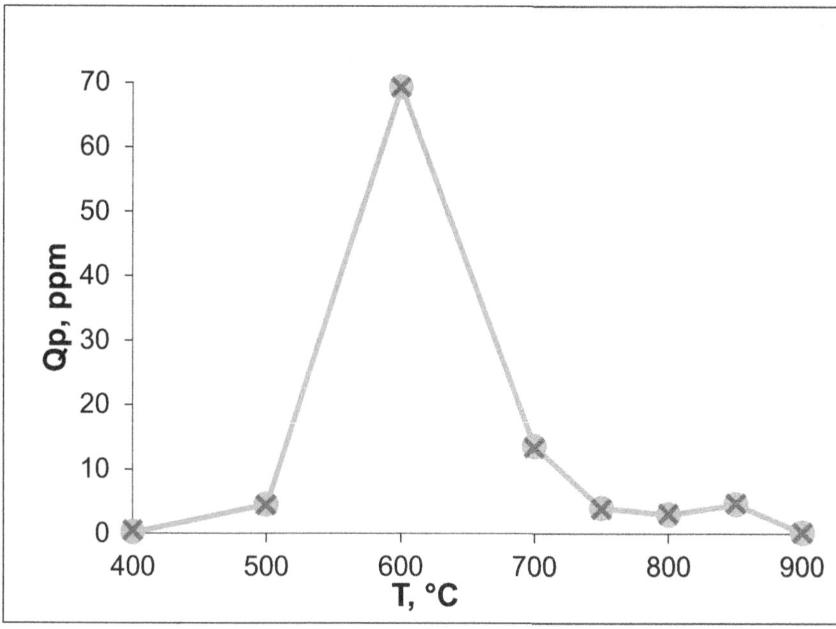

Fig. 3. DTDS step points. The DTDS points obtained by the experiment are black, the calculated ones are grey.

According to the results shown in Fig. 3, the experimental and theoretical amounts of hydrogen extracted at each stage coincidence almost perfectly.

The conducted comparison of the experimental data and theoretical results makes it possible to estimate the activation energies of hydrogen. The activation energies corresponding to gray points in Fig. 3 are shown in Table 1.

Table 1. The activation energies

Diffusion channel	Activation energy, eV
1	1,02
2	1,07
3	2,205
4	4,55

As a result, hydrogen can be divided into energy levels. This information allows further estimating the mechanical properties and residual resource of the material.

4 Conclusion

A new method for calculation the activation energies of hydrogen based on direct measurements was proposed.

The use of the DTDS method allow to obtain the results with a lower error, since the effect of background hydrogen flows absent.

The values of the activation energies are within the range of previously known values, although they are higher than for titanium manufactured by applying traditional methods.

The experimental and analytical obtained DTDS points coincidence almost perfectly. Thus, the method is verified.

New data on the concentration of hydrogen and binding energy for titanium were obtained.

Acknowledgements. The research is carried out under the financial support by Russian Science Foundation, grant 18-19-00160.

References

1. Wang, F., Williams, S., Colegrove, P., et al.: Metall. Mater. Trans. A **44**, 968 (2013). https://doi.org/10.1007/s11661-012-1444-6
2. Kirchheim, R.: Solubility, diffusivity and trapping of hydrogen in dilute alloys. Deformed and amorphous metals—II. Acta Metall. **30**(6), 1069–1078 (1982). https://doi.org/10.1016/0001-6160(82)90003-7
3. Woodtli, J., Kieselbach, R.: Damage due to hydrogen embrittlement and stress corrosion cracking. Eng. Fail. Anal. **7**(6), 427–450 (2000). https://doi.org/10.1016/S1350-6307(99)00033-3
4. Robertson, I.M., Sofronis, P., Nagao, A., et al.: Metall. Mater. Trans. A **46**, 2323 (2015). https://doi.org/10.1007/s11661-015-2836-1
5. Birnbaum, H.K., Sofronis, P.: Hydrogen-enhanced localized plasticity—a mechanism for hydrogen-related fracture. Mater. Sci. Eng., A **176**(1–2), 191–202 (1994). https://doi.org/10.1016/0921-5093(94)90975-X

6. Kissinger, H.E.: Reaction kinetics in differential thermal analysis. Anal. Chem. **28**(11), 1702–1706 (1957). https://doi.org/10.1021/ac60131a045
7. Belyaev, A.K., Polyanskiy, A.M., Polyanskiy, V.A.: Multichannel diffusion vs TDS model on example of energy spectra of bound hydrogen in 34CrNiMo6 steel after a typical heat treatment. Int. J. Hydrogen Energy **41**(20), 8627–8634 (2016). https://doi.org/10.1016/j.ijhydene.2016.03.198
8. Kuduzovic, A., Poletti, M.C., Sommitsch, C., Domankova, M., Mitsche, S., Kienreich, R.: Investigations into the delayed fracture susceptibility of 34CrNiMo6 steel, and the opportunities for its application in ultra-high-strength bolts and fasteners. Mater. Sci. Eng. **590**, 66–73 (2014). https://doi.org/10.1016/j.msea.2013.10.019
9. Polyanskiy, A.M., Polyanskiy, V.A., Yakovlev, Y.A.: Experimental determination of parameters of multichannel hydrogen diffusion in solid probe. Int. J. Hydrogen Energy **39** (30), 17381–17390 (2014). https://doi.org/10.1016/j.ijhydene.2014.07.080
10. Tapia-Bastidas, C.V., Atrens, A., Gray, E.M.: Thermal desorption spectrometer for measuring PPM concentrations of trapped hydrogen. Int. J. Hydrogen Energy **43**(15), 7600–7617 (2018). https://doi.org/10.1016/j.ijhydene.2018.02.161
11. Nickel, N.H., Brendel, K., Saleh, R.: Laser crystallization of hydrogenated amorphous silicon. Phys. Status Solidi **1**(5), 1154–1168 (2004). https://doi.org/10.1002/pssc.200304311

Author Index

D. A. Indeitsev and A. M. Krivtsov (Eds.): APM 2019, LNME, pp. 265–266, 2020.
https://doi.org/10.1007/978-3-030-49882-5

The manufacturer's authorised representative in the EU is Springer Nature Customer Service Centre GmbH, Europaplatz 3, 69115 Heidelberg, Germany. If you have any concerns regarding our products, please contact ProductSafety@springernature.com

Printed and bound by CPI Group (UK) Ltd, Croydon, CR0 4YY
30/09/2026
02225509-0007